T0091741

Ronald R. Yager, Janusz Kacprzyk, and Gleb Beliakov (Eds.)

Recent Developments in the Ordered Weighted Averaging Operators: Theory and Practice

Studies in Fuzziness and Soft Computing, Volume 265

Editor-in-Chief

Prof. Janusz Kacprzyk
Systems Research Institute
Polish Academy of Sciences
ul. Newelska 6
01-447 Warsaw
Poland
E-mail: kacprzyk@ibspan.waw.pl

Further volumes of this series can be found on our homepage: springer.com

Vol. 250. Xiaoxia Huang
Portfolio Analysis, 2010
ISBN 978-3-642-11213-3

Vol. 251. George A. Anastassiou
Fuzzy Mathematics: Approximation Theory, 2010
ISBN 978-3-642-11219-5

Vol. 252. Cengiz Kahraman, Mesut Yavuz (Eds.)
Production Engineering and Management under Fuzziness, 2010
ISBN 978-3-642-12051-0

Vol. 253. Badredine Arfi
Linguistic Fuzzy Logic Methods in Social Sciences, 2010
ISBN 978-3-642-13342-8

Vol. 254. Weldon A. Lodwick, Janusz Kacprzyk (Eds.)
Fuzzy Optimization, 2010
ISBN 978-3-642-13934-5

Vol. 255. Zongmin Ma, Li Yan (Eds.)
Soft Computing in XML Data Management, 2010
ISBN 978-3-642-14009-9

Vol. 256. Robert Jeansoulin, Odile Papini, Henri Prade, and Steven Schockaert (Eds.)
Methods for Handling Imperfect Spatial Information, 2010
ISBN 978-3-642-14754-8

Vol. 257. Salvatore Greco, Ricardo Alberto Marques Pereira, Massimo Squillante, Ronald R. Yager, and Janusz Kacprzyk (Eds.)
Preferences and Decisions,2010
ISBN 978-3-642-15975-6

Vol. 258. Jorge Casillas and Francisco José Martínez López
Marketing Intelligent Systems Using Soft Computing, 2010
ISBN 978-3-642-15605-2

Vol. 259. Alexander Gegov
Fuzzy Networks for Complex Systems, 2010
ISBN 978-3-642-15599-4

Vol. 260. Jordi Recasens
Indistinguishability Operators, 2010
ISBN 978-3-642-16221-3

Vol. 261. Chris Cornelis, Glad Deschrijver, Mike Nachtegael, Steven Schockaert, and Yun Shi (Eds.)
35 Years of Fuzzy Set Theory, 2010
ISBN 978-3-642-16628-0

Vol. 262. Zsófia Lendek, Thierry Marie Guerra, Robert Babuška, and Bart De Schutter
Stability Analysis and Nonlinear Observer Design Using Takagi-Sugeno Fuzzy Models, 2010
ISBN 978-3-642-16775-1

Vol. 263. Jiuping Xu and Xiaoyang Zhou
Fuzzy-Like Multiple Objective Decision Making, 2010
ISBN 978-3-642-16894-9

Vol. 264. Hak-Keung Lam and Frank Hung-Fat Leung
Stability Analysis of Fuzzy-Model-Based Control Systems, 2011
ISBN 978-3-642-17843-6

Vol. 265. Ronald R. Yager, Janusz Kacprzyk, and Gleb Beliakov (Eds.)
Recent Developments in the Ordered Weighted Averaging Operators: Theory and Practice, 2011
ISBN 978-3-642-17909-9

Ronald R. Yager, Janusz Kacprzyk,
and Gleb Beliakov (Eds.)

Recent Developments in the Ordered Weighted Averaging Operators: Theory and Practice

Editors

Prof. Ronald R. Yager
Machine Intelligence Institute
Iona College
New Rochelle, NY 10801
USA
E-mail: Yager@Panix.Com

Prof. Janusz Kacprzyk
Systems Research Institute
Polish Academy of Sciences
Ul. Newelska 6
01-447 Warsaw
Poland
E-mail: kacprzyk@ibspan.waw.pl

Prof. Gleb Beliakov
School of Information Technology
Deakin University
221 Burwood Highway
Burwood Victoria 3125
Australia
E-mail: gleb@deakin.edu.au

ISBN 978-3-642-17909-9 e-ISBN 978-3-642-17910-5

DOI 10.1007/978-3-642-17910-5

Studies in Fuzziness and Soft Computing ISSN 1434-9922

Typeset & Cover Design: Scientific Publishing Services Pvt. Ltd., Chennai, India.

Printed on acid-free paper

9 8 7 6 5 4 3 2 1

springer.com

Foreword

The area of aggregation operations is one of the most promising off-springs of fuzzy sets theory. Before Zadeh introduced fuzzy set connectives in 1965, there was a wide gap between logic and decision sciences. On the one hand, multiple-valued logics proposed many-valued extensions of the conjunctions and disjunctions for which the triangular norms and conorms today provide a natural setting. On the other hand, in decision sciences, the prototypical operation turned out to be the weighted average, whether for decision under uncertainty (the expected utility) or for multi-criteria decision-making, or yet in voting theory (the majority voting).

However, the setting of fuzzy sets theory, and soon after, the emergence of so-called fuzzy measures and integrals by Sugeno in 1974 led to the breaking of those borders. It was clear that while the triangular norms and co-norms, respectively, stood below the minimum and above the maximum, and the arithmetic means stood in-between. This led to the study of the family of averaging operations. All logical connectives could be viewed as aggregation operations of some sort, and the arithmetic mean could be viewed as an alternative set-theoretic operation. The Sugeno integral appeared as a natural ordinal family of weighted averaging operations based on the minimum and maximum. Another crucial step was the importance given to the notion of generalized quantifiers by Zadeh at the turn of the eighties, followed by several researchers including Ronald Yager. In multicriteria decision-making, it is natural to try and construct aggregation schemes computing to what extent most criteria are satisfied. This was achieved for additive aggregations by the introduction of the ordered weighted average (OWA) by Ronald Yager in 1988: instead of weighting criteria, the basic idea was to put weights on components of rating vectors after a preliminary ranking of the individual ratings. In this way it was possible to give importance weights to the fact of having prescribed numbers of criteria to be fulfilled. The other crucial contribution of Yager was to connect these weights (that sum up to 1 like for the usual averages) to fuzzy quantifiers like most, few, etc. This connective is a symmetric generalization of the arithmetic mean, the minimum and the maximum in the fuzzy set theory.

One important reason for the subsequent success of the ordered weighted averages (OWAs) is that scholar soon realised that they were nothing but the symmetric Choquet integrals. The emergence of Choquet integrals as a key aggregation operation in decision theory took place in the 1980's. Choquet had generalised the Lebesgue integral to non-additive set-functions in the 1950's for the purpose of modeling electric phenomena. He called a ``capacity" a monotone set-function employed by Sugeno later on, and independently named fuzzy measure. Ulrich Hoehle reintroduced the Choquet integral in 1982 in the mathematical framework of generalized probabilities (such as the Dempster-Shafer belief functions and possibility theory). Quite independently at the same time, David Schmeidler proposed a generalized framework for decision theory that could accommodate the Ellsberg paradox. Such paradoxes whereby people were repeatedly and consistently shown to violate the Savage Sure Thing Principle, a cornerstone of decision theory, justifying the systematic use of expected utility, had been extensively discussed in the seventies. What Schmeidler did was to reinvent the Choquet integral based on relaxed Savage axioms, especially that of the co-monotonioc additivity. This discovery led to numerous new axiomatic systems for decision criteria generalizing the expected utility akin to the Choquet integral. The OWA aggregation operator can be viewed as a discrete Choquet integral where values assigned to sets by the fuzzy measure involved depend only on the cardinality of these sets (these values depict a fuzzy quantifier).

Since then, the OWA operations have been the topic of active research, partly because of their intuitive meaning as a quantified aggregation, partly because of their computational simplicity (contrary to the Choquet Integrals, the specification of an OWA is of linear complexity).

There had been a first edited volume dedicated to OWA operations in the late 1990's and this new volume is a welcome addition that enables the reader to figure out where we stand and where we go on this topic. It shows that there is a small but active community of researchers that continue to dig the foundations of this aggregation operation, propose suitable extensions, explain how to derive the weights from data and linguistic information, and situate its role in the framework of social choice and group decision-making. The second part of the volume demonstrates a variety of possible applications in information fusion, image processing environmental engineering and the Web.

The editor sshould be congratulated for putting together such a collection of papers that gently introduces readers to the topic of the OWA operations while providing a survey of the latest developments as well as a collection of inspiring applications.

Toulouse, October 2010

Didier Dubois
Directeur de Recherche au CNRS
IRIT, CNRS and University of Toulouse

Preface

This volume is meant to present the state of the art of new developments, and some interesting and relevant applications of the OWA (ordered weighted averaging) operators. The OWA operators were introduced in the early 1980s by R.R. Yager as a conceptually simple, yet extremely powerful general aggregation operator. By an ingenious idea of first rearranging the weights from the highest to the lowest, and then using those rearranged weights in the well known weighted averaging scheme, a whole range of aggregation behaviors had been possible trough a proper selection of the weights (to be then rearranged, of course): from the pessimistic, safety first type minimum to the optimistic maximum, through all intermediate values including the mean value, and a linguistic quantifier driven aggregation exemplified by the aggregation of "most" values.

That generality of the OWA operators, combined with their intuitive appeal and conceptual similarity, have triggered much research both in the foundations and extensions of the OWA operators, and their relations to some other concepts like ordered statistics, and in their applications to a wide variety of problems in which the availability of tools for various aggregation operators is of a paramount importance. A notable example is here real life decision making in which a multitude of criteria, attributes, decision makers, decision making stages, etc. calls for an appropriate aggregation of pieces of evidence related to all those satisfactions of individual criteria, individual testimonies, results of particular stages, etc. The OWA operators have provided novel aggregation tools, and have opened new vistas and perspectives for the related fields of research and applications.

The papers included in Part I: Methods are concerned with more general issues related to the OWA operators. First, some extensions of the basic concept of the OWA operator are discussed, and the problem of how to properly select the weights of the OWA operators are dealt with. In the last papers of this section, first, the use of the OWA operators as an aggregation tools for an uniform representation of choice functions in group decision making and voting is presented. Then, the use of the linguistic OWA operators is shown to deal with the consensus reaching problem in group decision making under linguistic information (opinions of the individuals) when the linguistic term set is unbalanced.

M. Grabisch ("OWA Operators and Nonadditive Integrals") gives a survey on the relations between nonadditive integrals (the Choquet integral and the Sugeno integral) and the OWA operator and some of its relevant variants. The author then discusses some important behavioral indices for the OWA operator, notably the orness, veto and favor indices. Finally, the author propose the use of p-symmetric capacities for a natural generalization of the OWA operator.

V. Torra ("The WOWA Operator: A Review") is concerned with the WOWA (Weighted OWA) operator which was proposed as a generalization of both the OWA operator and the weighted mean. Formally, the WOWA is an aggregation operator that permits the aggregation of a set of numerical data with respect to two weighting vectors: one corresponding to that of the weighted mean and the other corresponding to the OWA. Some main definitions, issues and properties of the WOWA are discussed.

G. Beliakov and S. James ("Induced Ordered Weighted Averaging Operators") present the induced ordered weighted averaging operator (IOWA). Rather than reordering the input vector by the size of its arguments, the IOWA uses an auxiliary variable called the inducing vector, thus generalizing the standard OWA. The aggregation via the Induced OWA and its generalizations provide a useful framework for the modeling of many types of aggregation, including the nearest-neighbor rules. The authors introduce the IOWA and discuss many of its properties in relation to the associated inducing variable. They present then some of the generalizations discussed in the literature, as well as some important potential applications for aggregation where the ordering of the input vector is induced.

X. Liu ("A review of the OWA determination methods: classification and some extensions") discuss the problem of the determination of the appropriate OWA operators which is an important prerequisite for all applications of the OWA operators. The author gives a summary on the OWA determination methods with respect to the following classification: the optimization criteria methods, the sample learning methods, the function based methods, the argument dependent methods, and the preference methods. Some relationships between the methods in the same type and relationships between different types are provided. A uniform framework to view those OWA determination methods is discussed. Some extensions, problems and future research directions are outlined.

Sh.-M. Zhou, F. Chiclana, R.I. John, J.M. Garibaldi ("Fuzzification of the OWA operators for aggregating uncertain information with uncertain weights") generalize the source Yager's OWA operator and describe two novel uncertain OWA-type operators, namely the type-1 OWA operator and type-2 OWA operator. The type-1 OWA operator is meant to aggregate the type-1 fuzzy sets and the type-2 OWA operator is meant to aggregate the type-2 fuzzy sets. Therefore, the two new operators are capable of aggregating uncertain opinions or preferences with uncertain weights in decision making problems in soft settings. The authors indicate that not only the original Yager's OWA operator but also some existing operators of fuzzy sets, including the join and meet of the type-1 fuzzy sets, are special cases of the type-1 OWA operators. Then, the authors suggest new concepts of joinness and meetness of the type-1 OWA operators which can be considered as the extensions of the concepts of orness and andness in Yager's OWA operator, respectively. An attempt to alleviate the computational overhead involved in aggregating general type-2 fuzzy sets using the type-2 OWA is undertaken by using, an interval type-2 fuzzy sets oriented OWA operator. Examples are provided to illustrate the proposed tools.

G. Pasi and R.R. Yager ("A majority guided aggregation in group decision ma king") discuss the problem of majority modelling in the context of group (multi-expert) decision making, to the aim of defining a decision strategy which takes into account the individual opinions of the decision makers. The key concept of majority is concerned with the fact that what is often needed is an overall opinion which synthesizes the opinions of a majority of the experts. The reduction of the individual experts' opinions into a representative value (called the majority opinion) is usually performed through an aggregation process. The authors describe two distinct approaches to the definition and computation of a majority opinion within the fuzzy set theory, where majority can be expressed by a linguistic quantifier (such as most). First, they consider the case when the linguistic quantifiers are associated with aggregation operators, and a majority opinion is computed by aggregating the individual opinions. The Induced Ordered Weighted Averaging operators (IOWA) are used with a modified definition of their weighting vector. We then consider a second case where the concept of majority is modelled as a vague concept. Based on this interpretation, a formalization of a fuzzy majority opinion as a fuzzy subset is described.

J.L. García-Lapresta, B. Llamazares and T. Peña ("Generating OWA Weights from Individual Assessments") propose a method for generating the OWA weighting vectors from the individual assessments on a set of alternatives in such a way that these weights minimize the disagreement among individual assessments and the outcome provided by the OWA operator. For measuring that disagreement the authors aggregate distances between the individual and collective assessments by using a metric and an aggregation function. The Manhattan metric, the Chebyshev metric, and the arithmetic mean and maximum as the aggregation functions are employed. It is proven that the medians and the mid-range are the solutions for some cases considered. When a general solution is not available, the authors provide some mathematical programs for numerically solving the problem.

J. Kacprzyk, H. Nurmi and S. Zadrożny ("The role of the OWA operators as a unification tool for the representation of collective choice sets") consider various group decision making and voting procedures presented in the perspective of two kinds of aggregation of partial scores related to the individuals' (group's) testimonies with respect to alternatives and individuals. The authors show that the ordered weighthed averaging (OWA) operators can be viewed as a unique aggregation tool that – via the change of the order of aggregation, type of aggregation, etc. – can be used for a uniform and elegant formalization of basic group decision making, social choice and voting rules under fuzzy and nonfuzzy preference relations and fuzzy and nonfuzzy majority.

E. Herrera-Viedma, F.J. Cabrerizo, I.J. Pérez, M.J. Cobo, S. Alonso and F. Herrera ("Applying linguistic OWA operators in consensus models under unbalanced linguistic information") consider consensus reaching models in group decision making guided by different consensus measures which usually are obtained by aggregating similarities between the individuals' opinions. Most group decision making problem formulations based on linguistic approaches use symmetrically and uniformly distributed linguistic term sets to represent the opinions. However, there exist problems in which assessments need to be represented by unbalanced linguistic term sets, i.e., using term sets which are not uniformly and symmetrically distributed. The authors present different Linguistic OWA Operators (LOWAs) to compute the consensus measures in consensus models with unbalanced fuzzy linguistic information.

The papers in Part II: Applications show some more relevant applications of the OWA operators, mostly means, as powerful yet general aggregation operators. The applications concern many different application areas exemplified by environmental modeling, social networks, image analysis, financial decision making and water resource management.

G. Bordogna, M. Boschetti, A. Brivio, P. Carrara, M. Pagani and D. Stroppiana ("Fusion strategies based on the OWA operator in environmental applications" consider the modeling of ill-known environmental phenomena which is often done by means of multisource spatial data fusion. Generally, the fusion strategies have to cope with various kinds of uncertainty, related to the ill-defined knowledge of the phenomenon, a lack of classified data, different degrees of trust of the information sources, imprecision of the observed variables, etc. The authors discuss the advantages of modeling multisource spatial data fusion in the environmental field based on the OWA operator, and then overview two applications. The first application is aimed at defining an environmental indicator of anomaly at a continental scale based on a fusion of partial hints of the pieces of evidence of anomaly. The second application computes seismic hazard maps based on a consensual fusion strategy defined by an extended OWA operator that accounts for data imprecision, and the reliability of data sources. In particular, the proposed fusion function models consensual dynamics and is parameterized so as to consider a varying spatial neighborhood of the data to be fused.

J.M. Merigó and M. Casanovas ("Decision making with Dempster-Shafer theory using fuzzy induced aggregation operators") develop a new approach to decision making with the Dempster-Shafer theory of evidence when the available information is uncertain and can be assessed with fuzzy numbers. In this approach it is possible to represent the problem without losing relevant information so that the decision maker knows exactly which are the different alternatives and their consequences. To achieve this, the authors suggest the use of different types of fuzzy induced aggregation operators to be able to aggregate information considering all the different scenarios that could happen in the analysis. As a result, new types of fuzzy induced aggregation operators are obtained such as the belief structure – fuzzy induced ordered weighted averaging (BS-FIOWA) and the belief structure – fuzzy induced hybrid averaging (BS-FIHA) operator. Their main properties are discussed. Then, the approach is generalized by using the fuzzy induced generalized aggregation operators. An application of the new approach in a financial decision making problem on selection of financial strategies is presented.

H. Bustince, D. Paternain, B. De Baets, T. Calvo, J. Fodor, R. Mesiar, J. Montero and A. Pradera ("Two methods for image compression/reconstruction using OWA operators") address the problem of image compression by means of two alternative algorithms. In the first algorithm, the authors associate to each image an interval-valued fuzzy relation, and build an image which is n times smaller than the original one by using the two-dimensional OWA operators. The experimental results show that, in this case, the best results are obtained with the ME-OWA operators. In the second part of the work, the authors describe a reduction algorithm that replaces the image by several eigen fuzzy sets associated with it, and obtain these eigen fuzzy sets by means of an equation that relates the OWA operators used and the relation (image) considered. Finally, a reconstruction method is proposed based on an algorithm which minimizes a cost function built by means of two-dimensional OWA operators.

M. Brunelli, M. Fedrizzi, M. Fedrizzi ("OWA-based fuzzy m-ary adjacency relations in social network analysis") propose an approach to Social Network Analysis (SNA) based on fuzzy m-ary adjacency relations. In particular, the authors show that the dimensionality of the analysis can naturally be increased and interesting results can be derived. The fuzzy m-ary adjacency relations can be computed starting from the fuzzy binary relations and introducing OWA-based aggregations. The behavioral assumptions derived from the measure and the examination of individual propensity to connect with others suggest that the OWA operators can be considered to be particularly suitable for characterizing such relationships.

M. Zarghami and F. Szidarovszky ("Soft computing in water resources management by using OWA operator") introduce a new method to obtain the order weights of the OWA operator. The new method is based on the combination of fuzzy quantifiers and neat OWA operators. The fuzzy quantifiers are applied for soft computing in the modeling of the social preferences (an optimism degree of the decision maker, DM) while using the neat operators, the ordering of the inputs is not needed resulting in a better computation efficiency. The authors discuss one of the frequently-used ways to control water shortages is inter-basin water transfer (IBWT). Efficient decision making in this case is however a real challenge for the water authorities as these decisions should include multiple criteria, model uncertainty, and also a optimistic/pessimistic view (attitude) of the decision makers. The authors illustrate the theoretical results obtained by ranking four IBWT projects for the Zayanderud basin, Iran. The results demonstrate that by using the new method, more sensitive decisions can be obtained to deal with limited water resources, and also that this new method is more appropriate than other traditional MCDM methods in systems engineering since it takes the optimism/pessimism attitude into account in a quantifiable way. The comparison of the computational results with the current state of the projects shows an optimistic attitude of the real decision makers. The authors present a sensitivity analysis of how the rankings of the water projects depend on the optimism degree.

Q. Ji, P. Haase and G. Qi ("Combination of similarity measures in ontology matching using the OWA operator") provide a novel solution for ontology matching by using the ordered weighted average (OWA) operator to aggregate multiple values obtained from different similarity measures. They have implemented the solution in the ontology matching system FOAM. Using the similarity measures in FOAM, the authors analyze how to choose different OWA operators and compare their results with others.

We wish to thank all the contributors for their excellent work. All the contributions were anonymously peer reviewed by at least two reviewers, and we also wish to express our thanks to them. We hope that the volume will be interesting and useful to the entire intelligent systems research community, as well as other communities in which people may find the presented tools and techniques useful to formulate and solve their specific problems.

We also wish to thank Dr. Tom Ditzinger and Ms. Heather King from Springer for their multifaceted support and encouragement.

July 2010

Ronald R. Yager
Janusz Kacprzyk
Gleb Beliakov

Contents

Part I: Methods

Part II: Applications

Part I
Methods

OWA Operators and Nonadditive Integrals

Michel Grabisch

Abstract. We give a survey on the relations between nonadditive integrals (Choquet integral, Sugeno integral) and the OWA operator and its variants. We give also some behavioral indices for the OWA operator, as orness, veto and favor indices, etc. Finally, we propose the use of p-symmetric capacities for a natural generalization of the OWA operator.

1 Introduction

This paper offers a survey on the relations between the OWA operators (its classical definition and its variants) and the so-called fuzzy integrals (or more exactly nonadditive integrals).

Although the fact that the original OWA operator was a particular case of Choquet integral was discovered, with some surprise, only several years after its birth in 1988[1], their close relation appears rather obviously if one considers that both operators are linear up to a rearrangement of the arguments in increasing order. Later variants of the original definition, since still based on some rearrangement of the arguments, remain closely related to nonadditive integrals.

One may then consider that, since all OWA operators are more or less nonadditive integrals, these operators are no longer useful and there is no need to consider them any more. On the contrary, they provided useful and meaningful families of operators among the vast and unexplored realm of aggregation operators based on nonnaditive integrals (we refer the reader to some chapters of the recent monograph [10] for a detailed account of this

Michel Grabisch
Centre d'Economie de la Sorbonne, Université Paris I
106-112, Bd de l'Hôpital, 75013 Paris, France
e-mail: michel.grabisch@univ-paris1.fr

[1] Up to our knowledge, this was noticed by Murofushi and Sugeno in a 1993 paper [16].

R.R. Yager et al. (Eds.): Recent Developments in the OWA Operators, STUDFUZZ 265, pp. 3–15.
springerlink.com

question). In this respect, OWA operators provide aggregation operators with a clear interpretation. Moreover, by some crossfertilization process, many indices defined for nonadditive integrals can be applied to OWA operators to bring new insights into their behavioral properties.

Our survey will try to emphasize these issues. Due to size limitation, results are given without proofs. The reader is referred to the bibliography for more details.

2 Capacities and Nonadditive Integrals

Let us denote by $N := \{1, \ldots, n\}$ the index set of arguments to be aggregated (scores, utilities, etc.). For simplicity we consider here that scores to be aggregated lie in $[0, 1]$. Hence, all integrals will be defined for functions $f : N \to [0, 1]$, thus assimilated to vectors in $[0, 1]^n$.

In the whole paper, we use $\wedge, \vee$ for min and max.

Definition 1. A *capacity* [2] or *fuzzy measure* [20] on N is a mapping $\mu : 2^N \to [0, 1]$ satisfying

(i) $\mu(\emptyset) = 0$, $\mu(N) = 1$ (normalization)
(ii) $A \subseteq B$ implies $\mu(A) \leq \mu(B)$ (monotonicity).

Definition 2. A capacity μ on N is *symmetric* if for all $A, B \in 2^N$ such that $|A| = |B|$, we have $\mu(A) = \mu(B)$.

Definition 3. Let μ be a capacity on N. The *dual* (or *conjugate*) capacity $\overline{\mu}$ is a capacity on N defined by

$$\overline{\mu}(A) = 1 - \mu(\overline{A}), \quad \forall A \subseteq N$$

where $\overline{A} := N \setminus A$.

Definition 4. Let μ be a capacity on N. The *Möbius transform* [18] of μ is a mapping $m^\mu : 2^N \to \mathbb{R}$ defined by, for any $A \subseteq N$:

$$m^\mu(A) := \sum_{B \subseteq A} (-1)^{|A \setminus B|} \mu(B).$$

If there is no fear of ambiguity, we drop the superscript μ in m^μ. Given m^μ, it is possible to recover μ by the inverse transform (called *Zeta transform*):

$$\mu(A) = \sum_{B \subseteq A} m^\mu(B)$$

for any $A \subseteq N$.

Another interesting transform is the interaction transform [8, 9].

Definition 5. Let μ be a capacity on N. The *interaction transform* of μ is a mapping $I^\mu : 2^N \to \mathbb{R}$ defined by, for any $A \subseteq N$:

$$I^\mu(A) := \sum_{B \subset N \setminus A} \frac{(n-b-a)!b!}{(n-a+1)!} \sum_{K \subset A} (-1)^{a-k} v(K \cup B),$$

with a, b, k the cardinalities of A, B, K respectively.

For the following definition and for the rest of the paper we introduce the following notation: for any function $f : N \to [0,1]$, we denote $f(i)$ by f_i for all $i \in N$, thus assimilating f to a vector in $[0,1]^n$. Moreover, $f_{(i)}$ indicates the ith smallest value of f, that is, $(\cdot)$ indicates a permutation on N (nonnecessarily unique) such that

$$f_{(1)} \leq f_{(2)} \leq \cdots \leq f_{(n)}.$$

Definition 6. Let μ be a capacity on N and $f \in [0,1]^n$.

(i) The *Choquet integral* [2] of f w.r.t. μ is defined by

$$(C) \int f \, d\mu := \sum_{i=1}^{n} (f_{(i)} - f_{(i-1)}) \mu(A_i)$$

(ii) The *Sugeno integral* [20] of f w.r.t. μ is defined by

$$(S) \int f \, d\mu := \bigvee_{i=1}^{n} (f_{(i)} \wedge \mu(A_i)).$$

In the above expressions, $A_i := \{(i), \ldots, (n)\}$, and $f_{(0)} := 0$.

The Choquet integral can be equivalently written as follows:

$$(C) \int f \, d\mu = \sum_{i=1}^{n} f_{(i)} (\mu(A_i) - \mu(A_{i+1})) \tag{1}$$

with $A_{n+1} := \emptyset$. As we work on finite spaces and consider nonadditive integrals as aggregation operators, we prefer to use the notation $\mathcal{C}_\mu(f)$ instead of $(C) \int f \, d\mu$, and $\mathcal{S}_\mu(f)$ instead of $(S) \int f \, d\mu$.

3 Aggregation Operators and Main Classes of OWA Operators

As we work on $[0,1]$ only and with a fixed arity, the definition of an aggregation operator can be simplified as follows: a mapping $\mathsf{A} : [0,1]^n \rightarrow [0,1]$ is an *aggregation operator* (of n arguments) if it satisfies:

(i) $\mathsf{A}(0,\dots,0) = 0$, $\mathsf{A}(1,\dots,1) = 1$
(ii) A is nondecreasing with respect to each argument.

(for recent monographs on aggregation operators, see [1, 10]) The above defined Choquet and Sugeno integrals are aggregation operators.

Definition 7. An aggregation operator A on $[0,1]^n$ is *symmetric* if for any permutation σ on N and each $(x_1,\dots,x_n) \in [0,1]^n$, it holds

$$\mathsf{A}(x_1,\dots,x_n) = \mathsf{A}(x_{\sigma(1)},\dots,x_{\sigma(n)}).$$

Definition 8. Let A be an aggregation operator on $[0,1]^n$. Then the *dual* aggregation operator A^d is defined by:

$$A^d(x_1,\dots,x_n) = 1 - \mathsf{A}(1-x_1,\dots,1-x_n), \quad \forall (x_1,\dots,x_n) \in [0,1]^n.$$

Proposition 1. The Choquet integral w.r.t. a capacity μ is symmetric if and only if μ is symmetric. The same holds for the Sugeno integral.

Definition 9. Let $(w_1,\dots,w_n) \in [0,1]^n$ such that $\sum_{i=1}^n w_i = 1$ (*additive weight vector*). The *ordered weighted average* (OWA) [22] is an aggregation operator defined by, for all $(x_1,\dots,x_n) \in [0,1]^n$:

$$\mathsf{OWA}_w(x_1,\dots,x_n) := \sum_{i=1}^n w_i x_{(i)}$$

(recall that $(\cdot)$ means $x_{(1)} \leq \dots \leq x_{(n)}$).

Obviously, the OWA operator is symmetric.

Definition 10. Let $(w_1,\dots,w_n) \in [0,1]^n$ such that $\bigvee_{i=1}^n w_i = 1$ (*maxitive weight vector*). Then, for any $(x_1,\dots,x_n) \in [0,1]^n$

(i) The *ordered weighted maximum* [3] is an aggregation operator defined by

$$\mathsf{OWMax}_w(x_1,\dots,x_n) := \bigvee_{i=1}^n (w_i \wedge x_{(i)}),$$

with $w_1 \geq w_2 \geq \dots \geq w_n$.

(ii) The *ordered weighted minimum* [3] is an aggregation operator defined by

$$\mathsf{OWMin}_w(x_1, \dots, x_n) := \bigwedge_{i=1}^{n} ((1 - w_i) \vee x_{(i)}),$$

with $w_1 \leq w_2 \leq \cdots \leq w_n$.

In the case of the ordered weighted maximum, since

$$\bigvee_{i=1}^{n} (w_i \wedge x_{(i)}) = \bigvee_{i=1}^{n} \Big(\Big(\bigvee_{k=i}^{n} w_k \Big) \wedge x_{(i)} \Big)$$

the assumption $w_1 \geq w_2 \geq \cdots \geq w_n$ is not necessary, however it is useful in the next proposition. The same remark applies to the ordered weighted minimum.

Proposition 2. Let μ be a capacity. The following holds.

(i) $\mathcal{C}_\mu = \mathsf{OWA}_w$ if and only if μ is symmetric, with $w_i = \mu(C_{n-i+1}) - \mu(C_{n-i})$, $i = 2, \dots, n$, and $w_1 = 1 - \sum_{i=2}^{n} w_i$, where C_i is any subset of X with $|C_i| = i$ (equivalently, $\mu(A) = \sum_{j=0}^{i-1} w_{n-j}$, $\forall A, |A| = i$).

(ii) $\mathcal{S}_\mu = \mathsf{OWMax}_w$ if and only if μ is a symmetric capacity such that $\mu(A) = w_{n-|A|+1}$, for any $A \subseteq N$, $A \neq \emptyset$.

(iii) $\mathcal{S}_\mu = \mathsf{OWMin}_w$ if and only if μ is a symmetric capacity such that $\mu(A) = 1 - w_{n-|A|}$, for any $A \subsetneq N$.

Results (ii) and (iii) clearly show that the class of OWMin_w and OWMax_w coincide.

A weighted (hence nonsymmetric) version of OWA has been proposed by Torra [21].

Definition 11. Let $(w_1, \dots, w_n) \in [0,1]^n$ and $(p_1, \dots, p_n) \in [0,1]^n$ be two additive weight vectors (i.e., $\sum_{i=1}^{n} w_i = \sum_{i=1}^{n} p_i = 1$. Then for any $(x_1, \dots, x_n) \in [0,1]^n$, the *weighted OWA operator* is defined by

$$\mathsf{WOWA}_{w,p}(x_1, \dots, x_n) := \sum_{i=1}^{n} u_i x_{(i)},$$

where the weights $u_1, \dots, u_n$ are defined by

$$u_i = g\Big(\sum_{j \in A_i} p_j \Big) - g\Big(\sum_{j \in A_{i+1}} p_j \Big),$$

where $A_i := \{(i), \dots, (n)\}$ as before, and $g : [0,1] \to [0,1]$ is a nondecreasing function such that

- $g(0) := 0$, $g(\frac{i}{n}) := \sum_{j=n-i+1}^{n} w_j$ (hence $g(1) = 1$);
- g is linear if the points $(\frac{i}{n}, \sum_{j=n-i+1}^{n} w_j)$ lie on a straight line.

Note that the second condition happens if and only if $w_i = 1/n$ for all i. This entails that in this case, the weighted arithmetic mean with weight vector $(p_1, \ldots, p_n)$ is recovered. If $p_i = 1/n$ for $i = 1, \ldots, n$, then the usual OWA operator with weight vector $(w_1, \ldots, w_n)$ is recovered.

This complicated definition hides in fact a Choquet integral w.r.t. a distorted probability. Consider $g : [0, 1] \to [0, 1]$ a nondecreasing function such that $g(0) = 0$ and $g(1) = 1$, and a probability measure P defined on $(N, 2^N)$, with density $p_i = P(\{i\})$, $i = 1, \ldots, n$. Let us consider the *distorted probability* $\mu := g \circ P$, i.e.,

$$\mu(A) = g(\sum_{i \in A} p_i).$$

By the properties of g, clearly μ is a capacity on N. Using (1), the Choquet integral of any $x \in [0, 1]^n$ w.r.t. μ reads:

$$\begin{aligned}\mathcal{C}_{g \circ P}(f) = x_{(1)}(g(p_1 + \cdots p_n) - g(p_2 + \cdots + p_n)) + \cdots \\ \cdots + x_{(n-1)}(g(p_{n-1} + p_n) - g(p_n)) + x_{(n)} g(p_n) = \\ = \mathsf{WOWA}_{w,p}(x_1, \ldots, x_n). \quad (2)\end{aligned}$$

Note that the function g is however slightly restricted in the definition of $\mathsf{WOWA}_{w.p}$, hence the equivalence between the weighted OWA operators and the Choquet integral w.r.t. a distorded probability is not exact (the latter is more general).

4 Mathematical and Behavioral Properties of the OWA Operator

In the following, we restrict to the classical OWA operator. We begin by a result showing that the class of OWA operator is closed under duality.

Proposition 3. [7] Let OWA_w be an OWA operator with weight vector $(w_1, \ldots, w_n)$, with associated capacity μ. Then the dual capacity $\overline{\mu}$ corresponds also to an OWA operator $\mathsf{OWA}_{w'}$, with weight vector $w' = (w_n, \ldots, w_n)$, i.e., the OWA operator with the reversed weight vector. Moreover, for every $(x_1, \ldots, x_n) \in [0, 1]^n$

$$\mathrm{OWA}_{w_n, \ldots, w_1}(x_1, \ldots, x_n) = 1 - \mathrm{OWA}_{w_1, \ldots, w_n}(1 - x_1, \ldots, 1 - x_n),$$

i.e., it is the dual aggregation operator of OWA_w.

We turn now to the Möbius and interaction representations of OWA operators.

Proposition 4. [7] Let OWA_w be an OWA operator with weight vector $(w_1, \dots, w_n)$, and let μ be the associated capacity. Then its Möbius and interaction representations are given by:

$$m^\mu(A) = \sum_{i=1}^{|A|}(-1)^{|A|-i} w_{n-i+1}\binom{|A|-1}{i-1}, \tag{3}$$

$$I^\mu(A) = \frac{1}{n-|A|+1}\sum_{j=0}^{|A|-2}(-1)^j\binom{|A|-2}{j}[w_{j+1} - w_{n-|A|+j+2}], \tag{4}$$

for all $A \subset N$.

The interaction transform provides two meaningful indices describing the behavior of any aggregation function based on capacities. The first one is the Shapley value [19], obtained when A is a singleton:

$$\phi_i(\mu) := I^\mu(\{i\}) = \sum_{B \subseteq N\setminus i} \frac{b!(n-b-1)!}{n!}(\mu(B\cup i) - \mu(B)), \quad \forall i \in N.$$

The Shapley value $\phi_i(\mu)$ expresses the relative importance of dimension (let us call it criterion in this section) i in the aggregation. We find for the operator OWA_w that

$$\phi_i(\mu_w) = \frac{1}{n}, \quad \forall i \in N$$

where μ_w is the capacity corresponding to OWA_w. This result is immediate from the fact that μ_w is a symmetric capacity, and the Shapley value is constant for any symmetric capacity.

The second index provided by I^μ is the interaction index of Murofushi and Soneda [15] (also proposed by Owen as the covalue of μ [17]), obtained when A is a pair:

$$I_{ij}(\mu) := I^\mu(\{i,j\}) = \sum_{B\subseteq N\setminus\{i,j\}} \frac{b!(n-b-2)!}{(n-1)!}(\mu(B\cup\{i,j\}) - \mu(B\cup i) - \mu(B\cup j) + \mu(B)).$$

The interaction index describes the contribution to the overall score when two criteria i, j are both satisfied compared to the individual contribution of criteria i and j. We distinguish three cases [5]:

- positive interaction: in average, the simultaneous satisfaction of criteria i and j is more rewarding (for the overall score) than the sum of separate satisfactions of i and j. Criteria are said to be *complementary* in this case: the aggregation for i and j is of conjunctive type.
- negative interaction: in average, the sum of separate satisfactions of i and j is more rewarding than the simultaneous satisfaction of i and j. Criteria are said to be *substitutive*: the aggregation for i and j is of disjunctive nature.

- null interaction: in average, the sum of separate satisfactions of i and j equals the simultaneous satisfaction of i and j. Criteria are said to be *independent*: the aggregation for i and j is of additive nature.

The interaction index for the operator OWA_w with weight vector $(w_1, \ldots, w_n)$ is:

$$I_{ij}(\mu_w) = \frac{w_1 - w_n}{n-1}, \quad i, j = 1, \ldots, n, i \neq j.$$

Hence, as for the Shapley value, the interaction index is constant, and its value depends only on extreme weights w_1, w_n. If $w_1 > w_n$, then the interation is positive (the OWA operator is more of the min type), and if $w_1 < w_n$, the interaction is negative (the OWA is more of the max type).

The Shapley value and the interaction index provide two useful ways to describe the behavior of aggregation functions. In the rest of this paragraph, we present other indices describing behavior (see [10, Ch. 10] for a full description). The minimum operator is denoted by Min, and the maximum operator by Max.

Definition 12. [4] Let A be an aggregation function. The *orness value* of A is defined by:

$$\mathrm{orness}(\mathsf{A}) := \frac{\overline{\mathsf{A}} - \overline{\mathsf{Min}}}{\overline{\mathsf{Max}} - \overline{\mathsf{Min}}} = -\frac{1}{n-1} + \frac{n+1}{n-1}\overline{\mathsf{A}},$$

where, letting $x := (x_1, \ldots, x_n)$,

$$\overline{A} := \int_{[0,1]^n} \mathsf{A}(x)dx$$

is the mean value of A, and similarly for Min and Max.

It is shown in [11] that if A is the Choquet integral, then

$$\begin{aligned} \mathrm{orness}(\mathcal{C}_\mu) &= \sum_{\substack{K \subseteq N \\ 0 < |K| < n}} \frac{1}{(n-1)\binom{n}{|K|}} \mu(K) \\ &= \sum_{K \subseteq N} \frac{n - |K|}{(n-1)(|K|+1)} m^\mu(K). \end{aligned}$$

Applying this to the OWA operator yields

$$\mathrm{orness}(\mathsf{OWA}_w) = \frac{1}{n-1} \sum_{i=1}^{n} (i-1) w_i.$$

Next we introduce the veto and favor indices. The origin of these indices goes back to the idea of veto criterion and favor criterion proposed by the author [6]. A criterion i is a *veto* for the aggregation operator A if $\mathsf{A}(x_1,\ldots,x_n) \leq x_i$ for all $(x_1,\ldots,x_n) \in [0,1]^n$. Similarly i is a favor if $\mathsf{A}(x_1,\ldots,x_n) \geq x_i$ for all $(x_1,\ldots,x_n) \in [0,1]^n$. These condition being rarely satisfied (note that they are satisfied for min and max: every criterion is a veto for min, and a favor for max), Marichal proposed the concept of veto and favor indices, describing how close to a pure veto or favor effect is a criterion [11, 12].

Definition 13. Let A be an aggregation operator. The *veto* and *favor* indices of a criterion j w.r.t. A are defined by:

$$\text{veto}(\mathsf{A},j) := \frac{\overline{\mathsf{Max}(0_j x_{-j})} - \overline{\mathsf{A}(0_j x_{-j})}}{\overline{\mathsf{Max}(0_j x_{-j})} - \overline{\mathsf{Min}(0_j x_{-j})}} = 1 - \frac{n}{n-1}\overline{\mathsf{A}(0_j x_{-j})},$$

$$\text{favor}(\mathsf{A},j) := \frac{\overline{\mathsf{A}(1_j x_{-j})} - \overline{\mathsf{Min}(1_j x_{-j})}}{\overline{\mathsf{Max}(1_j x_{-j})} - \overline{\mathsf{Min}(1_j x_{-j})}} = \frac{n}{n-1}\overline{\mathsf{A}(1_j x_{-j})} - \frac{1}{n-1}.$$

The notation $\overline{\mathsf{A}(0_j x_{-j})}$ stands for the average of $\mathsf{A}(x)$ for all vector $x \in [0,1]^n$ whose jth component is 0 (and similarly for the others).

It is shown in [11, 12] that if A is the Choquet integral, then

$$\text{veto}(\mathcal{C}_\mu,j) = 1 - \sum_{K\subseteq N\setminus\{j\}} \frac{1}{(n-1)\binom{n-1}{|K|}}\mu(K)$$

$$\text{favor}(\mathcal{C}_\mu,j) = \sum_{K\subseteq N\setminus\{j\}} \frac{1}{(n-1)\binom{n-1}{|K|}}\mu(K\cup\{j\}) - \frac{1}{n-1}$$

and

$$\frac{1}{n}\sum_{i=1}^{n}\text{veto}(\mathcal{C}_\mu,j) = \text{andness}(\mathcal{C}_\mu),$$

$$\frac{1}{n}\sum_{i=1}^{n}\text{favor}(\mathcal{C}_\mu,j) = \text{orness}(\mathcal{C}_\mu).$$

The following proposition shows that it is only necessary to consider one of the two indices.

Proposition 5. For any aggregation operator A and criterion $j \in N$, it holds

(i) $\text{veto}(\mathsf{A}^d,j) = \text{favor}(\mathsf{A},j)$ and $\text{favor}(\mathsf{A}^d,j) = \text{veto}(\mathsf{A},j)$.
(ii) $\text{veto}(\mathsf{A},j) + \text{favor}(\mathsf{A},j) = 1 + \frac{n\phi_j(\mathsf{A})-1}{n-1}$.

Applying this to the OWA operator yields

$$\text{favor}(\mathsf{OWA}_w,j) = \frac{1}{n-1}\sum_{i=1}^{n}(i-1)w_i = \text{orness}(\mathsf{OWA}_w).$$

A related notion, although orthogonal in some sense, is the notion of k-conjunctivity and k-disjunctivity [12]. An aggregation operator A is k-conjunctive (respectively, k-disjunctive) for some $k \in \{1,\dots,n\}$ if $\mathsf{A}(x_1,\dots,x_n) \leq x_{(k)}$ (respectively, $\mathsf{A}(x_1,\dots,x_n) \geq x_{(n-k+1)}$). Note that OWA operators such that $w_i = 1$ for some $i \in N$ are both $(n-i+1)$-disjunctive and i-conjunctive.

As for veto and favor, it is more interesting to introduce indices to measure to which degree an aggregation operator is close to k-disjunctiveness or k-conjunctiveness.

Definition 14. Let A be an aggregation operator and k be an integer, $1 \leq k < n$. The *k-conjunctiveness* and *k-disjunctiveness* indices of A are defined by:

$$\mathrm{conj}_k(\mathsf{A}) := \frac{n-k+1}{n-k}\frac{1}{\binom{n}{k}}\sum_{\substack{K\subseteq N\\|K|=k}}\overline{\mathsf{A}(0_K x_{-K})},$$

$$\mathrm{disj}_k(\mathsf{A}) := \frac{n-k+1}{n-k}\frac{1}{\binom{n}{k}}\sum_{\substack{K\subseteq N\\|K|=k}}\overline{\mathsf{A}(1_K x_{-K})} - \frac{1}{n-k}.$$

Similarly to Definition 13, $\overline{\mathsf{A}(0_K x_{-K})}$ stands for the average of $\mathsf{A}(x)$ for all vectors $x \in [0,1]$ whose all components in K are 0.

Considering the Choquet integral, we have the following result [12]:

Proposition 6. For any capacity μ on N and $1 \leq k < n$, we have

$$\mathrm{conj}_k(\mathcal{C}_\mu) = 1 - \frac{1}{n-k}\sum_{j=0}^{n-k}\frac{1}{\binom{n}{j}}\sum_{\substack{J\subseteq N\\|J|=j}}\mu(J)$$

$$\mathrm{disj}_k(\mathcal{C}_\mu) = \frac{1}{n-k}\sum_{j=k}^{n}\frac{1}{\binom{n}{j}}\sum_{\substack{J\subseteq N\\|J|=j}}\mu(J) - \frac{1}{n-k}.$$

5 A Generalization of OWA: The p-Symmetric Choquet Integral

Proposition 2 shows the equivalence of the OWA operator with symmetric Choquet integrals, themselves being bijectively related to symmetric capacities. Hence, the idea of total symmetry in the weights is the basis of the OWA operator. The weighted OWA operator presented in Definition 11 is an attempt to escape total symmetry by introducing weights invidually on criteria. The weighted OWA operator can then be considered as the crossover of the original OWA operator and the good old weighted arithmetic mean.

Perhaps a generalization closer to the spirit of OWA would be to keep this idea of symmetry while just weakening it gradually. We think that this idea is well captured by the concept of p-symmetric capacities, proposed by Miranda and Grabisch [14, 13]. The basic concept for defining p-symmetric capacities is the one of subset of indifference.

Definition 15. Given a subset A of N, we say that A is a **subset of indifference** for a capacity μ over N if $\forall B_1, B_2 \subseteq A$, $|B_1| = |B_2|$, and $\forall C \subseteq N \setminus A$, we have

$$\mu(B_1 \cup C) = \mu(B_2 \cup C).$$

In words, inside a subset of indifference A, all is symmetric in the usual sense. It follows that for a symmetric capacity, any subset is a subset of indifference. However, observe that if A is a subset of indifference, then any subset of A is also a subset of indifference. Therefore, only the maximal subsets of indifference matter. On the other hand, observe that trivially any singleton is a subset of indifference for any capacity. From these two observations, it follows that for any capacity, the universe N can be partitioned into maximal subsets of indifference, say $B_1, \ldots, B_p$. This partition $\{B_1, \ldots, B_p\}$ is called the *basis* of μ, and leads to the following definition.

Definition 16. A capacity μ on N is said to be *p-symmetric* if its basis has p blocks (subsets).

Clearly, a capacity which is symmetric in the usual sense is 1-symmetric, and its basis is $\{N\}$. A capacity which has no symmetry property is n-symmetric, and its basis is $\{\{1\}, \ldots, \{n\}\}$.

The number of coefficients which are necessary to define a p-symmetric capacity depends on the basis. Specifically, if the basis is $\{B_1, \ldots, B_p\}$, then we need $(|B_1| + 1) \times \cdots \times (|B_p| + 1) - 2$ coefficients.

The "p-symmetric OWA" would be then the Choquet integral w.r.t. a p-symmetric capacity. Its expression is given by the following proposition.

Proposition 7. Let μ be a p-symmetric capacity on N with basis $\{B_1, \ldots, B_p\}$. Then, for all $x = (x_1, \ldots, x_n) \in [0, 1]^n$, the Choquet integral w.r.t. μ is given by

$$\mathcal{C}_\mu(x) = \sum_{i=1}^{p} \mathcal{C}_{\mu_{|B_i}}(x_{|B_i}) + \sum_{B \not\subseteq B_j, \forall j} m^\mu(B) \bigwedge_{i \in B} x_i,$$

where $\mu_{|B_i}$ is the restriction of μ to B_i, i.e.,

$$\mu_{|B_i}(C) := \mu(C), \ \forall C \subseteq B_i,$$

and $x_{|B_i}$ is the restriction of x to B_i.

(Note: the above expression is slightly simpler than the one given in [14]). An important observation is that for $i = 1, \ldots, p$, $\mu_{|B_i}$ is a symmetric non-normalized capacity on B_i, therefore $\mathcal{C}_{\mu_{|B_i}}$ is a classical OWA operator on B_i, with nonnegative weights $w_{i_1}, \ldots, w_{i_{|B_i|}}$ satisfying $\sum_{j=1}^{|B_i|} w_{i_j} = \mu(B_i)$.

References

[1] Beliakov, G., Pradera, A., Calvo, T.: Aggregation Functions: a Guide for Practitioners. Springer, Heidelberg (2007)
[2] Choquet, G.: Theory of capacities. Annales de l'Institut Fourier 5, 131–295 (1953)
[3] Dubois, D., Prade, H., Testemale, C.: Weighted fuzzy pattern matching. Fuzzy Sets & Systems 28, 313–331 (1988)
[4] Dujmović, J.J.: Weighted conjunctive and disjunctive means and their application in system evaluation, pp. 147–158. Univ. Beograd. Publ, Elektrotechn. Fak (1974)
[5] Grabisch, M.: The application of fuzzy integrals in multicriteria decision making. European J. of Operational Research 89, 445–456 (1996)
[6] Grabisch, M.: Alternative representations of discrete fuzzy measures for decision making. Int. J. of Uncertainty, Fuzziness, and Knowledge Based Systems 5, 587–607 (1997)
[7] Grabisch, M.: Alternative representations of OWA operators. In: Yager, R., Kacprzyk, J. (eds.) The Ordered Weighted Averaging Operators: Theory, Methodology, and Practice, pp. 73–85. Kluwer Academic, Dordrecht (1997)
[8] Grabisch, M.: k-order additive discrete fuzzy measures and their representation. Fuzzy Sets and Systems 92, 167–189 (1997)
[9] Grabisch, M.: The interaction and Möbius representations of fuzzy measures on finite spaces, k-additive measures: a survey. In: Grabisch, M., Murofushi, T., Sugeno, M. (eds.) Fuzzy Measures and Integrals — Theory and Applications, pp. 70–93. Physica Verlag, Heidelberg (2000)
[10] Grabisch, M., Marichal, J.-L., Mesiar, R., Pap, E.: Aggregation functions. Encyclopedia of Mathematics and its Applications, vol. 127. Cambridge University Press, Cambridge (2009)
[11] Marichal, J.-L.: Tolerant or intolerant character of interacting criteria in aggregation by the Choquet integral. Eur. J. of Operational Research 155(3), 771–791 (2004)
[12] Marichal, J.-L.: k-intolerant capacities and Choquet integrals. Eur. J. of Operational Research 177(3), 1453–1468 (2007)
[13] Miranda, P., Grabisch, M.: p-symmetric bi-capacities. Kybernetika 40(4), 421–440 (2004)
[14] Miranda, P., Grabisch, M., Gil, P.: p-symmetric fuzzy measures. Int. J. of Uncertainty, Fuzziness, and Knowledge-Based Systems 10(suppl.), 105–123 (2002)
[15] Murofushi, T., Soneda, S.: Techniques for reading fuzzy measures (III): interaction index. In: 9th Fuzzy System Symposium, Sapporo, Japan, pp. 693–696 (1993) (in Japanese)
[16] Murofushi, T., Sugeno, M.: Some quantities represented by the Choquet integral. Fuzzy Sets & Systems 56, 229–235 (1993)
[17] Owen, G.: Multilinear extensions of games. Management Sci. 18, 64–79 (1972)
[18] Rota, G.C.: On the foundations of combinatorial theory I. Theory of Möbius functions. Zeitschrift für Wahrscheinlichkeitstheorie und Verwandte Gebiete 2, 340–368 (1964)

[19] Shapley, L.S.: A value for n-person games. In: Kuhn, H.W., Tucker, A.W. (eds.) Contributions to the Theory of Games, Vol. II. Annals of Mathematics Studies, vol. 28, pp. 307–317. Princeton University Press, Princeton (1953)

[20] Sugeno, M.: Theory of fuzzy integrals and its applications. PhD thesis, Tokyo Institute of Technology (1974)

[21] Torra, V.: The weighted OWA operator. Int. J. of Intelligent Systems 12, 153–166 (1997)

[22] Yager, R.R.: On ordered weighted averaging aggregation operators in multicriteria decision making. IEEE Trans. Systems, Man & Cybern. 18, 183–190 (1988)

The WOWA Operator: A Review

Vicenç Torra

Abstract. The WOWA operator (Weighted OWA) was proposed as a generalization of both the OWA and the Weighted mean. Formally, it is an aggregation operator that permits the aggregation of a set of numerical data with respect to two weighting vectors: one corresponding to the one of the weighted mean and the other corresponding to the one of the OWA. In this chapter we review this operator as well as some of its main results.

1 Introduction

Aggregation operators [20, 21] permit us to combine data provided from several sources and return a single datum that is of better quality and, therefore, gives more accurate information. Several aggregation operators have been defined in the literature. Differences in the operators are based on the differences between the data, and the properties of these data.

A common classification of the operators is related to the nature of the data. In this way, we can distinguish between numerical data, categorical data, and also between data in other terms as e.g. partitions, (fuzzy) clusters, dendrograms, sequences.

In this chapter we will review some results about the WOWA operator. This operator, introduced in [13] and [14], was defined for numerical data.

From a practical point of view, this operator was defined to encompass in a single operator the advantages of the weighted mean and of the OWA operator. Informally, the weighted mean permits us to weight the information sources, and the OWA permits us to represent a compensation degree, or, alternatively, to give importance to the data according to their values.

Vicenç Torra
IIIA-CSIC, Campus UAB s/n, 08193 Bellaterra,
Catalonia, Spain
e-mail: vtorra@iiia.csic.es

R.R. Yager et al. (Eds.): Recent Developments in the OWA Operators, STUDFUZZ 265, pp. 17–28.
springerlink.com

From a mathematical point of view, the operator is a generalization of both the weighted mean and the OWA. That is, particular parametrizations of the operator lead to either the weighted mean or the OWA. In addition, it has also been proven [15] that the operator is a particular case of the Choquet integral [4].

In this paper we review some of the main results on this operator. In Section 2 we review the WOWA operator as well as other aggregation operators that are related to the WOWA. In Section 3 we discuss some generalizations of the operator. In Section 4 we review some learning approaches for the parameters of this operator.

2 The WOWA Operator and Other Aggregation Operators

Aggregation operators are functions that combine N different data into a single datum. We use $\mathbb{C}$ from $\mathbb{C}$onsensus or $\mathbb{C}$ombination to represent them. Then, in general, it is assumed that the aggregation of $a_1, \ldots, a_N$ in a given domain D is $\mathbb{C}(a_1, \ldots, a_N)$, also in this domain D. That is, $\mathbb{C} : D^N \to D$.

In some cases it is useful to represent in an explicit way where the data come from. That is, which is the information source that has supplied each data. We will use $X = \{x_1, \ldots, x_N\}$ to represent the set of information sources. Then, we will use f to represent the relationship between x_i and the supplied value a_i. That is, $f(x_i) = a_i$ represents that x_i supplies a_i. Using this notation, we have that the aggregation of the data supplied by the information sources in X is $\mathbb{C}(f(x_1), \ldots, f(x_N))$, or, with an abuse of notation, $\mathbb{C}(f)$.

There exist a few different definitions on what an aggregation function is. In general it is usual to require monotonicity and unanimity or idempotency (for at least a few elements in the domain). We consider aggregation operators as functions $\mathbb{C}$ satisfying:

- **Unanimity or idempotency:** $\mathbb{C}(a, \ldots, a) = a$ for all a in D
- **Monotonicity:** $\mathbb{C}(a_1, \ldots, a_N) \geq \mathbb{C}(a'_1, \ldots, a'_N)$ when $a_i \geq a'_i$

Some require unanimity only in the boundaries of D. In particular, if $D = [0, 1]$, unanimity is only required for 0 and 1. So, $\mathbb{C}(0, \ldots, 0) = 0$ and $\mathbb{C}(1, \ldots, 1) = 1$. This is the case of [2]. In this case, t-norms and t-conorms are aggregation functions. In this case, the term *mean operators* is used to name functions that satisfy unanimity for all a in D.

In addition, in some circumstances the symmetry condition is also required to aggregation operators. This property, which is formalized below, implies that there is no distinguished data.

- **Symmetry:** For any permutation π on $\{1, \ldots, N\}$ it holds that

$$\mathbb{C}(a_1, \ldots, a_N) = \mathbb{C}(a_{\pi(1)}, \ldots, a_{\pi(N)})$$

2.1 Arithmetic Mean, Weighted Mean and OWA Operator

The arithmetic mean, the weighted mean and the OWA operator are some of the most well known aggregation operators. Both the weighted mean and the OWA combine a set of data with respect to a weighting vector. The arithmetic mean does not include any parameter.

The weighting vector in the weighted mean permits us to take into account some *a prior* knowledge, following artificial intelligence jargon. about the information sources. We give below the definitions of the weighting vector, the arithmetic mean and the OWA operator.

Definition 1. Let $A = (a_1, \ldots, a_N)$ be N data in $\mathbb{R}$. Then, we define a weighting vector, the arithmetic mean ($AM : \mathbb{R}^N \to \mathbb{R}$) of A, and the weighted mean (WM) of A with respect to a weighting vector as follows:

- A vector $v = (v_1 \ldots v_N)$ is a *weighting vector* of dimension N if and only if $v_i \in [0,1]$ and $\sum_i v_i = 1$.
- AM is an *arithmetic mean*, if $AM(a_1, ..., a_N) = (1/N)\sum_{i=1}^{N} a_i$.
- WM is the *weighted mean* with respect to a weighting vector $\mathbf{p}$, if $WM_{\mathbf{p}}(a_1, ..., a_N) = \sum_{i=1}^{N} p_i a_i$.

The OWA (Ordered Weighting Averaging) operator has a definition similar to the one of the weighted mean. It is as follows:

Definition 2. [22, 23] Let $\mathbf{w}$ be a weighting vector of dimension N; then, a mapping OWA: $\mathbb{R}^N \to \mathbb{R}$ is an *Ordered Weighting Averaging (OWA) operator* of dimension N if

$$OWA_{\mathbf{w}}(a_1, ..., a_N) = \sum_{i=1}^{N} w_i a_{\sigma(i)},$$

where $\{\sigma(1), ..., \sigma(N)\}$ is a permutation of $\{1, ..., N\}$ such that $a_{\sigma(i-1)} \geq a_{\sigma(i)}$ for all $i = \{2, ..., N\}$ (i.e., $a_{\sigma(i)}$ is the ith largest element in the collection $a_1, ..., a_N$).

The weighted mean and the OWA operator are similar operators as both are a linear combination of the values with respect to the weights. Nevertheless, the ordering step that takes place in the OWA operator makes a fundamental difference. This difference makes different the interpretation of the weights in both operators.

In the weighted mean, the weight is attached to the information source. Due to this, weights correspond to the importance of the information sources. E.g., when the data correspond to sensors, the weight might correspond to the reliability of the corresponding sensor; and when the data correspond to the evaluation of some criteria (or experts) in a multicriteria decision making problem, the weights correspond to the importance of the criteria (or of the experts).

In contrast, in the OWA operator, the weight is attached to the data, with respect to its relative position. Due to this, weights permit us to give more importance to e.g. low values, central values or high values. For example, we can give more importance to small distances (e.g., if we want to avoid a collision of a robot, is more importance a nearer object than a farther one), or permit some compensation (e.g., if a bad evaluation of a criteria can be compensated with a good one – or in the extreme case, if a single good criteria can override all the others).

The degree of compensation in the OWA operator is measured with the *orness* degree. This is a measure that evaluates in what extent the outcome of the aggregation is near to the maximum of the data being aggregated. The larger the outcome, the larger the orness and the larger the compensation. Note that the maximum compensation corresponds to assigning the largest value to the output of the function.

The orness is formally defined below and, in fact, the definition is valid for all aggregation operators $\mathbb{C}$ and for all parameterizations P. It results that for some of the operators, the orness does not depend on the particular parameterization selected, while for others the orness depends on the particular parameterization used. The weighted mean is an example of the former (i.e., the orness of the weighted mean is independent of the parameter used), and the OWA is an example of the latter (i.e., different parameters give different orness for the OWA).

Definition 3. Let $\mathbb{C}$ be an aggregation operator with parameters P; then, the *orness* of $\mathbb{C}_P$ is defined by

$$orness(\mathbb{C}_P) := \frac{AV(\mathbb{C}_P) - AV(\min)}{AV(\max) - AV(\min)}. \quad (1)$$

The orness of the aggregation operators reviewed above is as follows:

- $orness\,(AM) = 1/2$
- $orness(WM_{\mathbf{p}}) = 1/2$
- $orness(OWA_{\mathbf{w}}) = \frac{1}{N-1}\sum_{i=1}^{N}(N-i)w_i$

From the orness of the OWA we can infer that its maximum orness is 1 when $w_1 = 1$ and $w_i = 0$ for all $i \neq 1$ (note that in this case the OWA corresponds to the maximum), and that the minimum ornes is 0 when $w_N = 1$ and $w_i = 0$ for all $i \neq N$ (note that in this case the OWA corresponds to the minimum).

2.2 The WOWA Operator

Due to the fact that in some applications it is of interest to assign weight to information sources and also to the compensation degree (or the relative importance of values), a generalization of both weighted mean and OWA was proposed. This generalization is the WOWA operator. WOWA operator stands for Weighted Ordered Weighted Averaging operator. Formally, it is also a linear combination of values $a_1, \ldots, a_N$ with respect to weights. Nevertheless, these weights are computed taking into account two weighting vectors. One of the weighting vectors has the interpretation of the ones in the weighted mean, and the other has the interpretation of the ones in the OWA. We use here $\mathbf{p}$ to represent the weights with the interpretation used in the weighted mean, and $\mathbf{w}$ to represent the weights with the interpretation used in the OWA. Note that although we use here different letters $\mathbf{w}$ and $\mathbf{p}$, both weighting vectors have the same mathematical properties.

Definition 4. [13, 14] Let **p** and **w** be two weighting vectors of dimension N; then, a mapping WOWA: $\mathbb{R}^N \to \mathbb{R}$ is a *Weighted Ordered Weighted Averaging (WOWA) operator* of dimension N if

$$WOWA_{\mathbf{p},\mathbf{w}}(a_1, ..., a_N) = \sum_{i=1}^{N} \omega_i a_{\sigma(i)},$$

where σ is defined as in the case of OWA (*i.e.*, $a_{\sigma(i)}$ is the ith largest element in the collection $a_1, ..., a_N$), and the weight ω_i is defined as

$$\omega_i = w^*(\sum_{j \leq i} p_{\sigma(j)}) - w^*(\sum_{j<i} p_{\sigma(j)}),$$

with w^* being a nondecreasing function that interpolates the points

$$\{(i/N, \sum_{j \leq i} w_j)\}_{i=1,...,N} \cup \{(0,0)\}.$$

The function w^* is required to be a straight line when the points can be interpolated in this way.

As stated above, this definition uses an interpolation method to build a function from the points in the set $\{(i/N, \sum_{j \leq i} w_j)\}_{i=1,...,N} \cup \{(0,0)\}$. The original definition used the interpolation method described in [17]. Other interpolations approaches have been used as e.g. linear interpolation. A discussion on the effects of different interpolation methods is given in [19].

For details on the WOWA operator, and about the meaning of the function see [20, 21].

Some extensions have been defined for this operator. One of them, the Linguistic WOWA (L-WOWA) operator [14], was given to deal with categorical data. L-WOWA operator can be seen as an extension of the L-OWA, in the same way that the WOWA operator is an extension of the OWA operator.

2.3 *The Choquet Integral*

The Choquet integral is an operator that generalizes the WOWA operator, as proven in [15]. As the WOWA generalizes the arithmetic mean, the weighted mean and the OWA, it can be said that all these functions belong to the same family of operators.

The basis of this integral, in comparison with the other mentioned operators, is that now the *weights* (or importances) are not of a single information source but to a set of them. While in the weighted mean, we have p_i as the weight of information source x_i, we can not consider the weight of e.g. the set $\{x_1, x_4\}$. Formally, in the weighted mean we have weights $p : X \to [0,1]$ such that $\sum_{x_i \in X} p(x_i) = 1$, and we use the notation $p_i = p(x_i)$. Thus, $p_i = p(x_i)$ is the importance of information source x_i.

In contrast, in the case of the Choquet integral we use functions μ over subsets of X. Then, $\mu(\psi)$ for $\psi \subseteq X$ is the importance of the elements in ψ taken together.

As in the case of the weighting vectors, $\mu(\psi) \in [0,1]$. These functions are known as fuzzy measures, and we review them below.

Definition 5. A fuzzy measure μ on a set X is a set function $\mu : \wp(X) \to [0,1]$ satisfying the following axioms:

(i) $\mu(\emptyset) = 0$, $\mu(X) = 1$ (boundary conditions)
(ii) $A \subseteq B$ implies $\mu(A) \leq \mu(B)$ (monotonicity)

That is, μ are set functions that satisfy monotonicity. Monotonicity means that the larger the set, the larger the measure, or, equivalently, the larger the set of criteria, the larger their importance. In addition, the maximum importance (equal to 1) is achieved for the whole set of criteria, and the minimum importance (equal to 0) is achived for the empty set.

Choquet integrals permit to aggregate values taking into account the importance expressed in the measures. The aggregation corresponds to the integral of a function with respect to the measure. The function corresponds to the data to be aggregated as expressed above with the expression $\mathbb{C}(f)$.

Definition 6. [4] Let μ be a fuzzy measure on X; then, the *Choquet integral* of a function $f : X \to \mathbb{R}^+$ with respect to the fuzzy measure μ is defined by

$$(C)\int f d\mu = \sum_{i=1}^{N} [f(x_{s(i)}) - f(x_{s(i-1)})]\mu(A_{s(i)}), \tag{2}$$

where $f(x_{s(i)})$ indicates that the indices have been permuted so that $0 \leq f(x_{s(1)}) \leq \cdots \leq f(x_{s(N)}) \leq 1$, and where $f(x_{s(0)}) = 0$ and $A_{s(i)} = \{x_{s(i)}, \ldots, x_{s(N)}\}$.

When no confusion exists, we can use $CI_\mu(a_1, \ldots, a_N) = (C)\int f d\mu$, where, $f(x_i) = a_i$, as before. There are alternative expressions for the Choquet integral that are equivalent to the one given above. The next proposition presents one of them.

Proposition 1. *Let μ be a fuzzy measure on X; then, the Choquet integral of a function $f : X \to \mathbb{R}^+$ with respect to μ can be expressed as*

$$(C)\int f d\mu = \sum_{i=1}^{N} f(x_{\sigma(i)})[\mu(A_{\sigma(i)}) - \mu(A_{\sigma(i-1)})], \tag{3}$$

where $\{\sigma(1), \ldots, \sigma(N)\}$ is a permutation of $\{1, \ldots, N\}$ such that $f(x_{\sigma(1)}) \geq f(x_{\sigma(2)}) \geq \cdots \geq f(x_{\sigma(N)})$, where $A_{\sigma(k)} = \{x_{\sigma(j)} | j \leq k\}$ (or, equivalently, $A_{\sigma(k)} = \{x_{\sigma(1)}, \ldots, x_{\sigma(k)}\}$ when $k \geq 1$ and $A_{\sigma(0)} = \emptyset$).

As stated above, the WOWA operator is a particular case of the Choquet integral. In particular, a WOWA operator corresponds to a Choquet integral with respect to a distorted probability. Distorted probabilities are a particular type of fuzzy measure. All Choquet integral with respect to this type of fuzzy measures are equivalent to a WOWA operator, and all WOWA operators with weights **p** and **w** are equivalent to a Choquet integral with respect to the distorted probability constructed from **p** and the function w^* constructed using the interpolation method in the definition of the WOWA (Definition 4).

In the next two definitions, we review the definition of a distorted probability.

Definition 7. Let $P : 2^X \to [0,1]$ be a probability distribution. Then, we say that a function f is strictly increasing with respect to P if and only if

$$P(A) > P(B) \text{ implies } f(P(A)) > f(P(B))$$

At this point it is relevant to state that as we suppose that X is a finite set, when there is no restriction on the function f, a strictly increasing function f with respect to P can be regarded as a strictly increasing function on $[0,1]$. Note that with respect to increasingness only the points in $\{P(A)|A \in 2^X\}$ are essential, the others are not considered by $f(P(A))$.

Definition 8. [1, 3] Let μ be a fuzzy measure. We say that μ is a distorted probability if there exists a probability distribution P and a strictly increasing function f with respect to P such that $\mu = f \circ P$.

3 Generalizations of the WOWA Operator

In a recent paper [11], we introduced an extension of distorted probabilities. This was motivated by the fact that distorted probabilities is only a small fraction [7, 11] of all possible fuzzy measures. m-dimensional distorted probabilities permits us, with an appropriate value m, to represent all fuzzy measures.

These measures, together with m-symmetric ones, permit us to naturally extend WOWA and OWA operators into m-dimensional WOWA and m-dimensional OWA. The m-dimensional ones with $m = |X|$ are equivalent to a Choquet integral with an unconstrained fuzzy measure. That is, a Choquet integral with an arbitrary fuzzy measure. Definitions and results are reviewed in this section.

3.1 m-Dimensional Distorted Probabilities

We start defining m-dimensional distorted probabilities, and then review two basic properties.

Definition 9. [11] Let $\{X_1, X_2, \cdots, X_m\}$ be a partition of X; then, we say that μ is an at most m-dimensional distorted probability if there exists a function f on $\mathbb{R}^m$ and probabilities P_i on $(X_i, 2^{X_i})$ such that:

$$\mu(A) = f(P_1(A \cap X_1), P_2(A \cap X_2), \cdots, P_m(A \cap X_m)) \quad (4)$$

where f on $\mathbb{R}^m$ is strictly increasing with respect to each variable.

We say that an at most m-dimensional distorted probability μ is an m-dimensional distorted probability if μ is not an at most $(m-1)$-dimensional.

The fact that all fuzzy measures can be represented as m-dimensional distorted probabilities follows from the next proposition, that is trivial from the above definition.

Proposition 2. *[11] Every fuzzy measure is an at most m-dimensional distorted probability with* $m = |X|$.

Note that for $n = |X|$, we are considering the following partition of X: $\{X_1 = \{x_1\}, \ldots, X_n = \{x_n\}\}$. So, $f(a_1, \ldots, a_n) = \mu(A)$ when $a_i = 1$ if and only if $x_i \in A$.

To complete the properties of these fuzzy measures, we have the following proposition that states that m-dimensional distorted probabilities define a family of measures with increasing complexity with respect to m. This means that increasing the value of m, the number of measures being representable increases. The following proposition establishes this property.

Proposition 3. *[11] Let* $\mathcal{M}_k$ *be the set of all fuzzy measures that are k-dimensional distorted probabilities and let* $\mathcal{M}_0$ *be the empty set. Then* $\mathcal{M}_{k-1} \subset \mathcal{M}_k$ *for all* $k = 1, 2, \ldots, |X|$.

3.2 *m*-Symmetric Fuzzy Measures

Symmetric fuzzy measures are those measures where the measure of a set depends only on the number of elements in the set. That is, $\mu(A) = f(|A|)$ for a function f ($|\cdot|$ stands for the cardinality of a set). It has been proven that an OWA operator corresponds to a Choquet integral with respect to the following symmetric fuzzy measure: $\mu(A) = \sum_{i=1}^{|A|} w_i$.

The concept of symmetric fuzzy measure has been extended to m-symmetric fuzzy measures [9, 8]. The definition is based on the *set of indifference*. Such set is defined by elements that do not affect the value of the measure. That is, the elements of a set are indistinguishable with respect to the fuzzy measure.

Definition 10. [9, 8] Given a subset A of X, we say that A is a set of indifference if and only if:

$$\forall B_1, B_2 \subseteq A, |B_1| = |B_2|,$$
$$\forall C \subseteq X \setminus A \quad \mu(B_1 \cup C) = \mu(B_2 \cup C)$$

In the case of $m = 2$, we have the following definition. Below is the general one.

Definition 11. [9, 8] Given a fuzzy measure μ, we say that μ is an at most 2-symmetric fuzzy measure if and only if there exists a partition of the universal set $\{X_1, X_2\}$, with $X_1, X_2 \neq \emptyset$ such that both X_1 and X_2 are sets of indifference. An at most 2-symmetric fuzzy measure is 2-symmetric if X is not a set of indifference.

Definition 12. [9, 8] Given a fuzzy measure μ, we say that μ is an at most m-symmetric fuzzy measure if and only if there exists a partition of the universal set $\{X_1, \ldots, X_m\}$, with $X_1, \ldots, X_m \neq \emptyset$ such that $X_1, \ldots X_m$ are sets of indifference.

It is clear from this definition that all fuzzy measures are m-symmetric for a large enough value m. This is stated in the next proposition.

Proposition 4. *Every fuzzy measure μ is an at most n-symmetric fuzzy measure for $n = |X|$.*

Definition 13. [9, 8] Given two partitions $\{X_1, \ldots, X_p\}$ and $\{Y_1, \ldots, Y_r\}$ on the finite universal set X, we say that $\{X_1, \ldots, X_p\}$ is coarser than $\{Y_1, \ldots, Y_r\}$ if the following holds:

$$\forall X_i \exists Y_j \text{ such that } Y_j \subseteq X_i$$

Definition 14. [9, 8] Given a fuzzy measure μ, we say that μ is m-symmetric if and only if the coarsest partition of the universal set in sets of indifference contains m non empty sets. That is, the coarsest partition is of the form: $\{X_1, \ldots, X_m\}$, with $X_i \neq \emptyset$ for all $i \in \{1, \ldots, m\}$.

It is known that symmetric fuzzy measures are a particular case of distorted probabilities. This is in relation to the fact that OWA operators are a particular case of WOWA operators. This relationship can also be established between m-symmetric fuzzy measures and m-dimensional distorted probabilities.

Proposition 5. *[10] Let μ be an m-symmetric fuzzy measure with respect to the partition $\{X_1, \ldots, X_m\}$. Then, μ is an m-dimensional distorted probability.*

The reversal of this proposition is not true, as it is the case for 1-dimensional ones, where the OWA and the WOWA are also not equivalent. The next proposition characterizes one case in which m-dimensional distorted probabilities are m-symmetric fuzzy measures.

Proposition 6. *[10] Let μ be an m-dimensional distorted probability. If, $p_i(x_j) = p_i(x_k)$ for all $x_j, x_k \in X_i$ and for all $i = 1, \ldots, m$, then μ is an m-symmetric fuzzy measure.*

3.3 *m-Dimensional OWA and m-Dimensional WOWA*

The definition of m-dimensional operators relies on the well known fact that OWA operators are equivalent to Choquet integrals with respect to symmetric fuzzy measures. On the basis of this fact, m-symmetric fuzzy measures permit us to define the corresponding generalization of the OWA operator. This is defined below.

Definition 15. [10] The m-dimensional OWA is defined as the Choquet integral with respect to an m-symmetric fuzzy measure.

In a similar way, WOWA operators are equivalent to Choquet integrals with respect to distorted probabilities [15]. Therefore, a Choquet integral with an m-dimensional probability can be seen as a generalization of the WOWA operator. We give this definition below.

Table 1 Data for learning

u_{C_1}	u_{C_2}	$\dots$	u_{C_N}	$R_{\mathbb{C}}$	$u_{\mathbb{C}}$
a_1^1	a_2^1	$\dots$	a_N^1	p^1	b^1
a_1^2	a_2^2	$\dots$	a_N^2	p^2	b^2
$\vdots$	$\vdots$		$\vdots$	$\vdots$	$\vdots$
a_1^M	a_2^M	$\dots$	a_N^M	p^M	b^M

Definition 16. [10] The m-dimensional WOWA is defined as the Choquet integral with respect to an m-dimensional distorted probability.

Definitions 15 and 16 permits us to establish the following result that is a corollary of Proposition 5.

Corollary 1. *[10] An m-dimensional OWA is a particular case of an m-dimensional WOWA. In other words, a Choquet integral with respect to an m-symmetric fuzzy measure is a particular case of a Choquet integral with respect to an m-dimensional distorted probability.*

Thus, the same relationship that holds for OWA and WOWA, also holds for m-dimensional OWA and m-dimensional WOWA.

4 Learning Parameters for the WOWA Operator

Learning parameters for the WOWA operator corresponds to the process of determining its weighting vectors **p** and **w**. This problem was considered in [18] under a supervised environment. That is, it is assumed that we have a set of examples for which both the input data and the output data are known. Table 1 represents this situation. Under this assumption, we select the weights **p** and **w** that minimize the difference between the expected output and the real output.

Assuming that the difference between the expected outcome and the actual outcome is computed using the Euclidean distance, the problem can be formalized as follows.

$$\begin{aligned}
&\text{Minimize } D_{WOWA}(\mathbf{p} = (p_1, \dots, p_N), \mathbf{w} = (w_1, \dots, w_N)) = \\
&\qquad\qquad \Sigma_{j=1}^{M} (\Sigma_{i=1}^{N} WOWA_{\mathbf{p},\mathbf{w}}(a_1^j, \dots, a_N^j) - b^j)^2 \\
&\textit{Subject to} \\
&\qquad \Sigma_{i=1}^{N} p_i = 1 \\
&\qquad \Sigma_{i=1}^{N} w_i = 1 \\
&\qquad p_i \geq 0 \\
&\qquad w_i \geq 0
\end{aligned} \tag{5}$$

To solve this problem, [18] used an hybrid approach that bootstrapped from the optimal solution obtained for the weighted mean and the OWA operator (following [16]), and then applied the gradient descent as proposed in [5, 6]. This hybrid approach is needed because the optimization problem in the case of the WOWA is not quadratic and there is no easy way to compute its optimal solution.

These results assume that data is complete, that is, there is no missing data. In the case of such data, we developed an approach based on genetic algorithms. Our approach, as well as some experiments, is reported in [12].

Finally, we have also considered the process of learning m-dimensional distorted probabilities. An approach for this type of problems is described in [11].

5 Summary

In this chapter we have described our main results about the WOWA operator, and some of its extensions. In particular, we have described m-dimensional WOWA operators. In addition, we have presented a short overview about the process of learning the weights of this operator. [20, 21] presents more details and some examples.

Acknowledgements. Partial support by the Generalitat de Catalunya (2005 SGR 00446 and 2005-SGR-00093) and by the Spanish MEC (projects ARES – CONSOLIDER INGENIO 2010 CSD2007-00004 – and eAEGIS – TSI2007-65406-C03-02) is acknowledged.

References

1. Aumann, R.J., Shapley, L.S.: Values of Non-Atomic Games. Princeton University Press, Princeton (1974)
2. Beliakov, G., Pradera, A., Calvo, T.: Aggregation Functions: A Guide for Practitioners. Springer, Heidelberg (2007)
3. Chateauneuf, A.: Decomposable measures, distorted probabilities and concave capacities. Mathematical Social Sciences 31, 19–37 (1996)
4. Choquet, G.: /54) Theory of capacities. Ann. Inst. Fourier 5, 131–295 (1953/1954)
5. Filev, D., Yager, R.R.: Learning OWA operator weights from data. In: Proc. of the 3rd IEEE Int. Conf. on Fuzzy Systems (IEEE WCCI), vol. 1, pp. 468–473 (1994)
6. Filev, D.P., Yager, R.R.: On the issue of obtaining OWA operator weights. Fuzzy Sets and Systems 94, 157–169 (1998)
7. Honda, A., Nakano, T., Okazaki, Y.: Distortion of fuzzy measures. In: Proc. of the SCIS/ISIS conference (2002)
8. Miranda, P., Grabisch, M.: p-symmetric fuzzy measures. In: Proc. of the IPMU 2002 Conference, pp. 545–552 (2002)
9. Miranda, P., Grabisch, M., Gil, P.: p-symmetric fuzzy measures. Int. J. of Unc., Fuzz. and Knowledge Based Systems 10, 105–123 (2002)
10. Narukawa, Y., Torra, V.: On n-dimensional distorted probabilities and p-symmetric fuzzy measures. In: IPMU 2004, Perugia, Italy, July 4-9, vol. 2, pp. 1279–1284 (2004) (ISBN 88-87242-54-2)
11. Narukawa, Y., Torra, V.: Fuzzy measure and probability distributions: distorted probabilities. IEEE Trans. on Fuzzy Systems 13(5), 617–629 (2005)
12. Nettleton, D., Torra, V.: A comparison of active set methods and genetic algorithm approaches for learning weighting vectors in some aggregation operators. Int. J. of Intel. Syst. 16(9), 1069–1083 (2001)

13. Torra, V.: Weighted OWA operators for synthesis of information. In: Proc. of the 5th IEEE Int. Conf. on Fuzzy Systems, pp. 966–971 (1996)
14. Torra, V.: The weighted OWA operator. Int. J. of Intel. Syst. 12, 153–166 (1997)
15. Torra, V.: On some relationships between the WOWA operator and the Choquet integral. In: Proc. of the IPMU 1998 Conference, pp. 818–824 (1998)
16. Torra, V.: On the learning of weights in some aggregation operators. Mathware and Soft Computing 6, 249–265 (1999)
17. Torra, V.: The WOWA operator and the interpolation function W*: Chen and Otto's interpolation method revisited. Fuzzy Sets and Systems 113(3), 389–396 (2000)
18. Torra, V.: Learning weights for Weighted OWA operators. In: Proc. of the IEEE Int. Conf. on Industrial Electronics, Control and Instrumentation (IECON 2000), pp. 2530–2535 (2000)
19. Torra, V., Lv, Z.: On the WOWA operator and its interpolation function. Int. J. of Intel. Systems (2009) (in press)
20. Torra, V., Narukawa, Y.: Modeling decisions: information fusion and aggregation operators. Springer, Heidelberg (2007)
21. Torra, V., Narukawa, Y.: Modelització de decisions: fusió d'informació i operadors d'agregació. UAB Press (2007)
22. Yager, R.R.: On ordered weighted averaging aggregation operators in multi-criteria decision making. IEEE Trans. on Systems, Man and Cybernetics 18, 183–190 (1988)
23. Yager, R.R., Kacprzyk, J. (eds.): The Ordered Weighted Averaging Operators: Theory and Applications. Springer, Heidelberg (1997)

Induced Ordered Weighted Averaging Operators

Gleb Beliakov and Simon James

1 Introduction

Since the introduction of the ordered weighted averaging operator [18], the OWA has received great attention with applications in fields including decision making, recommender systems [8, 21], classification [10] and data mining [16] among others. The most important step in the calculation of the OWA is the permutation of the input vector according to the size of its arguments. In some applications, it makes sense that the inputs be reordered by values different to those used in calculation. For instance, if we have a number of mobile sensor readings, we may wish to allocate more importance to the reading taken from the sensor closest to us at a given point in time, rather than the largest reading.

The idea of using an auxiliary variable to re-order the inputs had its first inception in the image compression work of Mitchell and Estrakh [12] and a follow-up application which sorted the inputs by *fuzzy ranks* [13]. In these applications, the arguments were sorted by a function of their values rather than the values themselves. Yager and Filev then formally defined the Induced OWA (IOWA) in [24], denoting the auxiliary variable associated with each input as an inducing variable. The properties of the IOWA were also investigated.

In this chapter, the induced OWA as well of some of its applications will be presented. In Section 2, the necessary definitions and background will be provided, leading to an overview of the Induced OWA, its properties and generalizations, in Section 3. Section 4 focuses on the inducing variable, and how certain choices may be appropriate in varying applications. A case-study of the induced aggregation framework and its use with fuzzy integrals to enhance nearest-neighbor approximation is given in Section 5, before we summarize in Section 6.

Gleb Beliakov · Simon James
School of Information Technology, Deakin University
221 Burwood Hwy, Burwood 3125, Australia
e-mail: gleb@deakin.edu.au, sjames@deakin.edu.au

R.R. Yager et al. (Eds.): Recent Developments in the OWA Operators, STUDFUZZ 265, pp. 29–47.
springerlink.com

2 Preliminaries

Although the OWA is studied in a number of topic fields, it is considered here within the framework of aggregation functions. Recent books concerning aggregation functions include [2, 7, 17]. The following definitions will be useful.

Definition 1. A function $f : [0,1]^n \to [0,1]$ is called an aggregation function if it is monotone non-decreasing in each variable and satisfies $f(0,0,\dots,0) = 0$, $f(1,1,\dots,1) = 1$.

Aggregation functions are classed depending on their behavior with respect to the inputs.

Definition 2. An aggregation function f is averaging if for every $\mathbf{x} = (x_1,\dots,x_n)$ it is bounded by

$$\min(x_1,\dots,x_n) \le f(x_1,\dots,x_n) \le \max(x_1,\dots,x_n).$$

Definition 3. A vector $\mathbf{w} = (w_1,\dots,w_n)$ is called a weighting vector if $w_i \in [0,1]$ and $\sum\limits_{i=1}^{n} w_i = 1$.

For weighted means, the weight w_i is some representation of the importance of the input x_i. The ordered weighted averaging function assigns its weights based on the magnitude of the inputs.

Definition 4. Given a weighting vector $\mathbf{w}$, the Ordered Weighted Averaging (OWA) function is

$$OWA_{\mathbf{w}}(x_1,\dots,x_n) = \sum_{i=1}^{n} w_i x_{\sigma(i)}, \tag{1}$$

where the $\sigma(.)$ notation[1] denotes the components of $\mathbf{x}$ being arranged in non-increasing order $x_{\sigma(1)} \ge x_{\sigma(2)} \ge \dots \ge x_{\sigma(n)}$.

The OWA is capable of expressing a number of order statistics such as the maximum function where $\mathbf{w} = (1,0,...,0)$ and the minimum for $\mathbf{w} = (0,...,0,1)$. It is also convenient for expressing the median $\mathbf{w}_k = 1$, for $n = 2k+1$ (n is odd) or $w_k = w_{k+1} = 0.5$ for $n = 2k$ (n is even) and $w_i = 0$ otherwise. Interestingly, it was found that the OWA operator is generalized by the Choquet integral, which is defined with respect to a fuzzy measure. The Choquet integral also requires a re-ordering step for its calculation, although generally the definition is provided with the inputs in non-decreasing order (whereas for OWA, inputs are ordered in non-increasing order). We provide the definitions for discrete fuzzy measures and the discrete Choquet integral below.

[1] It is often sufficient to simply write $x_{(i)}$, however here since we consider alternative orderings we will distinguish between non-increasing $\sigma(.)$, non-decreasing $\tau(.)$, and variable induced non-increasing $\eta(.)$ and non-decreasing $\theta(.)$.

Definition 5. Let $\mathcal{N} = \{1,2,\ldots,n\}$. A discrete fuzzy measure is a set function[2] $v : 2^{\mathcal{N}} \to [0,1]$ which is monotonic (i.e. $v(\mathcal{A}) \leq v(\mathcal{B})$ whenever $\mathcal{A} \subset \mathcal{B}$) and satisfies $v(\emptyset) = 0$ and $v(\mathcal{N}) = 1$.

Definition 6. The discrete Choquet integral with respect to a fuzzy measure v is given by

$$C_v(\mathbf{x}) = \sum_{i=1}^{n} x_{\tau(i)}[v(\{j|x_j \geq x_{\tau(i)}\}) - v(\{j|x_j \geq x_{\tau(i+1)}\})], \tag{2}$$

where $\tau(.)$ denotes a non-decreasing permutation of the input vector $\mathbf{x}$ such that $x_{\tau(1)} \leq \ldots \leq x_{\tau(n)}$ and $x_{\tau(n+1)} = \infty$ by convention.

A fuzzy measure is called additive if

$$v(\mathcal{A} \cup \mathcal{B}) = v(\mathcal{A}) + v(\mathcal{B})$$

for any $\mathcal{A}, \mathcal{B} \subset \mathcal{N}, \mathcal{A} \cap \mathcal{B} = \emptyset$.

The discrete Choquet integral with respect to an additive fuzzy measure is a weighted arithmetic mean.

A symmetric fuzzy measure satisfies

$$|\mathcal{A}| = |\mathcal{B}| \Rightarrow v(\mathcal{A}) = v(\mathcal{B}).$$

The discrete Choquet integral with respect to a symmetric fuzzy measure is an OWA function. The values of $v(\mathcal{A})$ are related to the weights of an OWA function by $v(A) = \sum_{i=1}^{|\mathcal{A}|} w_i$.

A related function to the Choquet integral is the Sugeno integral, which similarly, is defined with respect to a fuzzy measure. Sugeno integrals are often used for ordinal data as they are able to operate on finite ordinal scales that aren't necessarily numeric. The values of the fuzzy measure v must be commensurable with the input vales, i.e. if the inputs can only take the values $\{very\ small, small, mid-sized, large, very\ large\}$, then the values of the fuzzy measure must also be defined with these terms.

Definition 7. The Sugeno integral with respect to a fuzzy measure v is given by

$$S_v(\mathbf{x}) = \max_{i=1,\ldots,n} \min\{x_{\tau(i)}, v(H_i)\}, \tag{3}$$

where $\tau(.)$ denotes a non-decreasing permutation of the input vector, $x_{\tau(1)} \leq \ldots \leq x_{\tau(n)}$) and $H_i = \{(i),\ldots,(n)\}$.

[2] A set function is a function whose domain consists of all possible subsets of $\mathcal{N}$. For example, for $n = 3$, a set function is specified by $2^3 = 8$ values at $v(\emptyset)$, $v(\{1\})$, $v(\{2\})$, $v(\{3\})$, $v(\{1,2\})$, $v(\{1,3\})$, $v(\{2,3\})$, $v(\{1,2,3\})$.

3 Induced OWA

With the reordering step, the OWA is no longer a standard linear combination of weighted inputs, but rather a piecewise linear function, with its behavior differing on different parts of the domain. The induced OWA provides a more general framework for this reordering process. An inducing variable can be defined, on either numerical or ordinal spaces, which then dictates the order by which the arguments are permuted. After providing the definition, some important properties and generalizations are discussed.

3.1 Definition

The definition of IOWA is presented here as given by Yager and Filev in [24], in particular, their convention for ties is used.

Definition 8. Given a weighting vector $\mathbf{w}$ and an inducing variable $\mathbf{z}$ the Induced Ordered Weighted Averaging (IOWA) function is

$$IOWA_{\mathbf{w}}(\langle x_1, z_1\rangle, \ldots, \langle x_n, z_n\rangle) = \sum_{i=1}^{n} w_i x_{\eta(i)}, \tag{4}$$

where the $\eta(.)$ notation denotes the inputs $\langle x_i, z_i\rangle$ reordered such that $z_{\eta(1)} \geq z_{\eta(2)} \geq \ldots \geq z_{\eta(n)}$ and the convention that if q of the $z_{\eta(i)}$ are tied,
i.e. $z_{\eta(i)} = z_{\eta(i+1)} = \ldots = z_{\eta(i+q-1)}$,

$$x_{\eta(i)} = \frac{1}{q} \sum_{j=\eta(i)}^{\eta(i+q-1)} x_j, \tag{5}$$

An inducing variable can be based on any notion that associates a variable with each input x_i. Where x_i provides information to be aggregated, z_i provides some information about x_i, e.g. the importance, distance from the source, time displacement of the reading etc. The input pairs $\langle x_i, z_i\rangle$ may be two independent features of the same input, or can be related by some function, i.e. $z_i = f_i(x_i)$. It is conventional for inducing variables used with the IOWA to permute $\mathbf{z}$ in non-increasing order, while with fuzzy integrals the permutation will usually be non-decreasing. It is usually easy to reverse the permutation by using the reciprocal or negative of all z_i.

Example 1. For the weighting vector $\mathbf{w} = (0.6, 0.3, 0.1)$, and the input $\langle \mathbf{x}, \mathbf{z}\rangle = (\langle 0.2, 3\rangle, \langle 0.7, 2\rangle, \langle 0.05, 8\rangle)$, the aggregated value for the induced OWA is

$$IOWA_{\mathbf{w}}(\langle x, z\rangle) = 0.6(0.05) + 0.3(0.2) + 0.1(0.7) = 0.16\,.$$

Example 2. For the weighting vector $\mathbf{w} = (0.4, 0.3, 0.2, 0.1)$, and the input $\langle \mathbf{x}, \mathbf{z} \rangle = (\langle 0.8, 0.5 \rangle, \langle 0.2, 0.5 \rangle, \langle 0.9, 0.2 \rangle, \langle 0.9, 0.7 \rangle)$, the aggregated value for the induced OWA is

$$IOWA_{\mathbf{w}}(\langle x, z \rangle) = 0.4(0.9) + 0.3(\tfrac{0.8+0.2}{2}) + 0.2(\tfrac{0.8+0.2}{2}) + 0.1(0.9) = 0.7\,.$$

3.2 Properties

One immediately notices the similarity between the IOWA in Eq. (4) and the OWA defined in Eq. (1). With the exception of ties, the value obtained from an IOWA will be the same as that obtained from an OWA whose weights are permuted in a different order.

Proposition 1. *Given an inducing variable* $\mathbf{z}$ *where* $z_i = z_j \Leftrightarrow i = j$ *and a fixed input vector* $\mathbf{x}$,

$$IOWA_{\mathbf{w}}(\langle \mathbf{x}, \mathbf{z} \rangle) = OWA_{\mathbf{u}}(\mathbf{x}),$$

where $\mathbf{u}_{\pi(i)} = u_{\pi(1)}, \ldots, u_{\pi(n)} = \mathbf{w}$, *and* $\pi(i)$ *is some permutation of the values in the weighting vector* $\mathbf{u}$.

Although the properties and behavior of the IOWA will be largely dependent on the inducing variable $\mathbf{z}$, in most cases it will exhibit the properties which have shown to hold for the standard OWA, namely monotonicity, idempotency and symmetry. There are, however, certain choices for $\mathbf{z}$ such that symmetry and even monotonicity may be violated. We will discuss these properties with respect to the input vector $\mathbf{x}$.

Monotonicity: For a fixed vector $\mathbf{z}$, it will hold that

$$\mathbf{x} \leq \mathbf{y} \Rightarrow IOWA_{\mathbf{w}}(\langle \mathbf{x}, \mathbf{z} \rangle) \leq IOWA_{\mathbf{w}}(\langle \mathbf{y}, \mathbf{z} \rangle).$$

All values of the weighting vector $\mathbf{w}$ satisfy $w_i \geq 0$, hence an increase to any of the x_i cannot decrease the overall output. If we have $w_i > 0, \forall i$, the IOWA will be strictly monotone-increasing.

Remark 1. In some situations, however, the inducing variable $\mathbf{z}$ may change when $\mathbf{x}$ changes. As an example, suppose we have input pairs $\langle x_i, z_i \rangle_{i=1,\ldots,3}$ obtained from 3 observation stations. The z_i is the reliability of the reading x_i, based partially on the value of x as well as external information. We want to aggregate the x_i, giving preference to the most reliable readings. We hence use an IOWA which orders the observations from most to least reliable in accordance with the values z_i, and define a weighting vector with decreasing weights. For a given reading, the input vector has an induced order of $z_2 \geq z_1 \geq z_3$, so the value x_2 is allocated the largest weight w_1. Now suppose the value of x_2 increases to a value that is unusual at Station 2 and is hence less reliable. The value of z_2 decreases and the induced order is now $z_1 \geq z_2 \geq z_3$. This results in a smaller weight being given to the input x_2 and monotonicity is violated.[3]

Averaging / Idempotency: The positive and normalized (i.e. $\sum_{i=1}^{n} w_i = 1$) values of the weighting vector **w** ensure that the IOWA function will be both averaging and idempotent, i.e.,

$$\min(\mathbf{x}) \leq IOWA_{\mathbf{w}}(\langle \mathbf{x}, \mathbf{z} \rangle) \leq \max(\mathbf{x}),$$

$$IOWA_{\mathbf{w}}(\langle t, z_1 \rangle, \langle t, z_2 \rangle, ..., \langle t, z_n \rangle) = t.$$

Remark 2. In the study of aggregation functions, averaging behavior and idempotency are usually considered to be equivalent. Averaging behavior necessarily implies idempotency, however the property of idempotency is not sufficient for averaging behavior without monotonicity. As discussed previously in Remark 1, an IOWA-type function may not necessarily be monotone, however this does not cause either of these properties to be lost.

Symmetry: With respect to the input pairs $\langle \mathbf{x}, \mathbf{z} \rangle$, the initial indexing is unimportant, e.g. for $n = 2$,

$$IOWA_{\mathbf{w}}(\langle x_1, z_1 \rangle, \langle x_2, z_2 \rangle) = IOWA_{\mathbf{w}}(\langle x_2, z_2 \rangle, \langle x_1, z_1 \rangle).$$

With respect to the input vector **x**, however, certain choices of **z** may result in loss of the symmetry property. For instance, the z_i may be constant or a function that is somewhat dependant on the initial indexing, e.g. $z_i = f(i)$. As will be shown below, inducing variables that give a fixed calculation order result in weighted arithmetic means. If such instances arise in practice, it may make more sense to perceive the problem in terms of arithmetic means with a particular weight associated to each input, rather than defining a so-called inducing variable.

Homogeneity and Shift-Invariance: Standard OWA functions are both homogeneous and shift-invariant[4], i.e.

$$\lambda [OWA_{\mathbf{w}}(\mathbf{x})] = OWA_{\mathbf{w}}(\lambda \mathbf{x}), \forall \lambda \in [0,1] \quad \text{(homogeneity)},$$
$$OWA_{\mathbf{w}}(\mathbf{x}) + \lambda = OWA_{\mathbf{w}}(x_1 + \lambda, ..., x_n + \lambda), \forall \lambda \in [-1,1] \quad \text{(shift-invariance)}.$$

For the induced OWA, this will once again be somewhat dependent on the choice of inducing variable. The z_i may be specifically associated with each x_i, but needn't be a function of the actual values, i.e. $x_i = x_j$ does not necessarily imply

[3] The mean of Bajraktarevic [9], is another mean where the weights may vary depending on the value of x_i. It also fails the monotonicity condition, and is only considered an aggregation function for special cases.

[4] These properties of the OWA makes it easy to consider input vectors using an alternative scale to the unit interval. In these cases, λ takes values accordingly.

$z_i = z_j$. Given any input pair $\langle x_i, z_i \rangle$, let $z_{i+\lambda}$ be the value associated with the input $x_i + \lambda$ and $z_{\lambda i}$ be the value associated with the input $\lambda\ x_i$.

The induced OWA will be shift-invariant if:

$$z_i < z_j \Longleftrightarrow z_{i+\lambda} < z_{j+\lambda}, \forall i, j.$$

The induced OWA will be homogeneous if:

$$z_i < z_j \Longleftrightarrow z_{\lambda i} < z_{\lambda j}, \forall i, j (\lambda > 0).$$

In other words, provided λ does not change the relative ordering, IOWA will be stable for translations and homogenous. It should be noted that equalities also need to be preserved, i.e. $z_i = z_j \Leftrightarrow z_{\lambda i} = z_{\lambda j}$ (for homogeneity), which is implied by the above equivalence relations.

Duality: As with the standard OWA, the dual of an IOWA can be defined as the IOWA with respect to the reverse weighting vector, $\mathbf{w}_d = (w_n, w_{n-1}, ..., w_1)$. It is also possible to define the dual function of an IOWA by the inducing variable $\mathbf{z}_d$ which induces the reverse ordering to $\mathbf{z}$.

Special Cases: The IOWA generalizes many important aggregation functions. We provide some of its special cases here (for generalizations of the Induced OWA, see Section 3.3).

Standard OWA: The Induced OWA includes the standard OWA as a special case whenever the ordering of the inducing variable corresponds exactly with the order of the input variable, e.g. $z_i = x_i$. Clearly, the special cases of the OWA: the minimum, maximum and median, can be obtained through this choice and the appropriate selection of the weighting vector $\mathbf{w}$;

Reverse OWA: The reverse OWA is a function which orders the inputs in non-decreasing order $x_{(1)} \leq ... \leq x_{(n)}$. The inducing variable $\mathbf{z}$ can hence be chosen such that this order is achieved, e.g. $z_i = 1 - x_i$;

Weighted Arithmetic Mean: Any selection of $\mathbf{z}$ that maintains the order of the input components z_i when arranged in non-increasing order, e.g. $z_i = n + 1 - i$ will result in the weighted arithmetic mean $\sum_{i=1}^{n} w_i x_i$.

3.3 *Induced Generalized OWA*

Weighted quasi-arithmetic means are defined with the help of generating functions $g : [0,1] \to [-\infty, \infty]$ and generalize the weighted arithmetic mean. Many important families of means, including geometric, harmonic and power means have been found to correspond to special cases of the quasi-arithmetic mean. Similarly, the generalized OWA function was introduced to generalize the piecewise-linear OWA function. We give the definition here.

Definition 9. Let $g : [0,1] \to [-\infty,\infty]$ be a continuous strictly monotone function and let $\mathbf{w}$ be a weighting vector. The function

$$GenOWA_{\mathbf{w},g}(\mathbf{x}) = g^{-1}\left(\sum_{i=1}^{n} w_i g(x_{\sigma(i)})\right) \tag{6}$$

is called a generalized OWA (also known as ordered weighted quasi-arithmetic mean [3]). As for OWA, $x_{\sigma(i)}$ denotes the i-th largest value of $\mathbf{x}$.

Following from this, Chiclana et al. introduced the natural extension of the IOWA to the IOWG [4]. Of course, this corresponds to a special case of the Induced generalized OWA.

Definition 10. Given a weighting vector $\mathbf{w}$, an inducing variable $\mathbf{z}$ and a continuous strictly monotone function $g : [0,1] \to [-\infty,\infty]$, the Induced Generalized OWA (I-GenOWA) function is

$$I\text{-}GenOWA_{\mathbf{w},g}(\langle \mathbf{x},\mathbf{z}\rangle) = g^{-1}\left(\sum_{i=1}^{n} w_i g(x_{\eta(i)})\right) \tag{7}$$

As for IOWA, $\eta(.)$ notation denotes the inputs $\langle x_1, z_1\rangle$ reordered such that $z_{\eta(1)} \geq z_{\eta(2)} \geq \ldots \geq z_{\eta(n)}$ and Eq. (5) is employed for ties.

The induced generalized OWA was studied in [11]. With the I- GenOWA function, we can essentially use any quasi-arithmetic mean with respect to an order-inducing variable. Special cases of quasi-arithmetic mean include power means and harmonic means, however we will provide only the case for the geometric mean, since many studies (e.g. [4])have investigated its use and properties.

Definition 11. For a given weighting vector $\mathbf{w}$ and an inducing variable $\mathbf{z}$, the IOWG function is

$$IOWG_{\mathbf{w}}(\langle \mathbf{x},\mathbf{z}\rangle) = \prod_{i=1}^{n} x_{\eta(i)}^{w_i}. \tag{8}$$

where $\eta(.)$ notation denotes the inputs $\langle x_1, z_1\rangle$ reordered such that $z_{\eta(1)} \geq z_{\eta(2)} \geq \ldots \geq z_{\eta(n)}$. The same convention (Eq. (5)) is employed for ties as with IOWA and I- GenOWA.

3.4 Induced Fuzzy Integrals

It is well known that the OWA function is generalized by the Choquet integral, which also has a reordering step in its calculation. In [19], Yager made extensions to the induced Sugeno integral and in [22], the induced Choquet integral was presented.

Definition 12. The induced Choquet integral with respect to a fuzzy measure v and an order inducing variable $\mathbf{z}$ is given by

$$IC_v(\langle \mathbf{x}, \mathbf{z} \rangle) = \sum_{i=1}^{n} x_{\theta(i)}[v(\{j|z_j \geq z_{\theta(i)}\}) - v(\{j|z_j \geq z_{\theta(i+1)}\})], \tag{9}$$

where the $\theta(.)$ notation denotes the inputs $\langle x_1, z_1 \rangle$ reordered such that $z_{\theta(1)} \leq z_{\theta(2)} \leq \ldots \leq z_{\theta(n)}$ and $z_{\theta(n+1)} = \infty$ by convention.

The same convention can be used for ties, and note again that the calculation of the Choquet integral is usually performed by firstly arranging the arguments in non-decreasing order - in this case with respect to the inducing variable $\mathbf{z}$.

As the Choquet integral generalizes the OWA, motivation for extensions to the induced Choquet integral occur naturally in application. The fuzzy measure used in calculation still plays the same role, giving a weight to each coalition and allowing for interaction. Consider the final component in calculation of the standard Choquet integral, $x_{\tau(n)}v(j|x_j = x_{\tau(n)})$. The highest input is multiplied by the value of its corresponding singleton. This means that if $v(j)$ is high, a high score for x_j will be sufficient for an overall large score. In the case of the induced Choquet integral, however, this final term will be $x_{\theta(n)}v(j|z_j = z_{\theta(n)})$. If $v(j)$ is high in this instance, a high score for z_j results in the output being heavily influenced by the input x_j, whether it is small or large, i.e. $f(\langle \mathbf{x}, \mathbf{z} \rangle) \approx x_j$. Such examples arise naturally in nearest-neighbor rules as we will discuss later.

The Sugeno integral is similar to the Choquet integral, in that it is defined by a fuzzy measure v and is calculated with a reordering of the input vector. The Sugeno integral is based on operations of min and max and hence is capable of handling non-numerical inputs, provided they are taken from a finite ordinal scale.

Definition 13. The induced Sugeno integral with respect to a fuzzy measure v and an inducing variable $\mathbf{z}$ is given by

$$IS_v(\langle \mathbf{x}, \mathbf{z} \rangle) = \max_{i=1,\ldots,n} \min\{x_{\theta(i)}, v(H_{i,\mathbf{z}})\}, \tag{10}$$

where the $\theta(.)$ notation denotes the inputs $\langle x_1, z_1 \rangle$ reordered such that $z_{\theta(1)} \leq z_{\theta(2)} \leq \ldots \leq z_{\theta(n)}$, with $H_{i,\mathbf{z}} = \{\theta(i), \theta(i+1), \ldots, \theta(n)\}$. If q of the z_i are tied, i.e. $z_{\theta(i)} = z_{\theta(i+1)} = \ldots = z_{\theta(i+q-1)}$,

$$x_j \leq x_k \Rightarrow z_{\theta(j)} \leq z_{\theta(k)} \forall j,k = \theta(i), \ldots, \theta(i+q-1).$$

Note here that a different convention is employed for ties, namely that if the inducing variables are tied, the arguments within ties are reordered according to the relative values of x_i.

It might be noted that the circumstances under which an induced OWA could be applied may also allow for the application of the induced Choquet integral. The induced Sugeno integral, however, may be applied to situations which are quite different, where the semantics of an OWA function might not make sense. In particular, Sugeno integrals are capable of operating on values expressed linguistically e.g. $\{very\ good, good, fair, poor, very\ poor\}$. This is one reason why ties are dealt with differently, since with linguistic variables it might not make sense to take the average. The induced Sugeno integral has a tighter bound than that expressed by the averaging property. Yager noted in [19] that the bounds on IS_v are

$$\min\{\mathbf{x}\} \leq IS_v(\langle \mathbf{x}, \mathbf{z} \rangle) \leq x_{\theta(n)}.$$

In other words, the Sugeno integral is bounded from above by the input associated with the largest inducing variable input $z_{\theta(n)}$, as well as from below by the minimum argument value of $\mathbf{x}$.

4 Choices for the Inducing Variable

The process by which the arguments are reordered is clearly of fundamental concern when considering induced aggregation functions. The choice for the inducing variable $\mathbf{z}$ will clearly depend on the situation to be modeled. In many cases, we may wish to give weight to observations that are more similar or "closer" to an object of interest $\bar{\mathbf{x}}$, so we will let z_i represent some function of distance, i.e. $z_i = f(|x_i - \bar{x}_i|)$. In other cases, the z_i may be some representation of the reliability of the value x_i. If the z_i are constant, this essentially models a weighted arithmetic mean, however in many cases the accuracy of an observation may fluctuate. We consider some typical examples where induced aggregation functions can be useful, with particular focus on the choice of $\mathbf{z}$.

4.1 Standard Auxiliary Ordering

The inducing variable may simply be an attribute associated with the input $\mathbf{x}$ that is not considered in the actual aggregation process, but is informative about the object itself. For instance, consider the peer-review process for some journal and conference papers. Each reviewer allocates a score to each criterion, e.g. originality, relevance etc. Sometimes the reviewer also provides his/her evaluation of their own familiarity with the topic, e.g. very familiar, only marginally familiar etc. This last input, of course is not taken into account when aggregating the scores for submission, however could be taken into account in the weight allocation process. To give an overall score for each criterion, we can then use an IOWA where x_i is the score allocated by the i-th reviewer, z_i is the familiarity of the reviewer with the topic and $\mathbf{w}$ is a weighting vector with non-increasing weights such that the heavier weight is given to experts with more expertise on the given paper.

This allocation of weighting is different to providing the expert herself a weight based on her expertise, as the variability of $\mathbf{z}$ suggests that this may fluctuate depending on the paper she is marking.

Example 3. An editor for a journal considers two papers which have been evaluated by the same three reviewers.

	paper 1		paper 2	
	score (x)	expertise (z)	score (x)	expertise (z)
reviewer 1	70	8	92	4
reviewer 2	85	7	62	8
reviewer 3	76	4	86	5

Using the weighting vector $\mathbf{w} = (0.5, 0.3, 0.2)$, the score for the first paper is

$$IOWA_{\mathbf{w}}(\langle \mathbf{x}, \mathbf{z} \rangle) = 0.5(70) + 0.3(85) + 0.2(76) = 75.7\ .$$

The score for the second paper is

$$IOWA_{\mathbf{w}}(\langle \mathbf{x}, \mathbf{z} \rangle) = 0.5(62) + 0.3(86) + 0.2(92) = 75.2\ .$$

4.2 Nearest-Neighbor Rules

Nearest-neighbor methods and their variants have been popularly applied to classification and function approximation problems. The underlying assumption is that objects described similarly by their features will belong to the same class or have the same function output. For instance, consider a classification problem that requires an object $\bar{\mathbf{x}} = (x_1, x_2, ..., x_p)$ to be assigned to a class Y_1 or Y_2. Given a number of training data $\mathscr{D} = \{(\mathbf{x}_1, y_1), \ldots, (\mathbf{x}_K, y_K)\}$, we can identify the object $\mathbf{x}_j$ most similar to $\bar{\mathbf{x}}$ and allocate the same label. Extensions can be made such that the class labels of a number of the objects in $\mathscr{D}$ are aggregated to determine the class of $\bar{\mathbf{x}}$. Induced aggregation functions can be used to model this situation, with the weights and inducing variable often reflecting the similarity of each of the training data. Of course, the way similarity is calculated becomes very important.

The nearest-neighbor approach can also be used for function approximation. Suppose $\mathbf{x}_i \in [0,1]^p$ and $y_i \in [0,1]$. We assume the training data are generated by some function $f(\mathbf{x}_i) + \varepsilon_i = y_i$ where ε_i are random errors, and then approximate $y_{\bar{\mathbf{x}}}$ by aggregating the y_i of the most similar $\mathbf{x}_i$. The induced OWA and induced aggregation functions in general can be used in this context, where the input vector comprises the y_i values taken from $\mathscr{D}$ and $\mathbf{z}$ represents a similarity or distance function. For example, a standard representation of the nearest-neighbor model is,

$$y_{\bar{\mathbf{x}}} = IOWA_{\mathbf{w}}(\langle \mathbf{y}, \mathbf{z} \rangle) = \sum_{i=1}^{K} w_i y_{\eta(i)},$$

with $\mathbf{w} = (1, 0, ..., 0)$, and $z_i = \frac{1}{||\mathbf{x}_i - \bar{\mathbf{x}}||}$, the reciprocal of Euclidean distance between $\bar{\mathbf{x}}$ and $\mathbf{x}_i$ associated with y_i.

A function which approximates $y_{\bar{\mathbf{x}}}$ based on the k nearest-neighbors (known as the kNN method) could be calculated using the same function, with $w_i = \frac{1}{k}, i = 1, ..., k, w_i = 0$ otherwise. One can also consider weighted versions where the weights gradually decay, i.e. $w_1 \geq w_2 \geq ... \geq w_K$, e.g. $w_i = \frac{2(k+1-i)}{k(k+1)}$.

There exist alternative choices for the distance function, in particular, there are many suitable metrics depending on the spatial distribution of the data. Yager has studied the use of IOWA for modeling these situations in [19, 20, 22, 23] as well as in [24] with Filev.

Time-series smoothing can also be handled within the framework of induced aggregation operators. It is similar to a nearest-neighbor problem with a single-dimension variable, t. In time-series smoothing, we want to estimate or smooth the value $y_{\bar{t}}$ at a point in time $\bar{t}$ based on the values y_i obtained at previous times t_i. Induced OWA operators allow a simple framework for modeling this smoothing process. For instance, the 3-day simple moving average on a data set $\mathscr{D} = \{\langle t_1, y_1 \rangle, \ldots, \langle t_n, y_n \rangle\}$ can be modeled by $IOWA_{\mathbf{w}}(\bar{t})$, with $w = (\frac{1}{3}, \frac{1}{3}, \frac{1}{3}, 0, 0, ..., 0)$, $z_i = \frac{1}{\bar{t} - t_i}$.

4.3 Best-Yesterday Models

Extrapolation problems involve predicting future values based on previous observations. Weather prediction is an important example. In [24], Yager and Filev present the best-yesterday model for predicting stock market prices based on the opinions of multiple experts. We could aggregate their scores using a weighted mean, allocating a weight to the experts who seem more reliable, however an alternative is to use an IOWA operator, inducing the input vector based on the most accurate predictions from previous days. We consider an adapted example from [24] here.

Example 4. We have four experts who, daily, predict the next day's opening share price of the FUZ Company. Our data then consists of the predictions each day for each expert $i = 1, ..., 4$, $x_i(t)$ and the actual stock price of the FUZ Company each day $y(t)$. Our aggregated prediction could be the value obtained from the induced OWA

$$IOWA_{\mathbf{w}}(t) = \sum_{i=1}^{4} w_i x_{\eta(i)}(t),$$

where $z_i = -|x_i(t-1) - y(t-1)|$ and the weights w_i are non-increasing. This allocates more importance to the expert whose predictions were closest to the actual price yesterday.

Of course, we could order our experts by their accuracy for the past 2, 3, etc., days, or determine the weights using optimization techniques. In the Yager and Filev example, for instance, the fitted weighting vector was $\mathbf{w} = (0.2, 0.12, 0.08, 0.6)$ i.e., the best fitting weighted model gave more influence to the expert who was furthest from the mark the previous day.

4.4 Aggregation of Complex Objects

In [22], Yager presents an interesting application of induced aggregation functions: that of aggregating complex objects whose ordering is not easily defined. For instance, suppose we have a number of matrices $M_1, ..., M_n$ and we want some representative output $Agg(M_1, ..., M_n) = A$. Many basic aggregation functions and their extensions to matrix operations exist, e.g. the arithmetic mean, weighted arithmetic mean, however it would make no sense to apply a standard OWA in this case as no order as such exists among the M_i. In some situations, an OWA-type aggregation may be desired. We provide again another of Yager's examples.

Suppose we have a number of matrices M_i, each of which represents an expert's estimation of a probability distribution. We can measure the certainty of the expert's estimate with measures such as entropy, and induce the ordering from this measure. This allows us to then define a non-increasing weighting vector so that experts whose estimates are more certain can be given more importance. We note that in this example, the inducing variable $\mathbf{z}$ is some function of the values themselves, rather than an auxiliary input.

4.5 Group Decision Making

IOWA operators can be useful in group decision making (GDM) problems for modeling concepts such as consensus. In [5], the use of different inducing variables was considered for group decisions based on pair-wise preference matrices modeling multiple alternatives. The usual approach to varying weights is to have these reflect the importance of each expert. In the context of preference matrices, however it might also make sense to allocate importance to each input based on the consistency of each expert, inferred from how well their preferences satisfy transitivity etc. Another of the inducing variables presented took into account the overall preference for a particular alternative expressed by each expert.

The standard OWA is able to model majority concepts such as "most" or "80%" using weighting vectors based on linguistic quantifiers. In [15], it was proposed that consensus might be better achieved with inducing variables reflecting the *support* for each individual score. Consider the evaluations of 5 experts, $\mathbf{x} = (0.3, 0.1, 0.7, 0.9, 0.8)$. It makes sense that the score given by expert 5 is more representative of the group than say, expert 2. The support for evaluation x_i from x_j can be modeled simply using:

$$\mathrm{Sup}(x_i,x_j) = \begin{cases} 1, \text{ if } |x_i - x_j| < \alpha; \\ 0, \text{ otherwise,} \end{cases}$$

where α is a desired threshold. The inducing variable based on support can then be given by,

$$z_i = \sum_{j=1, j\neq i}^{n} \mathrm{Sup}(x_i,x_j).$$

In turn, weighting vectors with non-increasing weights can be specified such that experts with more support are allocated higher importance.

4.6 *Multiple Inducing Variables*

In [14], a generalization to multiple inducing variables was considered. Suppose we have N priorities and each of the inputs x_i are associated with N ratings or degrees of satisfaction with respect to these priorities. In this context, the order can be induced by some aggregation of these N scores, and a single inducing variable $\mathbf{z}^*$ can be considered as the vector of aggregated results.

Example 5. We want to measure the pollution levels at a beach, and we have multiple mobile sensors that report to a central computer for analysis. The reliability of the sensor readings depends somewhat on the time since they were transmitted, as well as the distance traveled in transmission and the local conditions as they were sent - for instance, varying water pressure, presence of animals etc. We hence decide to aggregate the pollution levels with an $IOWA_{\mathbf{w}}(\langle \mathbf{x}, \mathbf{z}^* \rangle)$ where ${\mathbf{z}^*}_i = f(z_{i1}, z_{i2}, z_{i3})$ is the aggregated inducing input associated with each pollution level input x_i.

In kNN classification and approximation methods, the distance between data points can be considered within this framework, where each dimension constitutes a separate inducing variable. It is noted in [14] that individual attributes may not be commensurable, in which cases the usual conceptions of distance cannot be used. The inducing variable could then be a weighted function of the distances in each dimension.

5 Case Study: Induced Choquet Integral for Function Approximation

In [1], the induced Choquet integral was used to enhance the performance of kNN-approximation. In nearest-neighbor approximation, the function values of the "closest" training data (usually calculated in some norm) are used to predict the value of an unknown datum. The implementation is quite simple, and with enough training data can obtain reasonable accuracy. One problem, however (particularly with sparse training data) is that neighbors may be skewed or correlated, contributing redundant information. Consider the data presented in Fig. 1(a). All the nearest neighbors happen to be on one side of the point in question, whereas there are plenty of neighbors on the other side whose votes are not counted. In Fig. 1(b), we see that when predicting local weather, we should take into account *where* the predictions are coming from as well as how far away. In these situations it is desirable to include information provided by the neighbors which are close to the point in question but also distributed all around it.

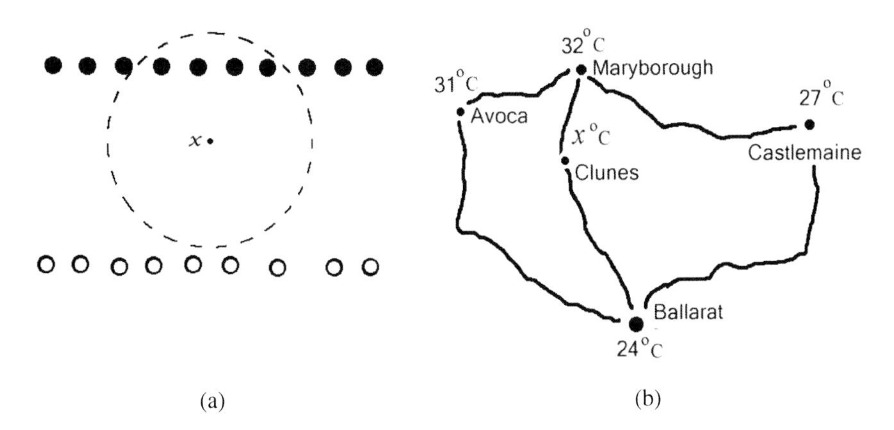

Fig. 1 (a) An example (from the area of remote sensing, the data are taken by an airplane flying over a region in two directions) illustrating the inadequacy of the kNN method. The value at x is determined exclusively by the data represented by filled circles, i.e. is extrapolated and not interpolated. (b) An example where data from nearest neighbors may be correlated. Cities that are close in proximity are likely to have similar weather, e.g. when predicting the weather for Clunes, we want to take into account the fact that Avoca and Maryborough are quite close together.

When considering kNN-approximation as modeled by the induced OWA, the weights $w_i,...,w_k$ represent the relative importance of the observation $(\mathbf{x}_{\eta(i)}, y_{\eta(i)})$ when we are aggregating. For standard *kNN*, these weights are equal, however decreasing weighting vectors can be used to reflect how close the data point $\mathbf{x}_{\eta(i)}$ is to the point $\bar{\mathbf{x}}$ in question. The main idea of the approach proposed in [1] is to replace the induced OWA with the induced Choquet integral so that the relative positioning of the data can be taken into account, not just the distance to the point in question.

Given a set of K training data, $\mathscr{D} = \{(\mathbf{x}_i, y_i)\}, i = 1,\ldots,K$, and an inducing vector given by $z_i = \frac{1}{||\mathbf{x}_i - \bar{\mathbf{x}}||}$, an unknown data point $\bar{\mathbf{x}}$ is assigned the value $y_{\bar{\mathbf{x}}} = IC_v(\langle y_1, z_1\rangle, ..., \langle y_k, z_k\rangle)$. The aim is to define a fuzzy measure, which reflects the importance of each of the k-nearest neighbors based on their proximity to $\bar{\mathbf{x}}$ as well as taking into account the degree to which they contribute redundant or complementary information. We approximate the importance of each of the k-nearest-neighbors by way of the Shapley value, which measures the average contribution of each of the inputs.

Definition 14. Let v be a fuzzy measure. The Shapley index for every $i \in \mathcal{N}$ is

$$\phi(i) = \sum_{\mathcal{A} \subseteq \mathcal{N} \setminus \{i\}} \frac{(n - |\mathcal{A}| - 1)!|\mathcal{A}|!}{n!} [v(\mathcal{A} \cup \{i\}) - v(\mathcal{A})].$$

The Shapley value is the vector $\phi(v) = (\phi(1), \ldots, \phi(n))$. It satisfies $\sum_{i=1}^{n} \phi(i) = 1$.

In turn, the interaction index for pairs, which verifies $I_{ij} < 0$ as soon as i, j are positively correlated (negative synergy) and $I_{ij} > 0$ for negatively correlated inputs (positive synergy), can be used to model redundancies in the data, for example, when two of the data are in the same position.

Definition 15. Let v be a fuzzy measure. The interaction index for every pair $i, j \in \mathcal{N}$ is

$$I_{ij} = \sum_{\mathcal{A} \subseteq \mathcal{N} \setminus \{i,j\}} \frac{(n - |\mathcal{A}| - 2)!|\mathcal{A}|!}{(n-1)!} \left(\sum_{\mathcal{B} \subseteq \{i,j\}} (-1)^{|\mathcal{B}|} v(\mathcal{A} \cup \mathcal{B}) \right) \tag{11}$$

Each of the $\phi(i)$ are calculated based on normalizing the vector of reciprocal Euclidean distances, while the interaction indices I_{ij} are approximated according to the angle, α_{ij} between the vectors $\mathbf{x}_i - \bar{\mathbf{x}}$ and $\mathbf{x}_j - \bar{\mathbf{x}}$. This is done using $I_{ij} = \max\{\cos(\alpha_{ij}), 0\}$, where the cosine can be calculated using the standard scalar product and Euclidean distance operations. From these values alone, a unique 2-additive measure v (possibly non monotonic) [6] can be defined. In order to ensure monotonicity, the actual values $\phi(i)$ and I_{ij} used for the aggregation process are chosen using optimization methods which minimize the residuals between the values determined from distance and angle calculations, and those which satisfy the properties of a monotonic fuzzy measure. See Fig. 2-4 for graphical comparisons of IOWA-based kNN approximation and IC-based kNN approximation. The test data

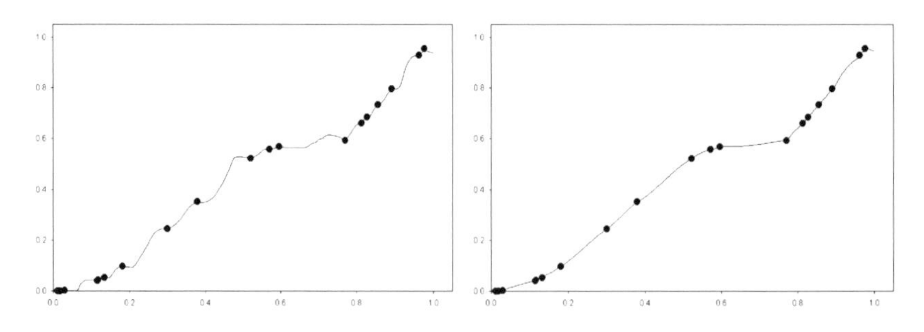

Fig. 2 Interpolation by IOWA-based kNN method (left) with k=3 (the best k for this data set) and IC-based kNN method (right) with $k = 15$ for the 1-dimensional data, test function $f_1(x) = \max(x\sin(\pi x), x^2)$.

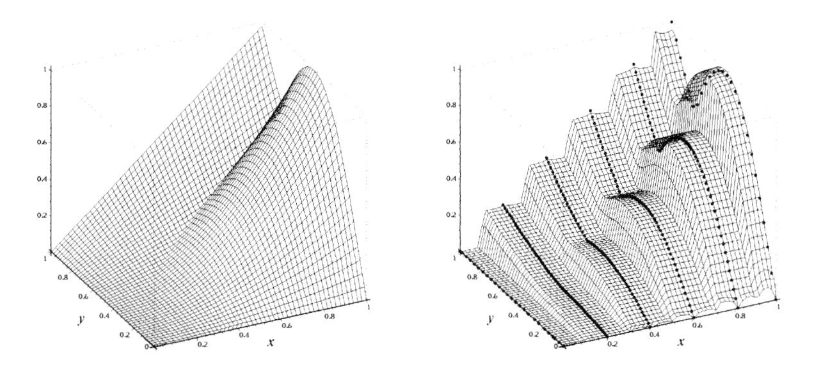

Fig. 3 The test function (left) and IOWA-based kNN method for $k = 5$. With standard kNN methods, accuracy is sensitive to the choice of k, increasing k does not always yield better results.

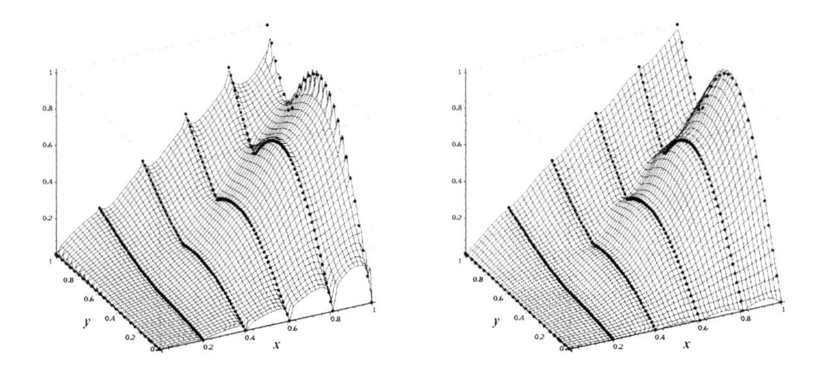

Fig. 4 Interpolation of 2D-data by the IOWA-based kNN method (left) and IC-based method (right) with k=50. The training data is structured similar to Fig. 1(a) and tested against the function in Fig. 3.

points are interpolated. The method was found to significantly reduce the root mean squared error (RMSE) for 1-, 2- and 3-dimensional test examples.

6 Summary

This chapter has presented the induced ordered weighted averaging function and many of its generalizations. This type of aggregation provides a framework that is useful for modeling various situations, which can benefit from the existing research concerning the OWA and similar aggregation functions. An important issue, central to the development of induced-order aggregation methods, is how best to choose the inducing variable. It seems that this line of research, in conjunction with the identification of weights, may yield valuable results for future applications.

References

1. Beliakov, G., James, S.: Using Choquet integrals for kNN approximation and classification. In: Feng, G.G. (ed.) IEEE International Conference on Fuzzy Systems (FUZZ-IEEE 2008), pp. 1311–1317 (2008)
2. Beliakov, G., Pradera, A., Calvo, T.: Aggregation Functions: A Guide for Practitioners. Springer, Heidelberg (2007)
3. Calvo, T., Kolesárová, A., Komorníková, M., Mesiar, R.: Aggregation operators: properties, classes and construction methods. In: Calvo, T., et al. (eds.) Aggregation Operators: New Trends and Applications, pp. 3–104. Physica-Verlag, Heidelberg (2002)
4. Chiclana, F., Herrera-Viedma, E., Herrera, F., Alonso, S.: Induced ordered weighted geometric operators and their use in the aggregation of multiplicative preference relations. International Journal of Intelligent Systems 19, 233–255 (2004)
5. Chiclana, F., Herrera-Viedma, E., Herrera, F., Alonso, S.: Some induced ordered weighted averaging operators and their use for solving group decision-making problems based on fuzzy preference relations. European Journal of Operational Research 182, 383–399 (2007)
6. Grabisch, M., Murofushi, T., Sugeno, M.: Fuzzy Measures and Integrals: Theory and Applications. Physica-Verlag, Heidelberg (2000)
7. Grabisch, M., Marichal, J.-L., Mesiar, R., Pap, E.: Aggregation Functions. In: Encyclopedia of Mathematics and its Applications, vol. 127, Cambridge University Press, Cambridge (2009)
8. Herrera-Viedma, E., Peis, E.: Evaluating the informative quality of documents in SGML format from judgements by means of fuzzy linguistic techniques based on computing with words. Information Processing and Management 39, 233–249 (2003)
9. Marques Pereira, R.A., Ribeiro, R.A.: Aggregation with generalized mixture operators using weighting functions. Fuzzy Sets and Systems 137, 43–58 (2003)
10. Mathiassen, H., Ortiz-Arroyo, D.: Automatic categorization of patent applications using classifier combinations. In: Corchado, E., Yin, H., Botti, V., Fyfe, C. (eds.) IDEAL 2006. LNCS, vol. 4224, pp. 1039–1047. Springer, Heidelberg (2006)
11. Merigó, J.M., Gil-Lafuente, A.M.: The induced generalized OWA operator. Information Sciences 179, 729–741 (2009)
12. Mitchell, H.B., Estrakh, D.D.: A modified OWA operator and its use in lossless DPCM image compression. Journal of Uncertainty, Fuzziness and Knowledged-based Systems 5, 429–436 (1997)
13. Mitchell, H.B., Estrakh, D.D.: An OWA operator with fuzzy ranks. International Journal of Intelligent Systems 13, 59–81 (1998)
14. Mitchell, H.B., Schaefer, P.A.: Multiple priorities in an induced ordered weighted averaging operator. International Journal of Intelligent Systems 15, 317–327 (2000)
15. Pasi, G., Yager, R.R.: Modeling the concept of majority opinion in group decision making. Information Sciences 176, 390–414 (2006)
16. Torra, V.: OWA operators in data modeling and reidentification. IEEE Transactions on Fuzzy Systems 12(5), 652–660 (2004)

17. Torra, V., Narukawa, Y.: Modeling Decisions: Information Fusion and Aggregation Operators. Springer, Heidelberg (2007)
18. Yager, R.R.: On ordered weighted averaging aggregation operators in multicriteria decision making. IEEE Trans. on Systems, Man and Cybernetics 18, 183–190 (1988)
19. Yager, R.R.: The induced fuzzy integral aggregation operator. International Journal of Intelligent Systems 17, 1049–1065 (2002)
20. Yager, R.R.: Using fuzzy methods to model nearest neighbor rules. IEEE Transactions on Systems, Man, and Cybernetics – Part B: Cybernetics 32(4), 512–525 (2002)
21. Yager, R.R.: Noble reinforcement in disjunctive aggregation operators. IEEE Transactions on Fuzzy Systems 11(6), 754–767 (2003)
22. Yager, R.R.: Induced aggregation operators. Fuzzy Sets and Systems 137, 59–69 (2003)
23. Yager, R.R.: Choquet aggregation using order inducing variables. International Journal of Uncertainty, Fuzziness and Knowledge-Based Systems 12(1), 69–88 (2004)
24. Yager, R.R., Filev, D.P.: Induced ordered weighted averaging operators. IEEE Transactions on Systems, Man, and Cybernetics – Part B: Cybernetics 20(2), 141–150 (1999)

A Review of the OWA Determination Methods: Classification and Some Extensions*

Xinwang Liu

Abstract. The OWA operator determination is an important prerequisite step for OWA operator applications. With the application of OWA operator in various areas, the OWA operator determination becomes an active topic in recent years. Based on recent developments, the paper give a summary on the OWA determination methods in classification way: the optimization criteria methods, the sample learning methods, the function based methods, the argument dependent methods and the preference methods. Some relationships between the methods in the same kind and the relationships between different kinds are provided. An uniform framework to connect these OWA determination methods together is also attempted. Some extensions, problems and future research directions are given with discussions.

Keywords: OWA operator, determination methods, optimization methods.

1 Introduction

The ordered weighted averaging (OWA) operator, which was introduced by Yager [68], has attracted much interest among researchers. It provides a general class of parameterized aggregation operators that include the min, max, $average$. The OWA operators have been used in a wide range of applications in the fields such as multicriteria and group decision making [2, 3, 16, 17, 24, 25, 27, 61], database query management and data mining [26, 52, 56, 57, 75, 84, 85], forecasting [80],

Xinwang Liu
School of Economics and Management, Southeast University
Nanjing, 210096, Jiangsu, P.R. China
e-mail: xwliu@seu.edu.cn

* The work is supported by the National Natural Science Foundation of China (NSFC) under project 70771025, and Program for New Century Excellent Talents in University of China NCET-06-0467.

R.R. Yager et al. (Eds.): Recent Developments in the OWA Operators, STUDFUZZ 265, pp. 49–90.
springerlink.com

data smoothing and data mining [56, 80], approximate reasoning [15], approximate reasoning, fuzzy system and control [32, 76, 83] and so on. The use of OWA operator is generally composed of the following three steps [65]:

1. Reorder the input arguments in descending order.
2. Determine the weights associated with the OWA operator by using a proper method.
3. Utilize the OWA weights to aggregate these reordered arguments.

It is clear that the actual type of aggregation performed by an OWA operator depends upon the form of the weight vector [14, 19–22, 71, 72, 74]. The weight vector determination is usually a prerequisite step in many OWA related applications, and it has become an active topic in recent years [1, 31, 36, 60, 65]. One of the appealing points of OWA operators is the concept of orness [68]. The orness measure can establish how "orlike" an operator is, which can be interpreted as the mode of decision making by conferring the semantic meaning to the weights used in the aggregation process [20, 22, 74, 76, 77]. A commonly used method for the OWA operator determination is to obtain the desired OWA operator under a given level of orness [19–22, 36, 46, 81], which is usually formulated as a constrained optimization problem. The objective to be optimized can be the (Shannon) entropy [19, 21, 36, 46], the variance [22, 31], the maximum dispersion [7, 60], the (generalized) Rényi entropy [41], the total square deviation or the Chi-square [59] or even the preemptive goal programming [5, 61]. O'Hagan [46] suggested the problem of constraint nonlinear programming with a maximum entropy procedure, and the solution is called a MEOWA (Maximum Entropy OWA) operator. Filev and Yager [19] further proposed a method to generate the MEOWA operator by an immediate parameter. Fullér and Majlender [21] transformed the maximum entropy model into a polynomial equation, which can be solved analytically. Liu and Chen [36] proposed general forms of the MEOWA operator with a parametric geometric approach, and discussed its aggregation properties. Besides the maximum entropy OWA operator, Fullér and Majlender [22] gives the minimal variability OWA operator problem in quadratic programming, and proposed an analytical method for solving it. Liu [31] gave this OWA operator generating method with the equidifferent OWA operator, which is an extension of [22]. A closely related work was done by Wang and Parkan [60]. They proposed a linear programming model with a minimax disparity approach to obtain the OWA operator under the desired orness level. Liu [33] proved the solution equivalence of the minimum variance problem and the minimax disparity problem. Amin and Emrouznejad [7] extended the minimax disparity approach. Majlender [41] proposed a maximum Rényi entropy OWA operator problem with an exponential objective function, which can include the maximum entropy and the minimum variance problem as special cases, and an analytical solution was proposed. Wang, Luo and Liu [58] proposed least squares deviation and Chi-square models to produce the OWA operator weights with a given orness degree. Recently, Liu [34] gave a more general form of OWA operator determination methods with a convex objective function, which can include the maximum entropy and minimum variance problems as special cases. Some properties were discussed and the solution equivalence to the minimax problem was also proved. From [34], we can see that the adjacent relationships of OWA operator elements can be changed by selecting different objective functions, and some function forms were proposed. The summarizations of some OWA operator determination methods were also given in [37, 65].

In spite of the optimization methods, generating OWA operator from empirical data is another classical technique for aggregation operator determination. From the point of view of applications, a particular operator has to be chosen to model correctly a particular situation [11]. This can be done by fitting an operator from a given class or family to some sort of empirical data: either observed or desired output of the system. It is also an active topic in recent years [9, 10, 20, 82]. This is also a commonly used technique for other aggregation operators [10, 54–56]. For a given empirical data set, the OWA operator determination can be formulated as an least square problem or some other criteria [9, 10]. Some OWA operator determination methods from examples are summarized by Beliakov, Pradera and Calvo recently [12].

Another important closely related topic is the OWA aggregation with Regular Increasing Monotone (RIM) quantifier, which was also proposed by Yager [70], various OWA operator weight series was proposed [69]. Based on the quantifier guided OWA aggregation method, Liu [30, 32] further analyzed the relationship between the OWA operator and the RIM quantifier with the generating function technique. The maximum entropy RIM quantifier and minimum variance RIM quantifier were proposed, and some properties of them were discussed [30, 35]. Xu [65] introduced a procedure for generating the symmetric OWA operator based on normal distribution. Yager [78] proposed a general form of symmetric OWA operator as the *centered OWA operator* with an additional monotonic condition. Sadiq and Tesfamariam [51] extended the OWA generating method from Gaussian distribution to the non-symmetric one with some ordinary probability density functions. Recently, Yager [79] proposed the OWA operator determination methods with the stress function method. An important advantage of the stress function based OWA determination method is that it can stress the places where the significant values of the OWA weighting vector to be generated. Some typical stress function shapes were discussed. These can be classified to the function based OWA operator determination methods.

In the above three kinds of OWA operator determination methods, a common feature is that the OWA weight elements and the aggregated elements are treated in a separate way. The weights obtained with these methods are used to aggregate the input data. However, the characteristics of the input data is not considered in the OWA determination. Next, we will give two other kinds of input related OWA operator determination methods: The argument dependent methods and the preference relation methods.

In [69, 81], Yager and Filev proposed the argument dependent method to generating OWA operator weights with power function of the input data. It is called BADD (BAsic Defuzzification Distribution) OWA operator. Unlike the ordinary OWA operator, the aggregated elements is neat that the input elements do not have to be ordered. It was used to aggregate the linguistic labels represented by partially ordered fuzzy numbers in fuzzy group decision making problems [43, 44]. In [38], Liu and Lou extended it to the weighted function average operator, which was called the Additive Neat OWA (ANOWA) operator. An orness measure for the weighted function average operator is proposed. Xu [64] proposed dependent OWA operator that can relieve the influence of unfair arguments on the aggregated results. Similarly, Wu, Liang and Huang [62] proposed an argument dependent approach based on normal distribution, which assigns very low weights to these false or biased opinions to relieve the influence of the unfair arguments. Peláez and Doña [47] proposed a majority additive-ordered weighting averaging operator based on the majority process. Boongoen and Shen [13] propose a cluster based argument dependent OWA operator called Clus-DOWA, which applies distributed structure of data or data clusters to determine its weight vector.

The preference based OWA determination methods utilize the preference information as inputs, that the preference relations between alternatives can be revealed. With these preference relation, the model to determine the most suitable OWA operator to these preference relations can be constructed. The preference based OWA operator methods can be seen as an extension of the empirical data, where the empirical data is replaced by the preference matrix of the experts. There is not systematic methods to obtain the OWA operator from the preference relation. Ahn [2] present a method for determining the OWA weights, when the preferences of some subset of alternatives over other subset of alternatives are specified in a holistic manner across all the criteria. Emrouznejad [18] proposed MP-OWA (The most preferred OWA) operator, where the preferences of alternatives across all the criteria are considered and based on the most popular criteria for all alternatives.

Comparing these OWA determination methods, they are proposed at different time, based on different ideas and also develops in different extent. Some methods and models are relatively mature, which are studied comprehensively and profoundly with some systematic results, such as the optimization based determination methods. While some others is just at their very beginning, there is neither clear ideas nor specific methods to deal with them, such as the preference based OWA determination methods. Some others are developed in the stage between these two extremes, such as the sample learning methods and the argument dependent methods.

Next, we will give a review on the existing OWA determination methods for different classification in a sequential way. Some connections between these methods and between different methods classes are proposed. Some extensions of these methods are provided, especially about the optimization methods. Some future problems and possible research directions are also pointed in a personal view. It should also be noted that, despite these OWA determination methods are reviewed in a classification way, there is not clear cut edges among different classes. Some other criteria and classification methods may also exist with different points of view.

2 The Classified Summarization of OWA Determination Methods

2.1 Preliminaries

An OWA operator of dimension n is a mapping $F_W : \mathbb{R}^n \to \mathbb{R}$ that has an associated weighting vector $W = (w_1, w_2, \ldots, w_n)$ having the properties

$$w_1 + w_2 + \cdots + w_n = 1; 0 \leqslant w_i \leqslant 1, \qquad i = 1, 2, \ldots, n$$

and such that

$$F_W(X) = F_W(x_1, x_2, \ldots, x_n) = \sum_{j=1}^{n} w_j y_j \tag{1}$$

with y_j being the jth largest of the x_i.

The degree of "orness" associated with this operator is defined as:

$$orness(W) = \sum_{j=1}^{n} \frac{n-j}{n-1} w_j \tag{2}$$

The *max, min* and *average* correspond to W^*, W_* and W_A respectively, where $W^* = (1, 0, \ldots, 0)$, $W_* = (0, 0, \ldots, 1)$ and $W_A = \left(\frac{1}{n}, \frac{1}{n}, \ldots, \frac{1}{n}\right)$, that is $F_{W_*}(X) = \min_{1 \leqslant i \leqslant n} \{x_i\}$, $F_{W^*}(X) = \max_{1 \leqslant i \leqslant n} \{x_i\}$ and $F_{W_A}(X) = \frac{1}{n} \sum_{i=1}^{n} x_i = A(X)$. Obviously, $orness(W^*) = 1$, $orness(W_*) = 0$ and $orness(W_A) = \frac{1}{2}$.

From (2), some properties about OWA operator are listed in the following.

Proposition 1. $0 \leqslant orness(W) \leqslant 1$.

Proposition 2. *[69, p.127] For OWA operator weighting vector* $W = (w_1, w_2, \ldots, w_n)$, $orness(W) = \alpha$, *then for the reverse order of* W, $\widetilde{W} = (w_n, w_{n-1}, \ldots, w_1)$, $orness(\widetilde{W}) = 1 - \alpha$.

Proposition 3. *If* $X = (x_1, x_2, \ldots, x_n)$ *is evenly distributed on* $[0, 1]$, *that is* $x_i = \frac{n-i}{n-1}$, *then* $F_W(X) = orness(W)$.

2.2 The Optimization Based Method

The optimization method is a commonly used technique for OWA determination. From the definition of OWA operator, the elements should be none negative and should sum to a unit, that is $w_i \geqslant 0$ and $\sum_{i=1}^{n} w_i = 1$. But this condition is not sufficiently to determine the OWA elements. An given orness level of OWA operator can be assigned. However, the OWA operator with given orness level still can not be uniquely determined unless in the two dimensional case. A commonly used technique is to let the OWA operator to satisfy an additional optimization criterion.

The maximum entropy OWA determination under given orness level can be formulated as follows [19, 46]:

$$\begin{aligned} \max \quad & -\sum_{i=1}^{n} w_i \ln w_i \\ \text{s.t.} \quad & \sum_{i=1}^{n} \frac{n-i}{n-1} w_i = \alpha, \qquad 0 \leqslant \alpha \leqslant 1 \\ & \sum_{i=1}^{n} w_i = 1. \\ & w_i \geqslant 0, i = 1, 2, \ldots, n. \end{aligned} \tag{3}$$

Various methods for its optimal solutions were discussed [19, 21, 36]. The optimal solutions has geometric form with $\frac{w_i+1}{w_i} = q$ that can be expressed as[36]:

$$w_i = \frac{q^{i-1}}{\sum_{j=0}^{n-1} q^j} \tag{4}$$

where q is the root solution of (5).

$$(n-1)\alpha q^{n-1} + \sum_{i=2}^{n} \left((n-1)\alpha - i + 1\right) q^{n-i} = 0 \tag{5}$$

Remark 1. *With the logarithmic function in the objective function of* (3)*, we must have* $w_i > 0 (i = 1, 2, \ldots, n)$*. Furthermore, the two special cases* $\alpha = 0$ *and* $\alpha = 1$*, which correspond to the unique OWA weight vectors* W_* *and* W^* *with zero value elements respectively, also can not be included in the problem.*

Another kind of OWA operator determination method is the minimum variance problem for the OWA operator which was proposed by Fullér and Majlender [22]:

$$\begin{aligned} \min D^2(W) = & \frac{1}{n} \sum_{i=1}^{n} w_i^2 - \frac{1}{n^2} \\ \text{s. t.} & \sum_{i=1}^{n} \frac{n-i}{n-1} w_i = \alpha, \quad 0 \leqslant \alpha \leqslant 1 \\ & \sum_{i=1}^{n} w_i = 1, \\ & w_i \geqslant 0, \qquad i = 1, 2, \ldots, n. \end{aligned} \tag{6}$$

Ahn and Park [4] also call it the least square OWA operator when the objective function is replaced with $\sum_{i=1}^{n} \left(w_i - \frac{1}{n}\right) = \sum_{i=1}^{n} w_i^2 - \frac{2}{n} + \frac{1}{n^2}$, but the same optimal solution keeps the same.

Liu [31] proved that the optimal solution of (6) is in equidifferent form and can be obtained in with the following algorithm:

Algorithm (1)

Step 1: Determine m with (7).

$$m = \begin{cases} [3\alpha(n-1)+2] & \text{if } 0 < \alpha < \frac{1}{3}; \\ n & \text{if } \frac{1}{3} \leqslant \alpha \leqslant \frac{2}{3}, \\ [3n - 3\alpha(n-1) - 1] & \text{if } \frac{2}{3} < \alpha < 1. \end{cases} \tag{7}$$

Step 2: Determine d with (8).

$$d = \begin{cases} \frac{6(2\alpha - 2n\alpha + m - 1)}{m(m^2-1)} & \text{if } 0 < \alpha < \frac{1}{3}, \\ \frac{6(1-2\alpha)}{n(n+1)} & \text{if } \frac{1}{3} \leqslant \alpha \leqslant \frac{2}{3}, \\ \frac{6(2\alpha - 2n\alpha + 2n - m - 1)}{m(m^2-1)} & \text{if } \frac{2}{3} < \alpha < 1. \end{cases} \tag{8}$$

Step 3: Determine $W = (w_1, w_2, \ldots, w_n)$ with (9).

$$\begin{aligned} &\text{Case 1: } 0 < \alpha < \tfrac{1}{3},\ w_i = \begin{cases} 0, & \text{if } 1 \leqslant i \leqslant n-m, \\ \frac{-dm^2+dm+2}{2m} + (i-n+m-1)d, & \text{if } n-m+1 \leqslant i \leqslant n. \end{cases} \\ &\text{Case 2: } \tfrac{1}{3} \leqslant \alpha \leqslant \tfrac{2}{3},\ w_i = \frac{-dn^2+dn+2}{2n} + (i-1)d, \quad i = 1, 2, \ldots, n. \\ &\text{Case 3: } \tfrac{2}{3} < \alpha < 1,\ w_i = \begin{cases} \frac{-dm^2+dm+2}{2m} + (i-1)d, & \text{if } 1 \leqslant i \leqslant m, \\ 0, & \text{if } n-m+1 \leqslant i \leqslant n. \end{cases} \end{aligned} \tag{9}$$

Liu [31, 36] gave the parametric forms of these two kinds of problems by replacing the constraint of a fixed orness level with a general constraint $\sum_{i=1}^{n} w_i x_i = c$.

A parameterized extension of these two kinds of problems is the OWA determination method with maximum Rényi entropy [41]:

$$\begin{aligned} \min H(W) = &\frac{1}{1-r} \log_2 \sum_{i=1}^{n} w_i^r \\ \text{s. t. } &\sum_{i=1}^{n} \frac{n-i}{n-1} w_i = \alpha, \quad 0 \leqslant \alpha \leqslant 1 \\ &\sum_{i=1}^{n} w_i = 1, \\ &w_i \geqslant 0, \qquad i = 1, 2, \ldots, n. \end{aligned} \tag{10}$$

The maximum entropy problem (3) and the minimum variance problem (6) correspond to the special cases of (10) with $r = 0$ and $r = 2$ respectively.

Liu [34] proposed a general model to obtain the desired OWA operator under a given orness level:

$$\begin{aligned} \min \quad & V_{OWA} = \sum_{i=1}^{n} F(w_i) \\ \text{s. t.} \quad & \sum_{i=1}^{n} \frac{n-i}{n-1} w_i = \alpha, \quad 0 \leqslant \alpha \leqslant 1 \\ & \sum_{i=1}^{n} w_i = 1, \\ & w_i \geqslant 0 \qquad i = 1, 2, \ldots, n. \end{aligned} \tag{11}$$

where $F(x)$ is a strictly convex function on $[0, 1]$, and it is at least second order differentiable.

Problems (3) and (6) become special cases of (11) with $F(x) = x \ln x$, $F(x) = x^2$ respectively. And (10) also becomes the special case of (11) with $F(x) = x^r$. It should be noted that the maximum entropy problem (3) is a maximum problem with an additional negative sign in the objective function.

The optimal solution of (11) is unique, and it can be expressed as $W = (w_1, w_2, \ldots, w_n)$ that [34]:

$$w_i = \begin{cases} g(\frac{n-i}{n-1}\lambda_1 + \lambda_2) & \text{if } i \in T \\ 0 & \text{otherwise} \end{cases} \tag{12}$$

where λ_1, λ_2 are determined by

$$\begin{cases} \sum\limits_{i \in T} \frac{n-i}{n-1} g\left(\frac{n-i}{n-1}\lambda_1 + \lambda_2\right) = \alpha \\ \sum\limits_{i \in T} g\left(\frac{n-i}{n-1}\lambda_1 + \lambda_2\right) = 1 \end{cases} \tag{13}$$

and $T = \{i | 1 \leqslant i \leqslant n, g\left(\frac{n-i}{n-1}\lambda_1 + \lambda_2\right) > 0\}$ with $g(x) = (F')^{-1}(x)$.

As $F(x)$ is convex, the monotonic increasing function $g(x) = (F')^{-1}(x)$ can be used to determine the relationship of the nonzero OWA operator elements, where the nonzero elements become an equidifferent series with function transformation $g(x)$.

Regarding all the entropy function (3), the variance function (6) and the Rényi entropy (10) can be seen as the dispersion indexes of the weight element distribution, and the convex objective function of (11) can be seen as a general dispersion index. The meaning of the OWA operator solution of this problem is to try to minimize the general dispersion index means to distribute the weight elements as evenly as possible.

Besides the dispersion measure with a separate function of the weight elements, Yager [69] also used a measure of entropy $1 - \max\limits_{i \leqslant i \leqslant n} w_i$ as an alternative form of the maximum entropy problem:

$$\begin{aligned}
\min \quad & \max_{i\leqslant i\leqslant n} w_i \\
\text{s. t.} \quad & \sum_{i=1}^{n} \frac{n-i}{n-1} w_i = \alpha, \quad 0 \leqslant \alpha \leqslant 1 \\
& \sum_{i=1}^{n} w_i = 1, \\
& w_i \geqslant 0 \qquad i = 1, 2, \ldots, n.
\end{aligned} \tag{14}$$

Recently, Wu *et al.* [63] proposed the linear programming model for minimizing the distance its vector from the vector of maximal entropy:

$$\begin{aligned}
\text{minimize} & \sum_{i=1}^{n} \left| w_i - \frac{1}{n} \right| \\
\text{s.t.} \quad & \sum_{i=1}^{n} \frac{n-i}{n-1} w_i = \alpha, \quad 0 < \alpha < 1 \\
& \sum_{i=1}^{n} w_i = 1, \\
& w_i \geqslant 0, \qquad i = 1, 2, \ldots, n.
\end{aligned} \tag{15}$$

A closely related OWA operator determination method with the optimization criteria is the minimax form in the objective function. The first minimax problem for OWA operator, called minimax disparity problem, was proposed by Wang and Parkan [60]. The objective is to minimize the maximum disparity, where the disparities between two adjacent weights are made as small as possible:

$$\begin{aligned}
\text{minimize} & \left\{ \max_{1\leqslant i\leqslant n-1} |w_i - w_{i+1}| \right\} \\
\text{s.t.} \quad & \sum_{i=1}^{n} \frac{n-i}{n-1} w_i = \alpha, \quad 0 < \alpha < 1 \\
& \sum_{i=1}^{n} w_i = 1, \\
& w_i \geqslant 0, \qquad i = 1, 2, \ldots, n.
\end{aligned} \tag{16}$$

The solution equivalence to the minimum variance problem (6) of Fullér and Majlender [22] was verified theoretically by Liu [33] with the dual theory of linear programming.

In [34], such conclusions are extended. The minimax problem corresponding to the maximum entropy problem (3) and the general optimization problem (11) can also be proposed in a similar way as (17) and (18).

$$\begin{aligned}
\min & \left\{\max_{1\leqslant i\leqslant n-1} |\ln(w_i) - \ln(w_{i+1})|\right\} \\
\text{s. t.} & \sum_{i=1}^{n} \frac{n-i}{n-1} w_i = \alpha, \quad 0 < \alpha < 1 \\
& \sum_{i=1}^{n} w_i = 1.
\end{aligned} \tag{17}$$

$$\begin{aligned}
\min & \quad \left\{\max_{1\leqslant i\leqslant n-1} |F'(w_i) - F'(w_{i+1})|\right\} \\
\text{s.t.} & \sum_{i=1}^{n} \frac{n-i}{n-1} w_i = \alpha, \quad 0 < \alpha < 1 \\
& \sum_{i=1}^{n} w_i = 1, \\
& w_i \geqslant 0, \qquad i = 1, 2, \ldots, n.
\end{aligned} \tag{18}$$

Both (3) and (17), (11) and (18) have the same optimal solution. Furthermore, (16) and (17) can be seen as the special case of (18) with $F(x) = x^2$ and $F(x) = x\ln(x)$. The general minimax problem for OWA operators tries to obtain the desired OWA weight vector under given orness level to minimize the maximum difference between the adjacent elements after a monotonic function transformation. Comparing the objective functions of the original optimization problem (11) and that of the minimax problem (18), the former minimizes the sum of $F(w_i)$ and the latter tries to minimize the maximum differences between the adjacent $F'(w_i)$s.

Contrary to the OWA determination methods with given orness value, Marchant [42] proposed the problems of the OWA operator determination to maximize the orness level with a fixed variance or entropy value. The analytical methods for these two problems are proposed with the Lagrange multiplier method.

The maximizing orness problem with a fixed entropy value for the OWA operator and the maximizing orness problem with a fixed variance value are shown in (19) and (20) respectively.

$$\begin{aligned}
\max & \sum_{i=1}^{n} \frac{n-i}{n-1} w_i \\
\text{s.t.} & \quad -\sum_{i=1}^{n} w_i \ln w_i = \beta, \quad 0 < \beta \leqslant \ln(n) \\
& \sum_{i=1}^{n} w_i = 1.
\end{aligned} \tag{19}$$

$$\max \sum_{i=1}^{n} \frac{n-i}{n-1} w_i$$
$$\text{s. t.} \frac{1}{n} \sum_{i=1}^{n} w_i^2 - \frac{1}{n^2} = \delta, \quad 0 \leqslant \delta \leqslant \frac{n-1}{n^2}$$
$$\sum_{i=1}^{n} w_i = 1,$$
$$w_i \geqslant 0, i = 1, 2, \ldots, n. \tag{20}$$

As shown in [36] and [31], the optimal solution of the maximum entropy problem (3) can be expressed with geometrical form and the optimal solution minimum variance problem (6) can be expressed with equidifferent form. Liu [29] further shows that the optimal solutions of their maximizing orness problems (19) and (19) also have the geometrical form or equidifferent form as (3) and (6). Furthermore, with a given entropy value or variance value, there are usually two geometric or equidifferent OWA operators with orness value of α and $1 - \alpha$ respectively. They are the optimal solutions of maximum problems (19), (20) (The optimal solutions are the OWA operator with oress level $\max\{\alpha, 1-\alpha\}$) and minimum problems (21), (22) (The optimal solutions are the OWA operator with oress level $\min\{\alpha, 1 - \alpha\}$) respectively.

$$\min \sum_{i=1}^{n} \frac{n-i}{n-1} w_i$$
$$\text{s.t.} \quad -\sum_{i=1}^{n} w_i \ln w_i = \beta, \quad 0 < \beta \leqslant \ln(n) \tag{21}$$
$$\sum_{i=1}^{n} w_i = 1.$$

$$\min \sum_{i=1}^{n} \frac{n-i}{n-1} w_i$$
$$\text{s. t.} \frac{1}{n} \sum_{i=1}^{n} w_i^2 - \frac{1}{n^2} = \delta, \quad 0 \leqslant \delta \leqslant \frac{n-1}{n^2}$$
$$\sum_{i=1}^{n} w_i = 1,$$
$$w_i \geqslant 0, i = 1, 2, \ldots, n. \tag{22}$$

These OWA operator are also the optimal solutions of (3) and (6) with orness level α respectively.

For the minimax disparity problem, Amin and Emrouznejad [7] also (16) give a more general case that

$$\begin{aligned} \text{minimize} & \left\{ \max_{i \neq j} |w_i - w_j| \right\} \\ \text{s.t.} & \sum_{i=1}^{n} \frac{n-i}{n-1} w_i = \alpha, \quad 0 < \alpha < 1 \\ & \sum_{i=1}^{n} w_i = 1, \\ & w_i \geqslant 0, \qquad i = 1, 2, \ldots, n. \end{aligned} \tag{23}$$

Amin [6] further discussed some its properties in an analytical way, but the analytical solution is still not expressed in a specific formula.

Recently, Wang, Luo and Liu [59] proposed the least squares deviation and chi-square χ^2 model as

$$\begin{aligned} \text{minimize} & \sum_{i=1}^{n-1} (w_i - w_j)^2 \\ \text{s.t.} & \sum_{i=1}^{n} \frac{n-i}{n-1} w_i = \alpha, \quad 0 < \alpha < 1 \\ & \sum_{i=1}^{n} w_i = 1, \\ & w_i \geqslant 0, \qquad i = 1, 2, \ldots, n. \end{aligned} \tag{24}$$

$$\begin{aligned} \text{minimize} & \sum_{i=1}^{n-1} \left(\frac{w_i}{w_{i+1}} + \frac{w_{i+1}}{w_i} - 2 \right)^2 \\ \text{s.t.} & \sum_{i=1}^{n} \frac{n-i}{n-1} w_i = \alpha, \quad 0 < \alpha < 1 \\ & \sum_{i=1}^{n} w_i = 1, \\ & w_i \geqslant 0, \qquad i = 1, 2, \ldots, n. \end{aligned} \tag{25}$$

Unlike the maximum entropy problem (3) or minimum variance problem (6) and their extension (11) or variance (16)-(22), in which the analytical solutions can be obtained, and the properties be observed in a very clear way. Despite some interesting properties both proved theoretically and observed from numerical examples, the analytical solutions of (23)-(93) are still not obtained in general way, that some profound properties can not be further analyzed or discussed.

In spite of the published results of OWA determination methods with optimization criteria, a moro general model than (11) can be proposed [28].

From (12), we can see the nonzero OWA element w_i is a linear function of $\frac{n-i}{n-1}$ with additional transformation $g(x)$. As $g(x)$ is monotonically increasing, we can only obtain the OWA operator with increasing or decreasing distributed elements (which means $w_1 \geqslant w_2 \geqslant \cdots \geqslant w_n$ if $\lambda_1 \geqslant 0$ or $w_1 \leqslant w_2 \leqslant \cdots \leqslant w_n$ if $\lambda_1 \leqslant 0$). However, other forms of the OWA operator are also needed in applications. A commonly used one is the symmetric OWA operator with $w_i = w_{n+1-i}$ [17, 65, 73, 78]. These OWA operators are usually non-monotonic and can not be obtained just by setting an expression in the objective function of (11). Here, we will extend (11) to a more general form (26) in the constraints, by which we can get more flexible OWA operator distribution as its optimal solution.

$$\begin{aligned}\min \quad & V_{OWA} = \sum_{i=1}^{n} F(w_i) \\ \text{s. t.} \quad & \sum_{i=1}^{n} w_i = 1 \\ & \sum_{i=1}^{n} h\left(\frac{n-i}{n-1}\right) w_i = \alpha \\ & w_i \geqslant 0.\end{aligned} \tag{26}$$

where $F(x)$ is a strictly convex function in $[0, 1]$ and it is at least two order differentiable. $h(x)$ is nonnegative and continuous on $[0, 1]$.

If $h(x) = x$, (26) becomes the OWA determination model under a given orness level (11). To keep the feasible domain nonempty, the feasible value of α may be not limited in $[0, 1]$ as the orness level be. The lower and upper bounds of the feasible α in (26) can be determined by the solutions of the following problems:

$$\begin{aligned}\min(\max) \quad & \alpha = \sum_{i=1}^{n} h\left(\frac{n-i}{n-1}\right) w_i \\ \text{s. t.} \quad & \sum_{i=1}^{n} w_i = 1 \\ & 0 \leqslant w_i \leqslant 1.\end{aligned} \tag{27}$$

In some cases, such as when $h(x)$ is monotonically increasing, the lower and upper bounds of α can be the minimum and maximum values of $h(x)$ on $[0, 1]$ respectively.

The optimal solution of (26) is unique, and it can be expressed as $W = (w_1, w_2, \ldots, w_n)$ such that

$$w_i = \begin{cases} g\left(h\left(\frac{n-i}{n-1}\right)\lambda_1 + \lambda_2\right) & \text{if } i \in T \\ 0 & \text{otherwise} \end{cases} \tag{28}$$

where λ_1, λ_2 are determined by

$$\begin{cases} \sum\limits_{i\in T} h\left(\frac{n-i}{n-1}\right) g\left(h\left(\frac{n-i}{n-1}\right)\lambda_1 + \lambda_2\right) = \alpha \\ \sum\limits_{i\in T} g\left(h\left(\frac{n-i}{n-1}\right)\lambda_1 + \lambda_2\right) = 1 \end{cases} \tag{29}$$

and $T = \{i|1 \leqslant i \leqslant n, g\left(h\left(\frac{n-i}{n-1}\right)\lambda_1 + \lambda_2\right) > 0\}$ with $g(x) = (F')^{-1}(x)$.

Besides the special case that (26) becomes (11) when $h(x) = x$, if we let (28) in the optimal solution of (26) with $g(x) = x$, the adjacent weights relationships should be determined by the shape of $h(x)$, with $(F')^{-1}(x) = g(x)$, $F(x) = \frac{1}{2}x^2$, (26) becomes:

$$\begin{aligned} \min \quad & V_{OWA} = \frac{1}{2}\sum_{i=1}^{n} w_i^2 \\ \text{s. t.} \quad & \sum_{i=1}^{n} w_i = 1 \\ & \sum_{i=1}^{n} h\left(\frac{n-i}{n-1}\right) w_i = \alpha \\ & w_i \geqslant 0. \end{aligned} \tag{30}$$

The unique optimal solution of (30) is

$$w_i = \begin{cases} h\left(\frac{n-i}{n-1}\right)\lambda_1 + \lambda_2 & \text{if } i \in T \\ 0 & \text{otherwise} \end{cases} \tag{31}$$

where λ_1, λ_2 are determined by

$$\begin{cases} \sum\limits_{i\in T} h\left(\frac{n-i}{n-1}\right)\left(h\left(\frac{n-i}{n-1}\right)\lambda_1 + \lambda_2\right) = \alpha \\ \sum\limits_{i\in T}\left(h\left(\frac{n-i}{n-1}\right)\lambda_1 + \lambda_2\right) = 1 \end{cases} \tag{32}$$

and $T = \{i|1 \leqslant i \leqslant n, h\left(\frac{n-i}{n-1}\right)\lambda_1 + \lambda_2 > 0\}$.

The optimal solution of (30) with (31) and (32) can also be expressed as

$$w_i = \begin{cases} h\left(\frac{n-i}{n-1}\right)\lambda_1 + \lambda_2 & \text{if } h\left(\frac{n-i}{n-1}\right)\lambda_1 + \lambda_2 \geqslant 0 \\ 0 & \text{otherwise} \end{cases} \quad (33)$$

where λ_1, λ_2 are determined by

$$\begin{cases} \sum_{i=1}^{n} h\left(\frac{n-i}{n-1}\right) w_i = \alpha \\ \sum_{i=1}^{n} w_i = 1 \end{cases} \quad (34)$$

By selecting the shape of $h(x)$ and the parameter α, we can obtain different OWA operator weight series corresponding with $h(x)$. Similar to the relationship between (11) and (18), there is also a minimax problems for (26) and (30). The optimal solution of (30) with (33) and (34) can be connected with the stress function method [79], which will be discussed in the function based methods.

2.3 *The Sample Learning Method*

Empirical fit is a very useful tool for aggregation operator determination because it has a direct quantitative interpretation. In most cases, the problem of choosing the operator is translated into some sort of regression problem such as least squares fit. We usually requires an aggregation operator with certain properties to some sort of empirical data. The data can be collected in an experiment, by questioning experts in the field or by conducting a mental experiment [9–12, 56].

Consider the problem of fitting an OWA operator to empirical data of the form:

$$\{(x_{1k}, x_{2k}, \cdots, x_{nk}), d_k\}, k = 1, 2, \ldots, K \quad (35)$$

There are K observations, and every observation has n observed arguments $(x_{1k}, x_{2k}, \cdots, x_{nk})$ and the observed aggregated value d^k. Our goal is to find an OWA operator weighting vector $W = (w_1, w_2, \ldots, w_n)^T$ to satisfy these observations and sum unit condition. For simplification, we will assume that the input arguments have been ordered, that is $x_{1k} \geqslant x_{2k} \geqslant \cdots \geqslant x_{nk}$.

A very special case is that when $K = 1$, we only have one sample data pair. It is obvious that the OWA operator can not be completely determined. This can be solved with an additional criteria where the sample data becomes the constraint in the optimization problem.

Yager [70] proposed a model to maximize the orness level under an aggregation observation that $\sum_{i=1}^{n} w_i x_i = d$. The problem can be formulated as

$$\begin{aligned} \text{maximize} \sum_{i=1}^{n} \frac{n-i}{n-1} w_i \\ \text{s.t.} \sum_{i=1}^{n} w_i x_i = d \\ \sum_{i=1}^{n} w_i = 1, \\ w_i \geqslant 0, \qquad i = 1, 2, \ldots, n. \end{aligned} \tag{36}$$

Liu [31, 36] also proposed parameterized extensions of the maximum entropy problem (19) and minimum variance problem (6) where the orness level constraint is replaced with a aggregation data example $\{(x_1, x_2, \cdots, x_n), d\}$. For (19), we have

$$\begin{aligned} \max \quad -\sum_{i=1}^{n} w_i \ln w_i \\ \text{s.t.} \quad \sum_{i=1}^{n} w_i x_i = d \\ \sum_{i=1}^{n} w_i = 1. \\ w_i \geqslant 0 \qquad i = 1, 2, \ldots, n. \end{aligned} \tag{37}$$

The objective function can be replaced any other objective function in the optimization models that

$$\begin{aligned} \min \quad V_{OWA} = \sum_{i=1}^{n} F(w_i) \\ \text{s. t.} \sum_{i=1}^{n} w_i x_i = d, \\ \sum_{i=1}^{n} w_i = 1, \\ w_i \geqslant 0 \qquad i = 1, 2, \ldots, n. \end{aligned} \tag{38}$$

It is natural that we will assume $\sum_{i=1}^{n} w_i x_i = c$ and $\sum_{i=1}^{n} w_i = 1$ should be independent.

In most cases, there are usually more than one observation data sets, the problem of choice of the OWA operator is translated into some sort of regression problem such as least squares fit. A commonly used technique is the least square method [11]. The goal is to find an OWA operator W, so that the generated operator approximates the data in the least squares sense:

$$\begin{aligned} & \min \sum_{k=1}^{K} \left(F_W(x_1^k, x_2^k, \ldots, x_n^k) - d^k\right)^2 \\ & \text{s.t.} \sum_{i=1}^{n} w_i = 1, \\ & \qquad w_i \geqslant 0 \qquad i = 1, 2, \ldots, n. \end{aligned} \tag{39}$$

As Beliakov [9] pointed out, because of the constraints in (39), its solution is not as simple as that of the traditional linear regression problem.

Filev and Yager [20] propose a nonlinear change of variables to transform the domain of w from the unit simplex to unrestricted domain.

$$w_i = \frac{e^{\lambda_i}}{\sum_{j=1}^{n} e^{\lambda_j}} \tag{40}$$

Then, an iterative procedure is developed to minimize the transformed (no longer quadratic) error function.

Beliakov [9] proposed another two alternatives for the solution of (39). One is to use the penalty function approach and add appropriate penalty for violating the restrictions to the expression in (39), which is subsequently minimized using standard descent algorithms. The other is to solve the restricted linear least squares problem directly, taking advantage of the linearity of (39) and the constraints, where the problem is formulated as linear nonnegative least squares problem with equality constraints. It shows that the third method performs better than the other two mentioned approaches in respect to speed and the quality of the solution, such as to avoid the local minimizers and handle high dimensional size problems. Similar approaches relying on quadratic programming was also proposed [8, 10, 11, 54, 56]. And they are shown to be numerically efficient ad stable with respect to rank deficiency [12]. Beliakov also develops a software package AOTool, which can be freely downloaded from http://www.deakin.edu.au/ gleb/aotool.html.

As an extension of the OWA determination method of singular data example with additional criteria. Liu, Yang and Fang [39] discussed the relationships between the optimization method and the sample learning method in a more systematic way.

If we regard the first constraint of (38) as an example of aggregated data set $X = (1, 1, \ldots, 1)$ and its aggregation value 1. A more general problem of the OWA determination method from small size examples can be proposed in the following.

For $K \leqslant n - 1$, there are usually infinity solutions for these linear equations of w_i, even the nonnegative constraints of w_i are considered. The OWA operator weight elements can not completely determined. A reasonable consideration is to make the OWA operator weights satisfy some criterion. This becomes the general OWA operator determination method (41).

$$\begin{aligned}
&\min \sum_{i=1}^{n} F(w_i) \\
&\sum_{i=1}^{n} w_i x_i^k = d^k, \quad k = 1, 2 \ldots, K \\
&\sum_{i=1}^{n} w_i = 1, \\
&w_i \geqslant 0 \qquad i = 1, 2, \ldots, n.
\end{aligned} \tag{41}$$

where F is a strictly convex function on $[0, 1]$, and it is at least two order differentiable.

Problem (41) includes some typical optimization method based OWA operator determination methods as special cases. The analytical solution was also proposed [39].

For problem (41), if the feasible domain is nonempty, then it has an unique optimal solution, and it can be expressed as $W = (w_1, w_2, \ldots, w_n)$ that

$$w_i = \begin{cases} (F')^{-1}(r_i) & \text{if } (F')^{-1}(r_i) > 0 \\ 0 & \text{otherwise} \end{cases} \tag{42}$$

where $r_i = -\sum_{i=1}^{K} x_i^k \lambda_k - \lambda_{K+1}$ and $\lambda_i (i = 1, , 2, \ldots, K + 1)$ are determined by

$$\begin{cases} \sum_{i=1}^{n} w_i x_i^k - d^k = 0 & k = 1, 2, \ldots, K \\ \sum_{i=1}^{n} w_i - 1 = 0 \end{cases} \tag{43}$$

If the number of examples is more than $n - 1$, problem (41) usually has no feasible solution. A commonly used technique is the least square method as (39). The minimum least square criterion can be adopted for (43), that is (42) and (43) can be transformed as

$$\begin{aligned}
&\min \sum_{k=1}^{K} \left(\sum_{i=1}^{n} w_i x_i^k - d^k \right)^2 \\
&\text{s. t.} \sum_{i=1}^{n} w_i - 1 = 0
\end{aligned} \tag{44}$$

where

$$w_i = \begin{cases} g(r_i) & \text{if } (F')^{-1}(r_i) > 0 \\ 0 & \text{otherwise} \end{cases}$$

and $r_i = -\sum_{k=1}^{K} \lambda_k x_i^k - \lambda_{K+1}$, $g(x) = (F')^{-1}(x)$ is a monotone increasing function.

Consider an special example of $g(x) = e^x$, as e^x is always positive, then (44) and (42) becomes

$$\begin{aligned} &\min \sum_{k=1}^{K} \left(\sum_{i=1}^{n} w_i x_i^k - d_k \right)^2 \\ &\text{s. t.} \sum_{i=1}^{n} w_i - 1 = 0 \end{aligned} \tag{45}$$

where w_i, $i = 1, 2, \ldots, n$ have the form as

$$w_i = e^{-\sum_{k=1}^{K} \lambda_i x_i^k - \lambda_{K+1}} \tag{46}$$

For the sack of easy manipulation, we can let $\lambda_i' = -\sum_{k=1}^{K} x_i^k \lambda_k - \lambda_{K+1}$ then $w_i = e^{\lambda_i'}$, with $\sum_{i=1}^{n} w_i - 1 = 0$, w_i can be expressed as

$$w_i = \frac{e^{\lambda_i'}}{\sum_{j=1}^{n} e^{\lambda_j'}} \tag{47}$$

This is just the OWA operator determination method from observations that was proposed by Filev and Yager [20]. That is the method of [20] becomes a special case of (44) with $g(x) = e^x$.

The OWA operator weight elements w_i in (44) is determined by the parameter λ_k, $k = 1, 2, \ldots, K + 1$. After a transformation of λ_k to r_i with

$$r_i = -\sum_{k=1}^{K} x_i^k \lambda_k - \lambda_{K+1}$$

As the observation examples are independent, so transformation matrix has the rank of n, the problem of determine w_i from (44) with parameter λ_k, $k = 1, 2, \ldots, K+1$ changes into the problem with parameter r_i, $i = 1, 2 \ldots, n$, that

$$\begin{aligned} &\min \sum_{k=1}^{K} \left(\sum_{i=1}^{n} w_i x_i^k - d^k \right)^2 \\ &\text{s. t.} \sum_{i=1}^{n} w_i - 1 = 0 \end{aligned} \tag{48}$$

where

$$w_i = \begin{cases} (F')^{-1}(r_i) & \text{if } (F')^{-1}(r_i) > 0 \\ 0 & \text{otherwise} \end{cases}$$

As $F(x)$ is a convex function, $(F')^{-1}$ is strictly increasing, if $(F')^{-1}$ can have the domain $[0, 1]$, (48) can further be transformed in a very simple way:

$$\min \sum_{k=1}^{K} \left(\sum_{i=1}^{n} w_i x_i^k - d^k \right)^2$$
$$\text{s. t.} \sum_{i=1}^{n} w_i - 1 = 0, \tag{49}$$
$$w_i \geqslant 0 \qquad i = 1, 2, \ldots, n.$$

This is just the ordinary least square OWA determination method. This means that when the number of observation examples is larger than the OWA operator dimension, or the feasible domain of w_i to satisfy these observation examples is empty, the criteria imposed on the form of OWA weights will take no effect.

In fact, Problem (41) can also be extended as a bi-level optimization problem

$$\min \sum_{i=1}^{n} F(w_i)$$
$$\min \sum_{k=1}^{K} \left(\sum_{i=1}^{n} w_i x_i^k - d^k \right)^2 \tag{50}$$
$$\sum_{i=1}^{n} w_i = 1$$
$$w_i \geqslant 0, i = 1, 2, \ldots, n.$$

When $K < n$, the lower level problem usually have infinite solutions that can make the objective function of the lower level $\sum_{k=1}^{K} \left(\sum_{i=1}^{n} w_i x_i^k - b^k \right)^2$ reaches its lower bound 0. The upper level objective function $\sum_{i=1}^{n} F(w_i)$ can be further be optimized within the optimal solutions of the lower level. The problem becomes the extended optimization model. However, if the number of observation examples is equal or more than n, the lower problem usually has an unique optimal solution, it is natural that the upper level objective will can take no influence on the OWA weight determination. The problem becomes the ordinary least square method.

2.4 The Function Based Methods

In [70], Yager proposed a method for obtaining the OWA weighting vectors via fuzzy linguistic quantifiers, especially the Regular Increasing Monotone (RIM) quantifier, which can provide information aggregation procedures guided by verbally expressed concepts and a dimension independent description of the desired aggregation.

Definition 1. *[70] A fuzzy subset Q of the real line is called a Regular Increasing Monotone(RIM) quantifier if $Q(0) = 0$, $Q(1) = 1$, and $Q(x) \geqslant Q(y)$ if $x > y$.*

Examples of this kind of quantifier are *all, most, many, there exists* [70].

With a RIM quantifier Q, the OWA weighting vector can be obtained as [70]:

$$w_i = Q\left(\frac{i}{n}\right) - Q\left(\frac{i-1}{n}\right) \tag{51}$$

The quantifier guided aggregation with OWA operator is

$$F_Q(X) = F_W(X) = \sum_{i=1}^{n} \left(Q\left(\frac{i}{n}\right) - Q\left(\frac{i-1}{n}\right)\right) x_i \tag{52}$$

With (51), the OWA operator and the corresponding RIM quantifier can be seen as the same aggregation operator in discrete and continuous cases respectively. With the generation of RIM quantifier [35] or the stress function [79], the properties and their generating methods of the OWA operator and the RIM quantifier can be corresponded each other [32, 34, 37, 76]. If we have known the solutions or the properties of one of them, the solution and the properties of the other form can also be anticipated, and vice versa.

In [69], Yager proposed various forms of OWA weighting vectors and the corresponding RIM quantifiers with piecewise linear membership functions. They include the Slide OWA (S-OWA), Step OWA and Window OWA operator and RIM quantifier forms. Here, the parameterized families of these RIM quantifiers are given.

The slide RIM quantifier corresponds to the Slide OWA(S-OWA) operator [69, 81]. A slide RIM quantifier can be defined as:

$$Q(x) = \begin{cases} 0 & \text{if} \quad x = 0, \\ \alpha + (1-\alpha-\beta)x & \text{if} \quad 0 < x < 1, \\ 1 & \text{if} \quad x = 1. \end{cases} \tag{53}$$

where $\alpha + \beta \leqslant 1$.

$$orness(Q) = \frac{1}{2}(1 + \alpha - \beta) \tag{54}$$

When $\alpha = 0$, this becomes the andlike S-OWA quantifier, and when $\beta = 0$, this becomes the orlike S-OWA quantifier, which was discussed in [69, p.134-136].

The corresponding OWA operator $W = (w_1, w_2, \ldots, w_n)$ in the discrete case is

$$w_i = \begin{cases} \frac{1}{n}(1-\alpha-\beta) + \alpha & \text{if} \quad i = 1, \\ \frac{1}{n}(1-\alpha-\beta) & \text{if} \quad i = 2, \ldots, n-1, \\ \frac{1}{n}(1-\alpha-\beta) + \beta & \text{if} \quad i = n. \end{cases} \tag{55}$$

$$orness(W) = \frac{1}{2}(1 + \alpha - \beta) \tag{56}$$

From (55), α, β can be seen as the parameters of rotation transformation.

A step RIM quantifier is a generator of the step OWA operator [69, p.136].

$$Q(x) = \begin{cases} 0 & \text{if } 0 \leqslant x \leqslant \gamma, \\ 1 & \text{if } \gamma < x \leqslant 1. \end{cases} \tag{57}$$

$$orness(Q) = 1 - \gamma \tag{58}$$

The step RIM quantifier can be interpreted as "at least γ percent" and the orness of the OWA operator maybe discontinuous.

The corresponding step OWA operator $W = (w_1, w_2, \ldots, w_n)$ is

$$w_i = \begin{cases} 1 & \text{if } i = k, \\ 0 & \text{if } i \neq k. \end{cases} \tag{59}$$

$$orness(W) = \frac{n-k}{n-1} \tag{60}$$

Obviously, a consistent RIM quantifier family can be obtained just by shifting γ on $[0, 1]$.

A window RIM quantifier corresponds to the window OWA operator [69, p.137], which disregards the top and bottom scoring elements.

$$Q(x) = \begin{cases} 0 & \text{if } 0 \leqslant x \leqslant \alpha, \\ \frac{x-\alpha}{\beta} & \text{if } \alpha < x \leqslant \alpha + \beta, \\ 1 & \text{if } \alpha + \beta < x \leqslant 1. \end{cases} \tag{61}$$

where $\alpha + \beta \leqslant 1$.

$$orness(Q) = 1 - \alpha - \frac{1}{2}\beta \tag{62}$$

The corresponding step OWA operator $W = (w_1, w_2, \ldots, w_n)$ is

$$w_i = \begin{cases} 0 & \text{if } i < k, \\ \frac{1}{m} & \text{if } k \leqslant i < k + m, \\ 0 & \text{if } i \geqslant k + m. \end{cases} \tag{63}$$

$$orness(W) = \frac{1}{n-1}\left(n - k - \frac{1}{2}(m-1)\right) \tag{64}$$

The $orness(W)$ is also discontinuous on $[0, 1]$.

Yager [78] extended this method to get the symmetrical OWA operator called centered OWA operator in any shape with the centering function.

$$w_i = \int_{\frac{j-1}{n}}^{\frac{j-1}{n}} f(x)\, dx$$

Where $f(x)$ is a centering function that $f(x) \geqslant 0$, $f(x) = f(1-x)$, $f(x)$ is increasing on $[0, 0.5]$ and $\int_0^1 f(x)dx = 1$. Various center function shapes and their corresponding OWA operators are illustrated.

Xu [65] proposed a method to generate OWA operator weights from the normal distribution

$$f(x) = \frac{1}{\delta\sqrt{2\pi}} e^{-\frac{(x-\mu)^2}{2\delta^2}} \tag{65}$$

that

$$w_i = \frac{e^{-(i-\mu_n)^2/2/\delta_n^2}}{\sum_{j=1}^{n} e^{-(j-\mu_n)^2/2/\delta_n^2}} \tag{66}$$

where $\mu_n = \frac{1+n}{2}$ and $\delta_n = \sqrt{\frac{1}{n}\sum_{i=1}^{n}(i-\mu_n)^2}$.

Both the centered OWA operator of Yager [78] and the OWA operator generated from normal distribution of Xu [65] are symmetrical that $w_i = w_{n-i+1}$.

Sadiq and Tesfamariam [51] also extended the method of Xu [65] to the non symmetric ones. A fractile index is introduced to locate the maximum value weight element with $\mu_n = \lambda(1+n)$, $\lambda \in [0,1]$ can be referred to as fractile or quantile representing the location of the maximum weight, which is assigned to the median ordinal position of the OWA operator. If $\lambda < 0.5$, a positively(skewed distribution can be generated, and if $\lambda > 0.5$, a negatively skewed distribution (leaning towards right) can be generated. In spite of the normal distribution, this method can also be extended to other distribution functions such as the inverse normal distribution and exponential distribution function.

Recently, Yager also [79] proposed the function based OWA operator aggregation from another view, the stress function method. An important advantage of the stress function method is that it allows the user very easily characterize the nature of the resulting OWA aggregation operator, which is accomplished by stressing the places where they want the significant values of the OWA weighting vector to be generated.

For example, the stress function tries to generate w_j with $F(j/n) - F((j-1)/n)$ in an approximate way, where $F'(x) = h(x)$ is the stress function. The basic form of this method is

$$w_j = \frac{h\left(\frac{j}{n}\right)}{\sum_{j=1}^{n} h\left(\frac{j}{n}\right)} \tag{67}$$

A general form of (67) is that

$$w_j = \frac{h\left(\xi_j\right)}{\sum_{j=1}^{n} h\left(\xi_j\right)} \tag{68}$$

where $x_j \in \left[\frac{j-1}{n}, \frac{j}{n}\right]$.

If $\xi_j = \frac{j}{n}$ then (68) becomes (67). Other typical alternatives for ξ_j selection are that $\xi_j = \frac{j-1}{n}$ or $\xi_j = \frac{j-0.5}{n}$ [79].

The following two methods can be regarded using the function of the weight element rather than a external analytical function for every weight elements. Filev and Yager [20] proposed a exponential smoothing method for OWA determination that

$$w_1 = \theta, w_2 = \theta(1-\theta), \ldots, w_{n-1} = \theta(1-\theta)^{n-1}, w_n = (1-\theta)^{n-1}$$

or

$$w_1 = \theta^{n-1}, w_2 = (1-\theta)\theta^{n-2}, \ldots, w_{n-1} = \theta(1-\theta), w_n = 1-\theta$$

When n is fixed, with different θ we can get the corresponding exponential OWA operator weights. θ can be seen as a parameter associated with the orness level.

Ahn [1] proposed a method of obtaining OWA operator with any orness level with the fact the some special forms of rank based weights can be result the constant level of orness. Such as For $W = (w_1, w_2, \ldots, w_n)$,

1. If $w_i^{(1)} = \frac{1}{n}\sum_{j=i}^{n} \frac{i}{j}$, then $orness(W^{(1)}) = \frac{3}{4}$
2. If $w_i^{(2)} = \frac{1}{n}\sum_{j=i}^{n} \frac{i}{n-j+1}$, then $orness(W^{(2)}) = \frac{1}{4}$.
3. If $w_i^{(3)} = \frac{1}{n}\sum_{j=1}^{n} \frac{n-i+1}{n-j+1}$, then $orness(W^{(3)}) = \frac{2}{3}$.
4. If $w_i^{(4)} = \frac{1}{n}\sum_{j=1}^{n} \frac{i}{n-j+1}$, then $orness(W^{(3)}) = \frac{1}{3}$.

For any two OWA operator W^1, and W^2 with constant level of ornress k_1, k_2, where $k_1 \neq k_2$, Lets us assume $k_1 \leqslant k_2$, then for any orness level $k \in [k_1, k_2]$, we can get an OWA operator $W = \beta W^1 + (1-\beta)W^2$ with constant level of orness k, where β is the solution of $\beta k_1 + (1-\beta)k_2 = k$.

Next, we will discuss the relationship between the general optimization model (30) and the stress function method (67). (67) can be seen as the solution of the following problem.

$$w_j = \lambda_1 h\left(\frac{j}{n}\right) \tag{69}$$

with $\sum_{i=1}^{n} w_i = 1$.

A variation of (67) is to let $\xi_j = \frac{j-1}{n-1}$ in (68), so that :

$$w_j = \frac{h\left(\frac{j-1}{n-1}\right)}{\sum_{j=1}^{n} h\left(\frac{j-1}{n-1}\right)} \tag{70}$$

(70) can be seen as the solution of the following problem:

$$w_j = \lambda_1 h\left(\frac{j-1}{n-1}\right) \tag{71}$$

with $\sum_{i=1}^{n} w_i = 1$.

On the other hand, the optimal solution of (30) is (33). From (33), we can see that w_i can be completely determined by the shape of $h(x)$. To make w_i be distributed in the shape of $h(x)$, it is obvious that we should have $T = \{i | 1 \leqslant i \leqslant n\}$, we will have

$$w_i = h\left(\frac{i-1}{n-1}\right)\lambda_1 + \lambda_2 \tag{72}$$

where λ_1, λ_2 is determined by

$$\begin{cases} \sum_{i=1}^{n} h\left(\frac{i-1}{n-1}\right)\left(h\left(\frac{i-1}{n-1}\right)\lambda_1 + \lambda_2\right) = \alpha \\ \sum_{i=1}^{n} h\left(\frac{i-1}{n-1}\right)\lambda_1 + n\lambda_2 = 1 \end{cases} \tag{73}$$

To make w_i in the shape of $h(x)$, we also should have $\lambda_1 \geqslant 0$.

Next, we will determine the bound of α to make w_i in the shape of $h(x)$. As $\lambda_1 \geqslant 0$, λ_1 monotonically increases with α [28], so the lower bound of α, α_L corresponds to $\lambda_1 = 0$, where the optimal solution is always $W_A = (\frac{1}{n}, \frac{1}{n}, \ldots, \frac{1}{n})$. In this case

$$\alpha_L = \sum_{i=1}^{n} h\left(\frac{i-1}{n-1}\right)\frac{1}{n} \tag{74}$$

Then, we will try to find the upper bound value of α that can keep the optimal solution distributed in $h(x)$. As $w_i = h\left(\frac{i-1}{n-1}\right)\lambda_1 + \lambda_2 (\lambda_1 \geqslant 0)$, so

$$\begin{aligned} \min_{1 \leqslant i \leqslant n}\{w_i\} &= \min_{1 \leqslant i \leqslant n}\left\{h\left(\frac{i-1}{n-1}\right)\lambda_1 + \lambda_2\right\} \\ &= \min_{1 \leqslant i \leqslant n}\left\{h\left(\frac{i-1}{n-1}\right)\right\}\lambda_1 + \lambda_2 \end{aligned} \tag{75}$$

Let $\min_{1 \leqslant i \leqslant n}\{w_i\} = 0$, we can find the corresponding value of λ_1 and λ_2 for the upper bound of α with

$$\alpha_U = \sum_{i=1}^{n} h\left(\frac{i-1}{n-1}\right)\left(h\left(\frac{i-1}{n-1}\right)\lambda_1 + \lambda_2\right) \tag{76}$$

where λ_1, λ_2 is determined by

$$\begin{cases} \min_{1 \leqslant i \leqslant n}\left\{h\left(\frac{i-1}{n-1}\right)\right\}\lambda_1 + \lambda_2 = 0 \\ \sum_{i=1}^{n} h\left(\frac{i-1}{n-1}\right)\lambda_1 + \lambda_2 = 1 \end{cases} \tag{77}$$

Thus, for any $\alpha \in [\alpha_L, \alpha_U]$, we can always find an OWA operator with the optimal solution of (30). (30) becomes (78) with an additional limitation on α.

$$\begin{aligned}
\min \quad & V_{OWA} = \frac{1}{2}\sum_{i=1}^{n} w_i^2 \\
\text{s. t.} \quad & \sum_{i=1}^{n} w_i = 1, \\
& \sum_{i=1}^{n} h\left(\frac{i-1}{n-1}\right) w_i = \alpha, \qquad \alpha \in [\alpha_L, \alpha_U] \\
& w_i \geqslant 0 \qquad i = 1, 2, \ldots, n.
\end{aligned} \tag{78}$$

The optimal solution of (78) can be expressed as

$$w_i = h\left(\frac{i-1}{n-1}\right)\lambda_1 + \lambda_2 \tag{79}$$

where λ_1, λ_2 is determined by

$$\begin{cases} \sum_{i=1}^{n} h\left(\frac{i-1}{n-1}\right)\left(h\left(\frac{i-1}{n-1}\right)\lambda_1 + \lambda_2\right) = \alpha \\ \sum_{i=1}^{n} h\left(\frac{i-1}{n-1}\right)\lambda_1 + n\lambda_2 = 1 \end{cases} \tag{80}$$

As the interval $[\alpha_L, \alpha_U]$ is dependent on $h(x)$. For easy comparison, we can select $\alpha = (1-t)\alpha_L + t\alpha_U$, where $t \in [0,1]$. t can be regard the parameter to control the distribution shape of the OWA operator between the average operator and the maximum curvature degree as $h(x)$.

It can be seen that (71) is a special case of (72) with $\lambda_2 = 0$. Comparing with (67) and (72), we can see that (67) obtains the OWA operator weight element w_j as the interpolation points of $\lambda_1 h(x)$ for $x_j = \frac{j}{n}$ with constraint $\sum_{j=1}^{n} w_j = 1$. While (72) obtains the OWA operator weight element w_j with the interpolation points of $\lambda_1 h(x) + \lambda_2$ as $x_j = \lambda_1 h(\frac{j-1}{n-1}) + \lambda_2$ with constraint $\sum_{j=1}^{n} w_j = 1$ and $\sum_{j=1}^{n} h(\frac{j-1}{n-1}) w_j = \alpha$. In he former case, the interpolation points are distributed on $\left[\frac{1}{n}, 1\right]$. In the latter case, the interpolation points are distributed on $[0,1]$, which can take into consideration the shape of $h(x)$ close to 0.

Another difference between (67) and (72) is that (67) only obtains OWA operator in the shape of $h(x)$ with scale parameter λ_1 as (69). However, $h(x)$ in (72) changes with both scale parameter λ_1 and vertical shift parameter λ_2. This makes the OWA operator solution always include the average operator as a special case with $\lambda_1 = 0$. So (72) can be seen as an extension of (67) with the optimization method.

From (78), for any nonnegative function $h(x) \in [0,1]$, we can always get an OWA operator that make it in the shape of $h(x)$. If $h(x)$ is selected as the normal distribution as (65) with $\mu = 0$, then we can get the symmetrical OWA operator weights similar to that of Xu [65]. Similar results to that of Yager [78] or Sadiq and Tesfamariam [51] can also be obtained if $h(x)$ is set to the corresponding centered function or the probability function respectively, and the distribution of the weight elements can be anticipated in an intuitive way. Another difference of between (78) and the other mentioned function based OWA determination methods is that in spite of using $h(x)$ to control the distribution of the OWA operator solution, we can also set α with different value that make the OWA operator solution with different curvature extent from the ordinary arithmetic average operator.

2.5 Argument Dependent Methods

In all of the proceeding we have assumed that the weights were fixed given constant values. The derived weights are associated with particular ordered positions of the aggregated arguments but have no connection with the aggregated arguments. In this section we shall generalize the concept of OWA aggregation by allowing the weights to be a function of the aggregated arguments.

Let $(x_1, x_2, \ldots, x_n)$ be the aggregated arguments set, and y_j be the jth largest element of the collection aggregated elements.

The first family of the argument dependent determination weights was proposed by Yager [69, 81].

$$w_i = \frac{y_i^\alpha}{\sum_{i=1}^n y_i^\alpha}$$

That

$$F(x_1, x_2, \ldots, x_n) = \frac{\sum_{i=1}^n y_i^{\alpha+1}}{\sum_{i=1}^n y_i^\alpha} = \frac{\sum_{i=1}^n x_i^{\alpha+1}}{\sum_{i=1}^n x_i^\alpha}$$

Yager [69] call it neat as the aggregated value is independent of the ordering. This operator is called BADD (BAsic Defuzzification Distribution) OWA operator.

With different $\alpha \in (-\infty, +\infty)$, we can make the aggregation value range among minimum, arithmetic average and maximum.

Xu [64] proposed the dependent OWA (DOWA) operator. Let $\mu = \frac{1}{n}\sum_{i=1}^n x_i$ be the average value of this argument set. The similarity between any argument x_j and the average value μ can be calculated as follows:

$$s(x_j, \mu) = 1 - \frac{|x_j - \mu|}{\sum_{i=1}^n |x_i - \mu|}$$

From this, a weight vector $W = (w_1, w_2, \ldots, , w_n)$ can be generated by applying the following:

$$w_j = \frac{s(y_j, \mu)}{\sum_{i=1}^n s(y_i, \mu)} \qquad j = 1, 2, \ldots, n$$

that

$$DOWA(x_1, x_2, \ldots, , x_n) = \sum_{i=1}^n w_i x_i$$

As an extension of the OWA determination method with normal distribution of [65], Wu, Liang and Huang [62] proposed an argument dependent approach based on normal distribution, which assigns very low weights to these false or biased opinions and can relieve the influence of the unfair arguments.

$$w_j = \frac{e^{-(y_j-\mu)^2/2\delta^2}}{\sum_{i=1}^{n} e^{-(y_i-\mu)^2/2\delta^2}}$$

where $\mu = \frac{1}{n}\sum_{i=1}^{n} y_j$, and $\delta = \sqrt{\frac{1}{n}\sum_{i=1}^{n}(y_j-\mu)^2}$.

Peláez and Doña [47] proposed a majority additive-ordered weighting averaging (MA-OWA) operator with based on the majority process, which use the cardinality of the elements to calculate the corresponding weights:

$$F_{MA} = (x_1, x_2, \ldots, x_n) = \sum_{j=1}^{n} w_j y_j = \sum_{j=1}^{n} f_j(y_1, y_2, \ldots, y_n) y_j$$

where $w_j \in [0,1]$ and $\sum_{j=1}^{n} w_j = 1$.

$$w_j = f_j(y_1, y_2, \ldots, y_n) = \frac{1}{\prod_{k=g_j}^{n} h_k(y_1, y_2, \ldots, y_n)}$$

where g_j where g_j is a function that indicate when the y_j element is used in the aggregation process. The h_k is a function that indicates the number of elements in each step in the aggregation process. It can also be extended to the quantified majority OWA operator called QM-OWA operator [48].

The weights of the MA-OWA operator are calculated as follows [49].

Let δ_i be the cardinality for the element i with $\delta_i > 0$, then

$$w_i = f_i(y_1, \ldots, y_n) = \frac{\gamma_i^{\delta_{\min}}}{\theta_{\delta_{\max}} \cdot \theta_{\delta_{\max}-1} \cdots \theta_{\delta_{\min}+1} \cdot \theta_{\delta_{\min}}} + \frac{\gamma_i^{\delta_{\min}+1}}{\theta_{\delta_{\max}} \cdot \theta_{\delta_{\max}-1} \cdots \theta_{\delta_{\min}+1}} + \cdots + \frac{\gamma_i^{\delta_{\max}}}{\theta_{\delta_{\max}}}$$

where

$$\gamma_i^k = \begin{cases} 1 & \text{if } \delta_i \geqslant k, \\ 0 & \text{otherwise.} \end{cases}$$

and

$$\theta_i = \begin{cases} (\text{number of item with cardinality} \geqslant i) + 1 & \text{if } i \neq \delta_{min}, \\ \text{number of item with cardinality} \geqslant i & \text{otherwise.} \end{cases}$$

Boongoen and Shen [13] propose a cluster based argument dependent OWA operator called Clus-DOWA, which applies distributed structure of data or data clusters to determine its weight vector.

With the clustering algorithm to a set of values $(x_1, x_2, \ldots, , x_n)$, the reliability of each value r_i can be directly estimated from the distance to its nearest cluster d_i recorded during the clustering process.

$$r_j = 1 - \frac{d_j}{\sum_{i=1}^{n} d_i}$$

thus

$$w_j = \frac{r_j}{\sum_{i=1}^{n} r_i}$$

and

$$Clus - DOWA(x_1, x_2, \ldots, x_n) = \sum_{i=1}^{n} w_i x_i$$

It is obvious that for these argument dependent weight determination methods, the aggregation property is determined by the weight generating function $w_j = f_j(y_1, y_2, \ldots, y_n)$, and the orness value of such argument dependent weights is variant for the same w_j. To solve such in consistent, Liu and Lou [38] give a general form of the additive argument dependent OWA (ADOWA) operator with

$$w_j = \frac{f(x_j)}{\sum_{i=1}^{n} f(x_i)}$$

That

$$ADOWA(x_1, x_2, \ldots, , x_n) = \frac{\sum_{i=1}^{n} x_i f(x_j)}{\sum_{i=1}^{n} f(x_i)} \tag{81}$$

which can include the BADD OWA operator as a special case with $f(x) = x^r$.

An alternative orness measure for such argument dependent weights and their determination methods were proposed.

$$orness(f) = \frac{\int_a^b (x - a) f(x) dx}{(b - a) \int_a^b f(x) dx} \tag{82}$$

Two function determination methods with the maximum entropy and minimum variance principle were proposed respectively. For example, the function $f(x)$ determination with maximum entropy principle can be formulated as

$$\begin{aligned} &\max - \int_a^b f(x) \ln f(x) dx \\ &\text{s.t.} \quad \int_a^b (1 - x) f(x) dx = \alpha, \qquad 0 < \alpha < 1 \\ &\qquad \int_0^1 f(x) dx = 1. \end{aligned} \tag{83}$$

The optimal solution is an exponential function:

$$f(x) = \frac{\lambda e^{\lambda x}}{e^{\lambda} - 1} \tag{84}$$

where λ is the root of the equation $\frac{e^{\lambda} - \lambda - 1}{\lambda(e^{\lambda} - 1)} = \alpha$.

And for the function $f(x)$ determination with minimum variance principle, we have

$$\min D^2(f(x)) = \frac{1}{b-a}\int_a^b f^2(x)dx - \frac{1}{(b-a)^2}$$
$$\text{s. t.} \int_a^b xf(x)dx = a + (b-a)\alpha, \quad 0 < \alpha < 1 \tag{85}$$
$$\int_a^b f(x) = 1, \quad f(x) \geqslant 0.$$

we can get the optimal solution in the following three cases:

1. If $\alpha \in (0, \frac{1}{3}]$, then

$$f(x) = \begin{cases} \frac{6\alpha(b-a)-2(x-a)}{9\alpha^2(b-a)^2}, & \text{if } a \leqslant x < a + 3\alpha(b-a), \\ 0, & \text{if } a + 3\alpha(b-a) \leqslant x \leqslant b. \end{cases} \tag{86}$$

2. If $\alpha \in (\frac{1}{3}, \frac{2}{3}]$, then

$$f(x) = \frac{(12\alpha - 6)(x-a) + (4-6\alpha)(b-a)}{(b-a)^2}, \qquad a \leqslant x \leqslant b. \tag{87}$$

3. If $\alpha \in (\frac{2}{3}, 1)$, then

$$f(x) = \begin{cases} 0, & \text{if } a \leqslant x < a + 3\left(\alpha - \frac{2}{3}\right)(b-a), \\ \frac{2(x-a)-(6\alpha-4)(b-a)}{9(1-\alpha)^2(b-a)^2}, & \text{if } a + 3\left(\alpha - \frac{2}{3}\right)(b-a) \leqslant x \leqslant b. \end{cases} \tag{88}$$

Like the case of $f(x) = x^r$, the ADOWA (81) with (84) can also range minimum, arithmetic average and maximum. Some properties of it are discussed [38].

It can be seen that the ADOWA maximum entropy problem (83) and the minimum variance problem (85) and their optimal solutions are very similar to that of the OWA operator cases with (3) and (6).

2.6 *The Preference Methods*

Besides the OWA operator determination purely consider the property OWA operator itself, the OWA determination can also be connected with the preference information in decision making. The preference based OWA determination methods utilize the preference information as inputs, that the preference relations between alternatives can be revealed. With these preference relation, the model to determine the most suitable OWA operator to these preference relations can be constructed. Despite the differences on the problems formulation on the first sight, the preference relation method can also be seen as a special form of the sample learning method where the sample data are provided as the preference matrix of the decision maker.

Ahn [2] proposed a OWA determination method from preference relations. The method tries to estimate the OWA weights in the direction of minimizing deviations implied by the preference relations, thus as consistent as possible with a priori preference relations.

For a decision matrix

$$\begin{array}{c} \\ A_1 \\ A_2 \\ \vdots \\ A_m \end{array} \begin{array}{c} \begin{array}{cccc} C_1 & C_2 & \cdots & C_n \end{array} \\ \begin{bmatrix} a_{11} & a_{12} & \cdots & a_{1n} \\ a_{21} & a_{22} & \cdots & a_{2n} \\ \vdots & \vdots & \vdots & \vdots \\ a_{m1} & a_{m2} & \cdots & a_{mn} \end{bmatrix} \end{array}$$

Where the set $A = \{A_1, A_2, \ldots, , A_m\}$ corresponds to a set of alternatives and the set $C = \{C_1, C_2, \ldots, , C_n\}$ corresponds to the set of multiple criteria considered. $a_{ij}, i = 1, \ldots, m, j = 1, \ldots, n$ indicates a consequence (or outcome, payoff, value, etc.) for selecting alternative A_i when the state of nature is C_j. The set of OWA operator $W = (w_1, w_2, \ldots, w_n)$ can be determined by considering the decision-makers holistic judgments between alternatives. If the decision maker indicates that alternative A_i is preferred to alternative A_j, then we should have

$$\sum_{k=1}^{n}(b_{ik} - b_{jk})w_k > 0$$

In which b_{ik} and b_{jk} are the reordered arguments of the arguments $a_{i1}, \ldots, a_{in}$ and $a_{j1}, \ldots, , a_{jn}$ respectively. Let $\Theta \in A \times A$ denote the set of ordered pairs (i, j) where i designates a preferred alternative from a paired comparison involving i and j.

Thus, the goal of analysis is to determine the solution W^* for which the conditions such as $(b_{ik} - b_{jk})w_k \geqslant \varepsilon$ for every a priori ordered pair $(i, j) \in \Theta$ are violated as minimally as possible in which ε is a small arbitrary. With the auxiliary variables δ_{ij} in $\sum_{k=1}^{n}(b_{ik} - b_{jk})w_k + \delta_{ij}\varepsilon > 0$ for every ordered pair $(i, j) \in \Theta$ and minimize the sum of auxiliary variables in the objective as shown in the following

$$\begin{aligned} \min \quad & \sum_{(i,j)\in\Theta} \delta_{ij} \\ \text{s.t.} \quad & \sum_{k=1}^{n}(b_{ik} - b_{jk})w_k + \delta_{ij} \geqslant \varepsilon, \\ & w_k \geqslant 0, \\ & \sum_{k=1}^{n} w_i = 1, \\ & w_k, \delta_{ij} \geqslant 0 \qquad \text{for all } (i, j) \in \Theta, \varepsilon > 0, k = 1, 2, \ldots, n. \end{aligned} \tag{89}$$

As remarked by Ahn [2], the objective function of (90) can also be replaced by the criteria in the optimization model such the entropy, the variance, the minimax disparity, or some other forms of entropy. If the maximum entropy is adopted, then the problem should be

$$\begin{aligned}
\max \quad & -\sum_{k=1}^{n} w_k \ln(w_k) \\
\text{s.t.} \; & \sum_{k=1}^{n} (b_{ik} - b_{jk}) w_k + \delta_{ij} \geqslant \varepsilon, \\
& w_k \geqslant 0, \\
& \sum_{k=1}^{n} w_i = 1, \\
& w_k, \delta_{ij} \geqslant 0 \text{ for all} (i,j) \in \Theta, \varepsilon > 0, k = 1, 2, \ldots, n.
\end{aligned} \tag{90}$$

Emrouznejad [18] proposed an OWA determination method by taking account of the preference information to the alternatives of the decision maker. The method is first to obtain an preference matrix from a give scale set $S = \{s_1, s_2, \ldots, s_r\}$ that $s_1 < s_2 < \ldots, s_r$. Then for each criteria, the number of each scale that is given by alternatives are counted. The most popular scale for each criteria can be obtained. Then the OWA operator according to this most popular scale vector can be obtained, which is called MP-OWA (Most Preferred Ordered Weighted Averaging).

Let $S_{ij} \in S$ be scale value of alternative $A_i (i = 1, 2 \ldots, n)$ for criteria $C_j (j = 1, 2 \ldots, m)$. Then for each criteria, the number of each scale for all the alternatives can be summarized. Let N_{kj} be the number of scale s_k is given to criteria C_j by all alternatives.

A vector in which its elements are the most popular scale for each criteria can be constructed

$$\begin{aligned}
V =& [V_1, V_2, \ldots, V_m] \\
=& [\max_{1 \leqslant k \leqslant r} \{N_{k1}\}, \max_{1 \leqslant k \leqslant r} \{N_{k2}\}, \ldots, \max_{1 \leqslant k \leqslant r} \{N_{km}\}]
\end{aligned}$$

The operator MP-OWA (Most Preferred Ordered Weighted Averaging) can be obtained with

$$W = (w_1, w_2, \ldots, w_m) = \left(\frac{V_1}{\sum_k V_k}, \frac{V_2}{\sum_k V_k}, \ldots, \frac{V_m}{\sum_k V_k} \right)$$

Renaud, Levrat and Fonteix [50] proposed an OWA weights determination by parametric identification method. The aggregation n data is the product evaluation according an unique scale. The weights are not fixed by criterion but according to utility level. First, a learning sample is ranked by the decision-maker. Then, this ranked sample is used in order to determine the weights by parametric identification. A hypothesis of equipartition of the scores of each sample is used.

Llamazares [40] proposed determining OWA operator weights regarding the class of majority rule when individuals do not grade their preferences between the alternatives. Since the same majority rule can be generalized through a wide variety of OWA operators, a procedure to determine the best-suited OWA operators in order to extend simple, Pareto, and absolute special majorities were suggested. The best-suited OWA operators such as the arithmetic mean, the median, and the average of the jth and the $(m+1-j)$th order statistics were obtained.

Wang and Parkan [61] proposed a preemptive goal programming method to determine the OWA operator weights, which combines the ordinary OWA determination method with that of the group decision making:

$$\begin{aligned}
\min\quad & P_1 \sum_{i=1}^{m} h_k(\varepsilon_k^+ + \varepsilon_k^-) + P_2\delta \\
\text{s.t.}\quad & \sum_{i=1}^{n} \frac{n-i}{n-1} w_i - \varepsilon_k^+ + \varepsilon_k^- = \alpha_k, \quad 0 < \alpha_k < 1 \\
& w_i - w_{i+1} - \delta \leqslant 0, \\
& w_i - w_{i+1} + \delta \geqslant 0, \\
& \sum_{i=1}^{n} w_i = 1, \\
& w_i, \varepsilon_k^+, \varepsilon_k^- \geqslant 0, \qquad i = 1, 2, \ldots, n, j = 1, 2, \ldots, m.
\end{aligned} \tag{91}$$

Where h_k $k = 1, \ldots, , m$, are the relative importance weights of experts, satisfying $\sum_{j=1}^{m} h_k = 1$. It can also be extended to the maximum entropy and minimum variance approaches such that

$$\begin{aligned}
\min\quad & P_1 \sum_{i=1}^{m} h_k(\varepsilon_k^+ + \varepsilon_k^-) + P_2 \sum_{i=1}^{n} w_i \ln w_i \\
\text{s.t.}\quad & \sum_{i=1}^{n} \frac{n-i}{n-1} w_i - \varepsilon_k^+ + \varepsilon_k^- = \alpha_k, \quad 0 < \alpha_k < 1 \\
& \sum_{i=1}^{n} w_i = 1, \\
& w_i, \varepsilon_k^+, \varepsilon_k^- \geqslant 0, \qquad i = 1, 2, \ldots, n, j = 1, 2, \ldots, m.
\end{aligned} \tag{92}$$

$$
\begin{aligned}
\min \quad & P_1 \sum_{i=1}^{m} h_k(\varepsilon_k^+ + \varepsilon_k^-) + P_2 \sum_{i=1}^{n} w_i^2 \\
\text{s.t.} \quad & \sum_{i=1}^{n} \frac{n-i}{n-1} w_i - \varepsilon_k^+ + \varepsilon_k^- = \alpha_k, \quad 0 < \alpha_k < 1 \\
& \sum_{i=1}^{n} w_i = 1, \\
& w_i, \varepsilon_k^+, \varepsilon_k^- \geqslant 0, \qquad i = 1, 2, \ldots, n, j = 1, 2, \ldots, m.
\end{aligned}
\tag{93}
$$

Amin [5] shows that the two stage goal programming model (91) can be integrated together that

$$
\begin{aligned}
\min \quad & \sum_{i=1}^{m} h_k(\varepsilon_k^+ + \varepsilon_k^-) \\
\text{s.t.} \quad & \sum_{i=1}^{n} \frac{n-i}{n-1} w_i - \varepsilon_k^+ + \varepsilon_k^- = \alpha_k, \quad 0 < \alpha_k < 1 \\
& w_i - w_{i+1} - \delta \leqslant 0, \\
& w_i - w_{i+1} + \delta \geqslant 0, \\
& \sum_{i=1}^{n} w_i = 1, \\
& w_i, \varepsilon_k^+, \varepsilon_k^- \geqslant 0, \qquad i = 1, 2, \ldots, n, j = 1, 2, \ldots, m.
\end{aligned}
\tag{94}
$$

Xu and Da [67] considered the situation where the weight information is available partially and suggested an approach to deal with it. The known weight information can be a constraint set H such as

1. A weak ranking: $\{w_i \geqslant w_j\}$.
2. A strick ranking: $\{w_i - w_j \geqslant \alpha_i\}$.
3. A rank with multiples: $\{w_i \geqslant \alpha_i w_j\}$.
4. An interval form: $\{\alpha_i \leqslant w_i \leqslant \alpha_i + \varepsilon_i\}$.
5. An rank of difference: $\{w_i - w_j \geqslant w_k - w_l, \text{for } j \neq k \neq l\}$.

Such weight information can be added into the optimization problem's constraints such as:

$$\begin{aligned}
\max \quad & -\sum_{i=1}^{n} w_i \ln w_i \\
\text{s.t.} \quad & \sum_{i=1}^{n} \frac{n-i}{n-1} w_i = \alpha, \qquad 0 \leqslant \alpha \leqslant 1 \\
& \sum_{i=1}^{n} w_i = 1, \\
& W \in H, \\
& w_i \geqslant 0, i = 1, 2, \ldots, n.
\end{aligned} \tag{95}$$

With the donations of (35), a sample learning model can also be proposed that [65]:

$$\begin{aligned}
\min \quad & J = \sum_{k=1}^{K} (e_k^+ + e_k^-) \\
\text{s.t.} \quad & \sum_{i=1}^{n} w_i x_i^k - d^k - \varepsilon_k^+ + \varepsilon_k^- = 0, \\
& w_i \in H, \\
& \sum_{i=1}^{n} w_i = 1, \\
& w_i, e_k^+, e_k^- \geqslant 0, \qquad i = 1, 2, \ldots, n, k = 1, 2, \ldots, K.
\end{aligned} \tag{96}$$

3 Comparison and Discussions

From the summarization of theses OWA operator determination methods, we can observed that:

1. The methods are proposed from different point of views, and there are some connections between them. I can be seen that the optimization method are mainly proposed in a theoretical way. However, the preference method closely connected to the real application problem. The objective function criteria are usually assigned in a general way. They are not connected to a specific decision problem and also there is no necessary to make the data follow these criteria. Some criteria even do not have concrete explanations of them. But the preference method is associated with the application data directly. Either the preference data or the OWA determination process have a specific meaning to a certain kind of decision problem. The other difference is that the optimization method have a very good mathematical structure and can even be solved

analytically, both the solution forms and solution properties can be discussed in a comprehensive way. There is also a problem structure and solution process for the optimization problems. Whereas the condition in the preference method is almost completely different. The data structure of the preference information can not be processed in the ordinary way. And the methods to deal with the preference information are also often based on some intuition ideas and there is not a uniform framework for them.

2. The developments of different kinds methods are unbalanced. Such unbalance includes both the different methods of the same class and also the methods in different classes. For example, in the optimization based methods, the extension of the maximum entropy OWA problem attracts many attention from the very beginning both in theory and applications. It is an very active in recent years. Several extension forms are provided. It can be extended to other methods and some other methods can be also be merged into it, such as the RIM quantifier determination, and the stress function method. But the developments are very unbalanced. We have the analytical solutions from the basic form (3) to a very general form (26), some of its properties can be clarified, and verify that the corresponding minimax problem also have the same optimal solution. However, we can not find the optimal solution of the minimax problem in an independent way. And for a variant of the basic form (16) and (24), we have not find their optimal solution and property discussion in publications. And even more, we can prove (24) also have some basic interesting properties as (3) and (85), we can even observe that the optimal solution behaves in the interpolation points of a piecewise three order polynomial function, but can not expressed it in an analytical way and further discuss its properties. Another example is that some optimization based OWA determination model have a relative comprehensive framework and theory of their properties, and also widely used in many applications. But the preference methods only have a few attempts with intuitive observations and ideas.
3. Despite the variety of these OWA determination methods, their problem background and problem formulation, their mathematical forms and their computing process, there are still many connections among these methods, either in theoretical or application points of view. It maybe difficult to build a uniform framework for these OWA operator methods, but we still can make some connections between them, that we can have a clear image of them and find other interesting topics for the future research.
4. The ideas and methods of these OWA determination methods are interactive and merged each other. For example, the optimization techniques are commonly used in many OWA determination methods, such as the sample learning methods and the preference methods. More specifically, the maximum entropy OWA determination problem have various extensions either in discrete or continuous forms. This makes some difficulties to classify these OWA determination methods. Another fact is that the ideas and methods of OWA

operator can also influence the research of the problems on other fields. For example, the RIM quantifier is primarily to determine the OWA operator weights from the linguistic quantifiers. However, from the quantifier guided OWA aggregation methods. It can be observed that the RIM quantifier can be regarded as the continuous case of the OWA operator. We can find the corresponding RIM solutions in the OWA research results. This promoted the research on the determination of RIM quantifier, which is an important topic in computing with words theory [37].

5. Despite the property discussions on some of these OWA determination methods are not included, and the relationship between the OWA operator and the RIM quantifier are also not discussed, there are also various extensions of the basic numerical form of this paper, such as the weighted OWA operator [32, 53], the geometric OWA operator [23, 66], the induced OWA operator [45, 82], the linguistic OWA operator [24] and many other forms. The research on the determination method of this basic form should also be useful for the determination of these extension or compound forms. So that the OWA operator weights and the corresponding weights determination can only be justified by the reasonability and connections to the real application problems.

4 Conclusions

The paper give a summary on the OWA determination methods in a classification way. It includes the optimization based methods, the sample learning methods, the function based methods, the argument dependent methods and the preference methods. Some connections between these methods are proposed and some extensions are also provided. A uniform framework for these OWA determination methods are also attempted to make.

Here, we mainly concern the OWA determination methods in the crisp numerical form. An important reason for the popular research about OWA operator is its ability to model some problems in a very simple and flexible way. We may encounter many different problems, which have various structures and different requirements. The listed OWA operator determination methods are proposed from different point of view. Some methods can be proposed in theoretical way, but some others are practical oriented. Furthermore, some methods are not completely understand either in theoretical or practical view, and even some are only existed in isolated way with intuitive ideas. Furthermore, we also do not have some principles on the OWA operator determination method selection in the face of practical problems, such as in what case does the maximum entropy or minimum variance criteria in OWA determination should be adopted? and which one? These are all problems deserve for further research in the crisp numerical OWA operator, even we disregard its various extension forms.

References

[1] Ahn, B.S.: On the properties of OWA operator weights functions with constant level of orness. IEEE Transactions on Fuzzy Systems 14(4), 511–515 (2006)
[2] Ahn, B.S.: Preference relation approach for obtaining OWA operators weights. International Journal of Approximate Reasoning 47(2), 166–178 (2008)
[3] Ahn, B.S., Park, H.: An efficient pruning method for decision alternatives of OWA operators. IEEE Transactions On Fuzzy Systems 16(6), 1542–1549 (2008)
[4] Ahn, B.S., Park, H.: Least-squared ordered weighted averaging operator weights. International Journal of Intelligent Systems 23(1), 33–49 (2008)
[5] Amin, G.R.: Note on A preemptive goal programming method for aggregating OWA operator weights in group decision making. Information Sciences 177(17), 3636–3638 (2007)
[6] Amin, G.R.: Notes on properties of the OWA weights determination model. Computers & Industrial Engineering 52(4), 533–538 (2007)
[7] Amin, G.R., Emrouznejad, A.: An extended minimax disparity to determine the OWA operator weights. Computers and Industrial Engineering 50(3), 312–316 (2006)
[8] Beliakov, G.: Methods of construction of OWA operators from data. In: IEEE International Conference on Fuzzy Systems, Melbourne, vol. 1 (2001)
[9] Beliakov, G.: How to build aggregation operators from data. International Journal of Intelligent Systems 18(8), 903–923 (2003)
[10] Beliakov, G.: Learning weights in the generalized OWA operators. Fuzzy Optimization and Decision Making 4, 119–130 (2005)
[11] Beliakov, G., Mesiar, R., Valaskova, L.: Fitting generated aggregation operators to empirical data. International Journal Of Uncertainty Fuzziness And Knowledge-Based Systems 12(2), 219–236 (2004)
[12] Beliakov, G., Pradera, A., Calvo, T.: Aggregation Functions: A Guide for Practitioners. Springer, Heidelberg (2007)
[13] Boongoen, T., Shen, Q.: Clus-DOWA: A new dependent OWA operator. In: IEEE International Conference on Fuzzy Systems. IEEE, Hong Kong
[14] Carbonell, M., Mas, M., Mayor, G.: On a class of monotonic extended OWA operators. In: The Sixth IEEE International Conference on Fuzzy Systems, Barcelona (1997)
[15] Chakraborty, C., Chakraborty, D.: A decision scheme based on owa operator for an evaluation programme: an approximate reasoning approach. Applied Soft Computing 5(1), 45–53 (2004)
[16] Cheng, C.H., Chang, J.R.: MCDM aggregation model using situational ME-OWA and ME-OWGA operators. International Journal of Uncertainty Fuzziness and Knowledge-Based Systems 14(4), 421–443 (2006)
[17] Chiclana, F., Herrera, F., Herrera-Viedma, E., Martínez, L.: A note on the reciprocity in the aggregation of fuzzy preference relations using OWA operators. Fuzzy Sets and Systems 137(1), 71–83 (2003)
[18] Emrouznejad, A.: MP-OWA: The most preferred OWA operator. Knowledge-Based Systems 21, 847–851 (2008)
[19] Filev, D., Yager, R.R.: Analytic properties of maximum entropy OWA operators. Information Sciences 85(1-3), 11–27 (1995)
[20] Filev, D., Yager, R.R.: On the issue of obtaining OWA operator weights. Fuzzy Sets and Systems 94(2), 157–169 (1998)
[21] Fullér, R., Majlender, P.: An analytic approach for obtaining maximal entropy OWA operator weights. Fuzzy Sets and Systems 124(1), 53–57 (2001)

[22] Fullér, R., Majlender, P.: On obtaining minimal variability OWA operator weights. Fuzzy Sets and Systems 136(2), 203–215 (2003)
[23] Herrera, F., Herrera-Viedma, E., Chiclana, F.: A study of the origin and uses of the ordered weighted geometric operator in multicriteria decision making. International Journal of Intelligent Systems 18(6), 689–707 (2003)
[24] Herrera, F., Herrera-Viedma, E., Verdegay, J.: Direct approach processes in group decision making using linguistic OWA operators. Fuzzy Sets and Systems 79(2), 175–190 (1996)
[25] Herrera-Viedma, E., Cordón, O., Luque, M., Lopez, A.G., Muñoz, A.M.: A model of fuzzy linguistic IRS based on multi-granular linguistic information. International Journal of Approximate Reasoning 34(2-3), 221–239 (2003)
[26] Kacprzyk, J., Zadrozny, S.: Computing with words in intelligent database querying: standalone and internet-based applications. Information Sciences 134(1), 71–109 (2001)
[27] Kacprzyk, J., Zadrożny, S., Fedrizzi, M., Nurmi, H.: On group decision making, consensus reaching, voting and voting paradoxes under fuzzy preferences and a fuzzy majority: A survey and some perspectives. In: Bustince, H., Herrera, F., Montero, J. (eds.) Fuzzy Sets and Their Extensions: Representation, Aggregation and Models. Springer, Heidelberg (2008)
[28] Liu, X.: Parameterized OWA operator determination with optimization criteria: A general model, submitted to Information Sciences
[29] Liu, X.: The relationships between two kinds of variability optimization and orness optimization problems for OWA operator with their RIM quantifier extensions. International Journal of General Systems (2008) (accpeted)
[30] Liu, X.: On the properties of equidifferent RIM quantifier with generating function. International Journal of General Systems 34(5), 579–594 (2005)
[31] Liu, X.: On the properties of equidifferent OWA operator. International Journal of Approximate Reasoning 43(1), 90–107 (2006)
[32] Liu, X.: Some properties of the weighted OWA operator. IEEE Transactions on Systems, Man and Cybernetics, Part B 36(1), 118–127 (2006)
[33] Liu, X.: The solution equivalence of minimax disparity and minimum variance problems for OWA operator. International Journal of Approximate Reasoning 45(1), 68–81 (2007)
[34] Liu, X.: A general model of parameterized OWA aggregation with given orness level. International Journal of Approximate Reasoning 48(2), 598–627 (2008)
[35] Liu, X.: On the properties of regular increasing monotone RIM quantifiers with maximum entropy. International Journal of General Systems 37(2), 167–179 (2008)
[36] Liu, X., Chen, L.: On the properties of parametric geometric OWA operator. International Journal of Approximate Reasoning 35(2), 163–178 (2004)
[37] Liu, X., Han, S.: Orness and parameterized RIM quantifier aggregation with OWA operators: A summary. International Journal of Approximate Reasoning 48(1), 77–97 (2008)
[38] Liu, X., Lou, H.: Parameterized additive neat OWA operators with different orness levels. International Journal of Intelligent Systems 21(10), 1045–1072 (2006)
[39] Liu, X., Yang, X., Fang, Y.: The relationships between two kinds of OWA operator determination methods. In: IEEE International Conference on Fuzzy Systems, Hong Kong
[40] Llamazares, B.: Choosing OWA operator weights in the field of social choice. Information Sciences 177(21), 4745–4756 (2007)

[41] Majlender, P.: OWA operators with maximal Rényi entropy. Fuzzy Sets and Systems 155(3), 340–360 (2005)
[42] Marchant, T.: Maximal orness weights with a fixed variability for OWA operators. International Journal of Uncertainty, Fuzziness Knowledge-Based Systems 14(3), 271–276 (2006)
[43] Marimin, M., Umano, M., Hatono, I., Tamura, H.: Linguistic labels for expressing fuzzy preference relations in fuzzy group decision making. IEEE Transactions on Systems, Man and Cybernetics, Part B 28(2), 205–218 (1998)
[44] Marimin, M., Umano, M., Hatono, I., Tamura, H.: Hierarchical semi-numeric method for pairwise fuzzy group decision making. IEEE Transactions on Systems, Man and Cybernetics, Part B 32(5), 691–700 (2002)
[45] Merigo, J.M., Gil-Lafuente, A.M.: The induced generalized OWA operator. Information Sciences 179(6), 729–741 (2009)
[46] O'Hagan, M.: Aggregating template or rule antecedents in real-time expert systems with fuzzy set. In: Grove, P. (ed.) Proc. 22nd Annu. IEEE Asilomar Conf. on Signals, Systems, Computers, CA (1988)
[47] Peláez, J.I., Doña, J.M.: Majority additive-ordered weighting averaging: a new neat ordered weighting averaging operator based on the majority process. International Journal of Intelligent Systems 18, 469–481 (2003)
[48] Peláez, J.I., Doña, J.M.: A majority model in group decision making using QMA-OWA operators. International Journal of Intelligent Systems 21(2), 193–208 (2006)
[49] Peláez, J.I., Doña, J.M., Gómez-Ruiz, J.A.: Analysis of OWA operators in decision making for modelling the majority concept. Applied Mathematics and Computation (New York) 186(2), 1263–1275 (2007)
[50] Renaud, J., Levrat, E., Fonteix, C.: Weights determination of OWA operators by parametric identification. Mathematics and Computers in Simulation 77(5-6), 499–511 (2008)
[51] Sadiq, R., Tesfamariam, S.: Probability density functions based weights for ordered weighted averaging (OWA) operators: An example of water quality indices. European Journal of Operational Research 182(3), 1350–1368 (2007)
[52] Sicilia, M.-A., García-Barriocanal, E., Sánchez-Alonso, S.: Empirical assessment of a collaborative filtering algorithm based on OWA operators. International Journal of Intelligent Systems 23, 1251–1263 (2008)
[53] Torra, V.: The weighted OWA operator. International Journal of Intelligent Systems 12(2), 153–166 (1997)
[54] Torra, V.: Learning weights for weighted OWA operators. In: 26th Annual Confjerence of the IEEE Industrial Electronics Society, vol. 4 (2000)
[55] Torra, V.: Learning weights for the quasi-weighted means. IEEE Transactions on Fuzzy Systems 10(5), 653–666 (2002)
[56] Torra, V.: OWA operators in data modeling and reidentification. IEEE Transactions on Fuzzy Systems 12(5), 652–660 (2004)
[57] Wang, J.W., Chang, J.R., Cheng, C.H.: Flexible fuzzy OWA querying method for hemodialysis database. Soft Computing 10(11), 1031–1042 (2006)
[58] Wang, Y., Luo, Y., Hua, Z.: Aggregating preference rankings using OWA operator weights. Information Sciences 177(16), 3356–3363 (2007)

[59] Wang, Y., Luo, Y., Liu, X.: Two new models for determining OWA operator weights. Computers and Industrial Engineering 52, 203–209 (2007)

[60] Wang, Y., Parkan, C.: A minimax disparity approach for obtaining OWA operator weights. Information Sciences 175(1), 20–29 (2005)

[61] Wang, Y., Parkan, C.: A preemptive goal programming method for aggregating OWA operator weights in group decision making. Information Sciences 177, 1867–1877 (2007)

[62] Wu, J., Liang, C.Y., Huang, Y.Q.: An argument-dependent approach to determining OWA operator weights based on the rule of maximum entropy. International Journal of Intelligent Systems 22(2), 209–221 (2007)

[63] Wu, J., Sun, B.-L., Liang, C.-Y., Yang, S.-L.: A linear programming model for determining ordered weighted averaging operator weights with maximal yager's entropy. Computers & Industrial Engineering (2009), doi:10.1016/j.cie.2009.02.001

[64] Xu, Z.S.: Dependent OWA operators. In: Torra, V., Narukawa, Y., Valls, A., Domingo-Ferrer, J. (eds.) MDAI 2006. LNCS (LNAI), vol. 3885, pp. 172–178. Springer, Heidelberg (2006)

[65] Xu, Z.S.: An overview of methods for determining OWA weights. International Journal of Intelligent Systems 20(8), 843–865 (2005)

[66] Xu, Z.S., Da, Q.L.: The ordered weighted geometric averaging operators. International Journal of Intelligent Systems 17(7), 709–716 (2002)

[67] Xu, Z.S., Da, Q.L.: Approaches to obtaining the weights of the ordered weighted aggregation operators. Journal of Southeast University 33, 94–96 (2003)

[68] Yager, R.R.: On ordered weighted averaging aggregation operators in multicriteria decision making. IEEE Transactions on Systems, Man and Cybernetics 18(1), 183–190 (1988)

[69] Yager, R.R.: Families of OWA operators. Fuzzy Sets and Systems 59(2), 125–143 (1993)

[70] Yager, R.R.: Quantifier guided aggregation using OWA operators. International Journal of Intelligent Systems 11(1), 49–73 (1996)

[71] Yager, R.R.: On the analytic representation of the Leximin ordering and its application to flexible constraint propagation. European Journal of Operational Research 102(1), 176–192 (1997)

[72] Yager, R.R.: Fuzzy modeling for intelligent decision making under uncertainty. IEEE Transactions on Systems, Man and Cybernetics, Part B 30(1), 60–70 (2000)

[73] Yager, R.R.: A hierarchical document retrieval language. Information Retrieval 3(4), 357–377 (2000)

[74] Yager, R.R.: On the valuation of alternatives for decision-making under uncertainty. International Journal of Intelligent Systems 17(7), 687–707 (2002)

[75] Yager, R.R.: Toward a language for specifying summarizing statistics. IEEE Transactions on Systems, Man and Cybernetics, Part B 33(2), 177–187 (2003)

[76] Yager, R.R.: OWA aggregation over a continuous interval argument with applications to decision making. IEEE Transactions on Systems, Man and Cybernetics, Part B 34(5), 1952–1963 (2004)

[77] Yager, R.R.: An extension of the naive bayesian classifier. Information Sciences 176(5), 577–588 (2006)

[78] Yager, R.R.: Centered OWA operators. Soft Computing 11(7), 631–639 (2007)

[79] Yager, R.R.: Using stress functions to obtain OWA operators. IEEE Transactions on Fuzzy Systems 15(6), 1122–1129 (2007)

[80] Yager, R.R.: Time series smoothing and OWA aggregation. IEEE Transactions on Fuzzy Systems 16(4), 994–1007 (2008)

[81] Yager, R.R., Filev, D.P.: Parameterized and-like and or-like OWA operators. International Journal of General Systems 22(3), 297–316 (1994)

[82] Yager, R.R., Filev, D.P.: Induced ordered weighted averaging operators. IEEE Transactions on Systems, Man and Cybernetics, Part B 29(2), 141–150 (1999)

[83] Yager, R.R., Kacprzyk, J.: The ordered weighted averaging operators—Theory and applications. Kluwer Academic Publishers, Dordrecht (1997)

[84] Zadrozny, S., Kacprzyk, J.: Computing with words for text processing: An approach to the text categorization. Information Sciences 176(4), 415–437 (2006)

[85] Zadrożny, S., Kacprzyk, J.: On tuning OWA operators in a flexible querying interface. In: Larsen, H.L., Pasi, G., Ortiz-Arroyo, D., Andreasen, T., Christiansen, H. (eds.) FQAS 2006. LNCS (LNAI), vol. 4027, pp. 97–108. Springer, Heidelberg (2006)

Fuzzification of the OWA Operators for Aggregating Uncertain Information with Uncertain Weights

Shang-Ming Zhou, Francisco Chiclana,
Robert I. John, and Jonathan M. Garibaldi

Abstract. Yager's ordered weighted averaging (OWA) operator has been widely applied in various domains. Yager's traditional OWA operator focuses exclusively on the aggregation of crisp numbers with crisp weights. However, uncertainty prevails in almost every process of real world decision making, and so there is a need to find OWA mechanisms to aggregate uncertain information. In this chapter, we generalise Yager's OWA operator and describe two novel uncertain operators, namely the *type-1 OWA operator* and *type-2 OWA operator*. The *type-1 OWA operator* is able to aggregate type-1 fuzzy sets, whilst the *type-2 OWA operator* is able to aggregate type-2 fuzzy sets. Therefore, the two new operators are capable of aggregating uncertain opinions or preferences with uncertain weights in soft decision making. This chapter also indicates that not only Yager's OWA operator but also some existing operators of fuzzy sets, including the *join* and *meet* of type-1 fuzzy sets, are special cases of the type-1 OWA operators. We further suggest the concepts of *joinness* and *meetness* of type-1 OWA operators, which can be considered as the extensions of the concepts-*orness* and *andness* in Yager's OWA operator, respectively. Given the high computing overhead involved in aggregating general type-2 fuzzy sets using the type-2 OWA operator, an interval type-2 fuzzy sets oriented OWA operator is also defined. Some examples are provided to illustrate the proposed concepts.

Shang-Ming Zhou · Francisco Chiclana · Robert I. John
The Centre for Computational Intelligence, Department of Informatics,
De Montfort University, Leicester, LE1 9BH, UK
e-mail: smzhou@ieee.org, chiclana@dmu.ac.uk, rij@dmu.ac.uk

Jonathan M. Garibaldi
The Intelligent Modelling and Analysis Research Group,
School of Computer Science, University of Nottingham,
Nottingham, NG8 1BB, UK
e-mail: jmg@cs.nott.ac.uk

R.R. Yager et al. (Eds.): Recent Developments in the OWA Operators, STUDFUZZ 265, pp. 91–109.
springerlink.com

1 Introduction

The aggregation operation is not only one of the most important steps in dealing with multi-expert decision making (i.e. group decision making), multi-criteria decision making and multi-expert multi-criteria decision making [1, 2, 3], but also necessary in many other application domains, such as database integration and knowledge discovery [4, 5], information fusion [6], etc. The objective of aggregation is to combine individual experts' preferences or criteria into an overall one in a proper way so that the final result of aggregation takes into account in a given fashion all the individual contributions [2]. Currently, at least 90 different families of aggregation operators have been studied [1, 2, 3, 7, 13, 8, 9]. Among these aggregation operators, the Ordered Weighted Averaging (OWA) operator proposed by Yager [1] is arguably the most widely used. However, the majority of the existing aggregation operators, including the OWA operator, focus on aggregating crisp numbers. But uncertainty prevails in almost every aspect of real world decision applications. For example, in practice, human experts perceive the distance, size, weight, likelihood, and other characteristics of physical and mental objects in a very natural way via linguistic terms, such as "very long","big", "very heavy", "good" etc., when they cannot provide exact numbers for expressing vague and imprecise opinions [25]. These uncertain opinions are widely characterised by type-1 fuzzy sets or type-2 fuzzy sets, where type-1 fuzzy sets are the traditional fuzzy sets proposed by Zadeh in 1965 [24], while type-2 fuzzy sets were proposed by Zadeh later in 1975 [25]. Hence, the problem arises as to how to effectively aggregate uncertain judgments for decision makers.

In order to tackle this issue, Zhou et al. have proposed a so-called *type-1 OWA* operator to aggregate uncertain information with uncertain weights via the OWA mechanism [26], where the aggregated uncertain objects and weights are modelled as type-1 fuzzy sets. Furthermore, *type-2 linguistic quantifiers* were suggested to induce linguistic weights for type-1 OWA operators. Other researchers have also proposed methods to aggregate uncertain objects or linguistic terms from different perspectives [12, 14, 15, 16, 17, 18, 19]. Among the exisiting uncertain aggregation operators, those in [17, 18, 19] are probably the most closely related to the efforts on type-1 OWA operators. Mitchell and Schaefer also applied Zadeh's Extension Principle to fuzzifying Yager OWA operator, but Mitchell and Schaefer's approach focused on ordering fuzzy sets during aggregation. It is known that real numbers can produce incontrovertible ordering, but there does not exist a standard ordering method which is able to yield consistent results for ordering fuzzy sets. Different methods for ordering fuzzy sets may lead to different aggregation results if the fuzzified OWA operators are dependent on the ordering of fuzzy sets. The proposed approach to type-1 OWA aggregation [26] is still based on ordering crisp values and so avoids ordering fuzzy sets.

Due to certain equivalence between OWA operators and Choquet integrals, fuzzified Choquet integrals would lead to the same mechanisms as the type-1 OWA operators. But the existing approaches [18, 19] only considered the cases of aggregating fuzzy sets with crisp weights, while the type-1 OWA operator aggregates fuzzy sets with fuzzy set weights. Another widely investigated fuzzified aggregation operator, the fuzzy weighted averaging operators [10, 11, 21], can also be used to aggregate fuzzy sets with fuzzy set weights. As Yager's OWA operator is a nonlinear aggregation operator, it is significantly different from the fuzzy weighted average operator (which is linear).

In this chapter, we extend the *orness* and *andness* of OWA operators, and propose the *joinness* and *meetness* of type-1 OWA operators. The *joinness* and *meetness* of type-1 OWA operators are type-1 fuzzy sets that indicate the linguistic expressions of the degree of compensation included in the uncertain aggregation process. We further show that some existing operators of fuzzy sets, including the well known *join* and *meet* operators of type-1 fuzzy sets, are special cases of type-1 OWA operators.

An alternative way of modelling uncertainty is via type-2 fuzzy sets [20]. Type-2 fuzzy sets offer the ability to model higher levels of uncertainty than type-1 fuzzy sets. In this chapter, we further extend Yager's OWA operator to the case of aggregating type-2 fuzzy sets, and proposed the named *type-2 OWA* operators. A procedure to perform type-2 OWA operations on interval type-2 fuzzy sets is also described.

The organisation of this chapter is as follows. Section 2 briefly reviews the type-1 OWA operator, then propose the definitions of the *joinness* and *meetness* of type-1 OWA operator. Some special cases of type-1 OWA operators are presented in Section 3. Section 4 defines type-2 OWA operators and provides examples of their use. Finally, Section 5 concludes this chapter with a brief summary of the contribution.

2 Type-1 OWA for Aggregating Type-1 Fuzzy Sets

2.1 Definition of Type-1 OWA Operator

The departure point for suggesting type-1 OWA operators is to aggregate the linguistic variables (modelled as type-1 fuzzy sets) used to express human opinions or preferences in soft decision making. Let $F(X)$ be the set of type-1 fuzzy sets defined on the domain of discourse X. Based on Zadeh's Extension Principle, we extend Yager's OWA operator by defining the type-1 OWA operator which aims at the aggregation of type-1 fuzzy sets [26].

Definition 1: *Given n linguistic weights* $\{W_i\}_{i=1}^n$ *in the form of type-1 fuzzy sets defined on the domain of discourse* $U \subseteq [0,\ 1]$, *a type-1 OWA operator is a mapping* Φ,

$$\begin{aligned} \Phi\colon F(X) \times \cdots \times F(X) &\longrightarrow F(X) \\ (A_1, \cdots, A_n) &\mapsto G \end{aligned} \tag{1}$$

$$\mu_G(y) = \sup_{\substack{\sum_{k=1}^{n} \bar{w}_i a_{\sigma(i)} = y \\ w_i \in U, a_i \in X}} \left(\mu_{W_1}(w_1) * \cdots * \mu_{W_n}(w_n) * \mu_{A_1}(a_1) * \cdots * \mu_{A_n}(a_n) \right) \quad (2)$$

that is associated with the $\{W_i\}_{i=1}^{n}$ to aggregate the type-1 fuzzy sets $\{A_i\}_{i=1}^{n}$, in which $*$ is a t-norm operator, $\bar{w}_i = w_i / \sum_{i=1}^{n} w_i$, and $\sigma : \{1, \cdots, n\} \rightarrow \{1, \cdots, n\}$ is a permutation function such that $a_{\sigma(i)} \geq a_{\sigma(i+1)}$, $\forall i = 1, \cdots, n-1$, i.e., $a_{\sigma(i)}$ is the ith largest element in the set $\{a_1, \cdots, a_n\}$.

From the above definition, it can be seen that the aggregating result $\Phi(A_1, \cdots, A_n) = G \in F(X)$ is a type-1 fuzzy set defined on X.

However, given the linguistic weights $\{W_i\}_{i=1}^{n} \subset F(U)$ and aggregated objects n type-1 fuzzy sets $\{A_i\}_{i=1}^{n} \subset F(X)$, one can not directly perform the type-1 OWA aggregation on $\{A_i\}_{i=1}^{n}$ using the equation (2) due to the so-called *over-partition* of input space [26], i.e., given the discretised domains of X and U: $\hat{X} = \{\hat{x_1}, \cdots, \hat{x_p}\}$ and $\hat{U} = \{\hat{u_1}, \cdots, \hat{u_k}\}$, the $\sum_{k=1}^{n} \bar{w}_i a_{\sigma(i)}$ with all the combinations of $(w_1, \cdots, w_n, a_1, \cdots, a_n)$ may produce another partition of X, where $w_i \in \hat{U}, a_i \in \hat{X}, i = 1, \cdots, n$, i.e.,

$$\overline{X} = \{\bar{x}_j\} = \left\{ \sum_{k=1}^{n} \bar{w}_i a_{\sigma(i)} \middle| w_i \in \hat{U}, a_i \in \hat{X}, i = 1, \cdots, n \right\} \quad (3)$$

The problem is that $\hat{X} \neq \overline{X}$, i.e., the two discretised versions of X may be different, and cardinality of the set $\overline{X}$ is greater than that of $\hat{X}$: $|\overline{X}| \geq |\hat{X}|$. In other words, there are many points in the $\overline{X}$ that lie between the neighouring points in the $\hat{X}$. $\overline{X}$ is referred to as *over-partition* of input space given the used $\hat{X}$. The consequence is that the fuzzy set, $\overline{G}$, generated on the $\overline{X}$ according to the Extension Principle is likely to be unreadable, because for some data points that are in the $\overline{X}$ but not in the $\hat{X}$, their membership grades may not be consistent with the membership grades of the corresponding nearest points in the $\hat{X}$. In [26], a procedure for correctly performing type-1 OWA operation under *over-partition* has been proposed. This procedure is described as follows.

Step 1: Initialisation

- Given the linguistic weights $\{W_i\}_{i=1}^{n} \subseteq F(U)$ for aggregating the objects $\{A_i\}_{i=1}^{n} \subseteq F(X)$;
- Given the discretised domains of linguistic weights, $\hat{U}$, and that of aggregated objects, $\hat{X}$.
- Let the initial $\overline{G} = (\overline{X}, \mu_{\overline{G}})$, where $\overline{X} = \{0\}$, and the $\mu_{\overline{G}}(\bar{x}) = 0$.

Step 2: Induce the initial result $\overline{G}$

- Select $w_1 \in \hat{U}, \cdots, w_n \in \hat{U}$, $a_1 \in \hat{X}, \cdots, a_n \in \hat{X}$:
- Normalise $(w_1, \cdots, w_n)$ as $\bar{w}_i = w_i / \sum_{i=1}^{n} w_i$

- Perform Yager's OWA operation: $\bar{y} = \phi_{\bar{w}_1,\cdots,\bar{w}_n}(a_1,\cdots,a_n)$
- Calculate μ_0:

$$\mu_0 = \mu_{W_1}(w_1) * \cdots * \mu_{W_n}(w_n) * \mu_{A_1}(a_1) * \cdots * \mu_{A_n}(a_n)$$

- If there exists $y_k \in \overline{X} : \bar{y} = y_k$, then update the potential membership grade $\mu_{\overline{G}}(y_k)$:

$$\mu_{\overline{G}}(y_k) \leftarrow max\left(\mu_{\overline{G}}(y_k), \mu_0\right)$$

  Otherwise, $\bar{y}$ is added to $\overline{X}$, and the $\mu_{\overline{G}}(\bar{y}) \triangleq \mu_0$.
- Go to ***Step 2***-1, and continue until all the weight vectors and aggregating points are selected.

Step 3: Induce the fuzzy set G on the $\hat{X}$:

$$\mu_G(\hat{x}) = \sup_{\bar{x}_j \in \Theta_{\hat{x}}} \left(\mu_{\overline{G}}(\bar{x}_j)\right)$$

In the following, let us give an example of aggregating two type-1 fuzzy sets by type-1 OWA operator given the linguistic weights.

Example 1. Supposing the domains $U = \{0.1, 0.8\}$ and $X = \{1.0, 2.0, 3.0\}$. Let the given linguistic weights on U be

$$W_1 = \begin{pmatrix} u_i \\ \mu_{W_1}(u_i) \end{pmatrix}_{u_i \in U} = \begin{pmatrix} 0.1 & 0.8 \\ 0.2 & 1.0 \end{pmatrix}; \; W_2 = \begin{pmatrix} 0.1 & 0.8 \\ 1.0 & 0.2 \end{pmatrix}$$

and the aggregated objects on X be

$$A_1 = \begin{pmatrix} 1.0 & 2.0 & 3.0 \\ 0.4 & 0.6 & 1.0 \end{pmatrix}; \; A_2 = \begin{pmatrix} 1.0 & 2.0 & 3.0 \\ 0.6 & 1.0 & 0.4 \end{pmatrix}$$

According to the described procedure (the t-norm $\star = product$ is used in the following):

- Calculate the $\overline{G}$ on the over-partition $\overline{X}$ as follows,

$$\overline{G} = \begin{pmatrix} 1.00 & 1.11 & 1.22 & 1.50 & 1.89 & 2.00 & 2.11 & 2.50 & 2.78 & 2.89 & 3.00 \\ 0.24 & 0.02 & 0.02 & 0.08 & 0.4 & 0.6 & 0.04 & 0.2 & 0.6 & 1.0 & 0.4 \end{pmatrix}$$

- The aggregating result obtained by the type-1 OWA is a type-1 fuzzy set induced as,

$$G = \begin{pmatrix} 1.0 & 2.0 & 3.0 \\ 0.4 & 1.0 & 0.4 \end{pmatrix}$$

2.2 *Joinness of Type-1 OWA Operator*

Given two type-1 fuzzy sets A and B, their meet ($A \sqcap B$) and join ($A \sqcup B$) are defined as follows [23, 25]:

$$\mu_{A \sqcap B}(z) = \sup_{\substack{x \wedge y = z \\ x \in D_A,\ y \in D_B}} \mu_A(x) * \mu_B(y) \tag{4}$$

$$\mu_{A \sqcup B}(z) = \sup_{\substack{x \vee y = z \\ x \in D_A,\ y \in D_B}} \mu_A(x) * \mu_B(y) \tag{5}$$

where $D_A, D_B \subseteq X$ represent the domains of A and B respectively; $*$ is a t-norm operator; $\wedge$ represents the minimum operation; and $\vee$ represents the maximum operation.

Interestingly, a type-1 OWA operator can be used to perform the join and meet operations by selecting appropriate linguistic weights in the forms of type-1 fuzzy sets. In the following, we define the degree of *joinness* and the degree of *meetness* associated with the linguistic weights of a type-1 OWA operator to characterise the degrees to which the aggregation is like the *join* and *meet* operation respectively.

Definition 2: *Given a type-1 OWA operator with n linguistic weights $\{W_i\}_{i=1}^n$ in the form of type-1 fuzzy sets on $U \subseteq [0,\ 1]$, its joinness is defined as follows,*

$$\mu_{joinness}(v) = \sup_{\substack{\frac{1}{(n-1)\sum_{i=1}^{n} w_i} \sum_{i=1}^{n} (n-i)w_i = v \\ w_i \in U}} \mu_{W_1}(w_1) * \cdots * \mu_{W_n}(w_n) \tag{6}$$

while its meetness is defined as follows,

$$\mu_{meetness}(v) = \sup_{\substack{1 - \frac{1}{(n-1)\sum_{i=1}^{n} w_i} \sum_{i=1}^{n} (n-i)w_i = v \\ w_i \in U}} \mu_{W_1}(w_1) * \cdots * \mu_{W_n}(w_n) \tag{7}$$

where $*$ is a t-norm operator.

The *joinness* and *meetness* of type-1 OWA operators can be considered as the extensions of *orness* and *andness* of Yager's OWA operator [1] respectively. The *joinness* and *meetness* are type-1 fuzzy sets that indicate the linguistic expressions of the degrees of compensations induced in their aggregations.

3 Special Cases of Type-1 OWA Operators

Interestingly, some existing aggregation operators can be implemented via type-1 OWA mechanism, as they are the special cases of type-1 OWA operators.

3.1 Yager's OWA Operator

Naturally, Yager's OWA operator is a special case of the proposed type-1 OWA operator: when the linguistic weights $\{W_i\}_{i=1}^n$ and the aggregated objects $\{A_i\}_{i=1}^n$ are singleton type-1 fuzzy sets, the type-1 OWA operator reduces to Yager's OWA operator.

3.2 Join and Join-Like Operators

The following theorem indicates that the join operator (5) of type-1 fuzzy sets [23, 25] can be obtained via a special case of type-1 OWA operator.

Theorem 1. *If the linguistic weights of a type-1 OWA operator are the singleton weights as:* $W_1 = \dot{1}; W_i = \dot{0}\ (i \neq 1)$, *i.e.,*

$$\mu_{W_1}(w) = \begin{cases} 1 & w = 1 \\ 0 & otherwise \end{cases} \tag{8}$$

$$\mu_{W_i}(w) = \begin{cases} 1 & w = 0 \\ 0 & otherwise \end{cases} \quad (i \neq 1) \tag{9}$$

then for any aggregated objects $\{A_i\}_{i=1}^n$,

$$\Phi(A_1, A_2, \cdots, A_n) = A_1 \sqcup A_2 \sqcup \cdots \sqcup A_n \tag{10}$$

Proof: In this special type-1 OWA operator, we can only consider the special weight vector $\hat{w}$ defined as $\hat{w}_1 = 1, \hat{w}_2 = 0, \cdots, \hat{w}_n = 0$ applying to the aggregation process. For all $w_1, \cdots, w_n \in U, a_1, \cdots, a_n \in X$, if the weight vector $w = (w_1, w_2, \cdots, w_n) \neq \hat{w}$, then $\mu_{W_1}(w_1) * \cdots * \mu_{W_n}(w_n) = 0$, so $\mu_{W_1}(w_1) * \cdots * \mu_{W_n}(w_n) * \mu_{A_1}(a_1) * \cdots * \mu_{A_n}(a_n) = 0$, which indicates that the aggregating point $\bar{a} \equiv \sum_{i=1}^{n} \bar{w}_i a_{\sigma(i)}$ does not lie in the support set of the final aggregating type-1 fuzzy set.

For $\hat{w}$, $\mu_{W_1}(\hat{w}_1) * \cdots * \mu_{W_n}(\hat{w}_n) = 1$ leading to $\mu_{W_1}(\hat{w}_1) * \cdots * \mu_{W_n}(\hat{w}_n) * \mu_{A_1}(a_1) * \cdots * \mu_{A_n}(a_n) = \mu_{A_1}(a_1) * \cdots * \mu_{A_n}(a_n)$, while, the aggregating point $\bar{a} \equiv \sum_{i=1}^{n} \bar{w}_i a_{\sigma(i)} = a_{\sigma(1)} = max(a_1, \cdots, a_n)$, hence $\Phi(A_1, A_2, \cdots, A_n) = A_1 \sqcup A_2 \sqcup \cdots \sqcup A_n$. ■

The *joinness* and *meetness* of this particular type-1 OWA operator, Φ, are *joinness* $(\{W_i\}_{i=1}^n) = \dot{1}$ and *meetness*$(\{W_i\}_{i=1}^n) = \dot{0}$, which further confirm that this particular type-1 OWA operator is the join operator of type-1 fuzzy sets. Figure 1 shows three type-1 fuzzy sets and their aggregation result by this particular case of type-1 OWA operator.

Furthermore, join-like type-1 OWA operators can be obtained by selecting appropriate linguistic weights. For example, if the first linguistic weight approaches to $\dot{1}$, and other linguistic weights approach to $\dot{0}$ as depicted in Figure 2a and Figure 2b, then the type-1 OWA operator shows a join-like type behaviour. Indeed, the

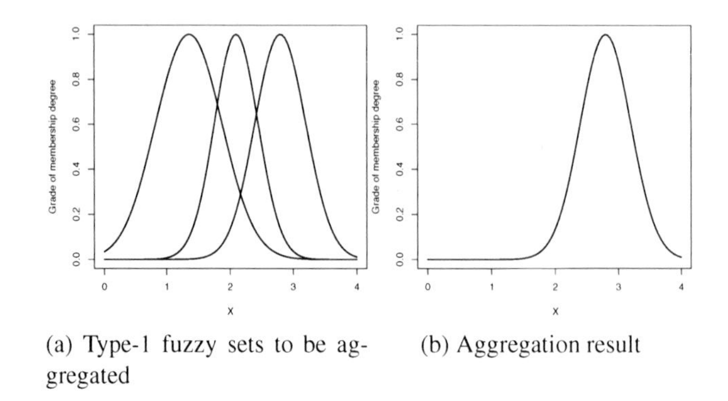

(a) Type-1 fuzzy sets to be aggregated

(b) Aggregation result

Fig. 1 Aggregation of type-1 OWA operator as join

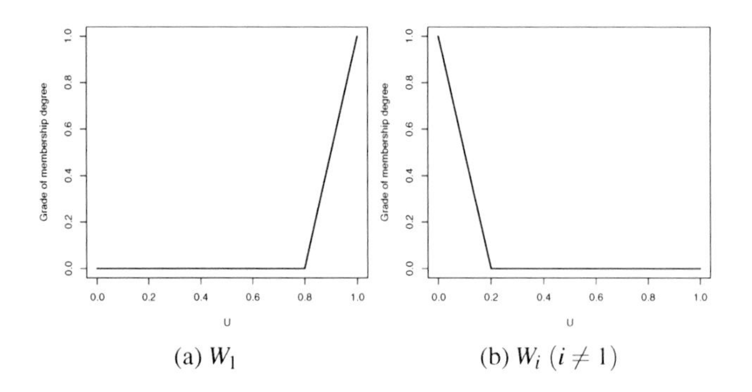

(a) W_1

(b) W_i $(i \neq 1)$

Fig. 2 Linguistic weights for a join-like operator

joinness of this type-1 OWA operator is illustrated in Figure 3a, which shows that the aggregation by this operator is very much like a join operation. Figure 3b delineates the aggregation results of three type-1 fuzzy sets by this operator, which shows its join-like behaviour.

3.3 Meet and Meet-Like Operators

The following theorem indicates that the meet operation (4) of type-1 fuzzy sets [23, 25] can also be obtained via a special case of type-1 OWA operation.

Theorem 2. *If the linguistic weights of type-1 OWA operator are the singleton weights as:* $W_i = \dot{0}$ $(i \neq n)$; $W_n = \dot{1}$, *then for any aggregated objects* $\{A_i\}_{i=1}^{n}$,

$$\Phi(A_1, A_2, \cdots, A_n) = A_1 \sqcap A_2 \sqcap \cdots \sqcap A_n \tag{11}$$

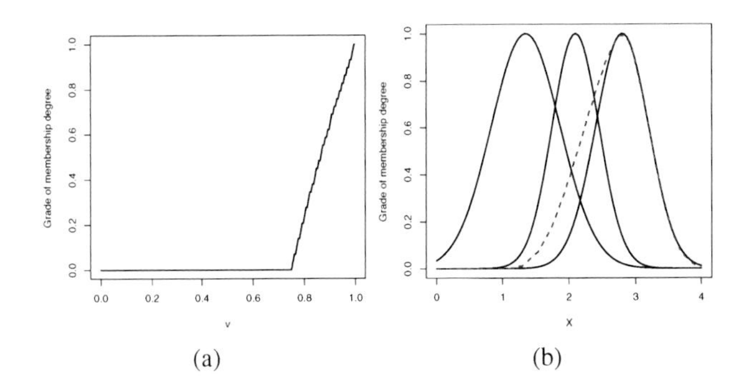

Fig. 3 (a)-*Joinness* of the join-like type-1 OWA operator with linguistic weights given in Figure 2; (b)-Aggregation result by this type-1 OWA operator– solid lines: fuzzy sets to be aggregated; dashed line: aggregation result

Proof: Similar to the Proof of Theorem 2. ■

The *joinness* and *meetness* of this particular type-1 OWA operator, Φ, are *joinness* $(\{W_i\}_{i=1}^n) = \dot{0}$ and *meetness* $(\{W_i\}_{i=1}^n) = \dot{1}$, which further confirm that this particular type-1 OWA operator is the meet operator of type-1 fuzzy sets. Figure 4 depicts the aggregation result of three type-1 fuzzy sets by this operator.

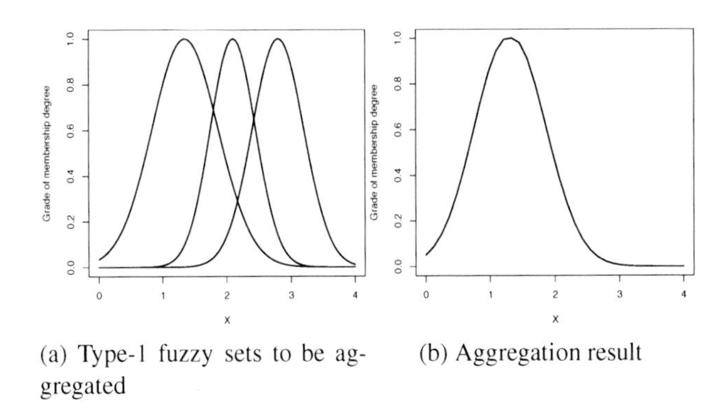

(a) Type-1 fuzzy sets to be aggregated

(b) Aggregation result

Fig. 4 Aggregation of type-1 OWA operator as meet

Moreover, meet-like operation can be achieved by type-1 OWA operator via selecting appropriate linguistic weights. For example, as depicted in Figure 5a and Figure 5b, the last linguistic weight is close to $\dot{1}$, and other linguistic weights close to $\dot{0}$. Figure 5c depicts the aggregation result of three type-1 fuzzy sets by this operator, which shows its meet-like behaviour.

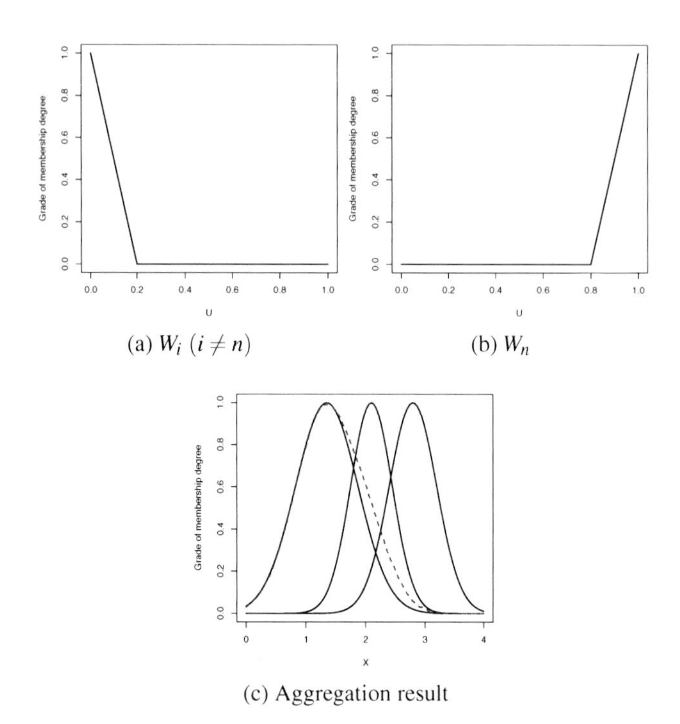

(a) W_i $(i \neq n)$ (b) W_n

(c) Aggregation result

Fig. 5 (a) and (b)- Linguistic weights for a meet-like operator; (c)-Solid lines represent aggregated fuzzy sets, dashed line represents aggregation result.

3.4 Mean and Mean-Like Operators

It is known that Yager's OWA operator reduces to the mean averaging operation when its associated weights are all equal to $1/n$. In type-1 OWA operation, when the linguistic weights are all chosen to be the singleton type-1 fuzzy sets $1/n$, then the associated type-1 OWA operator becomes the extended mean operation on type-1 fuzzy sets, i.e.,

$$\mu_{\Phi(A_1,\cdots,A_n)}(y) = \sup_{\substack{\frac{1}{n}\sum\limits_{i=1}^{n} a_i = y \\ a_i \in X}} \mu_{A_1}(a_1) * \cdots * \mu_{A_n}(a_n) \tag{12}$$

For instance, three identical weights in the form of singleton type-1 fuzzy sets as depicted in Figure 6a are used to aggregate type-1 fuzzy sets. Figure 6b depicts the aggregation result of three type-1 fuzzy sets by this operator, and Figure 6c shows the *joinness* of this OWA operator: $joinness(\{W_i\}_{i=1}^{n}) = meetness(\{W_i\}_{i=1}^{n}) = 0.5$, which indicate that the obtained type-1 OWA operator neither has the tendency to the join operation nor the meet operation.

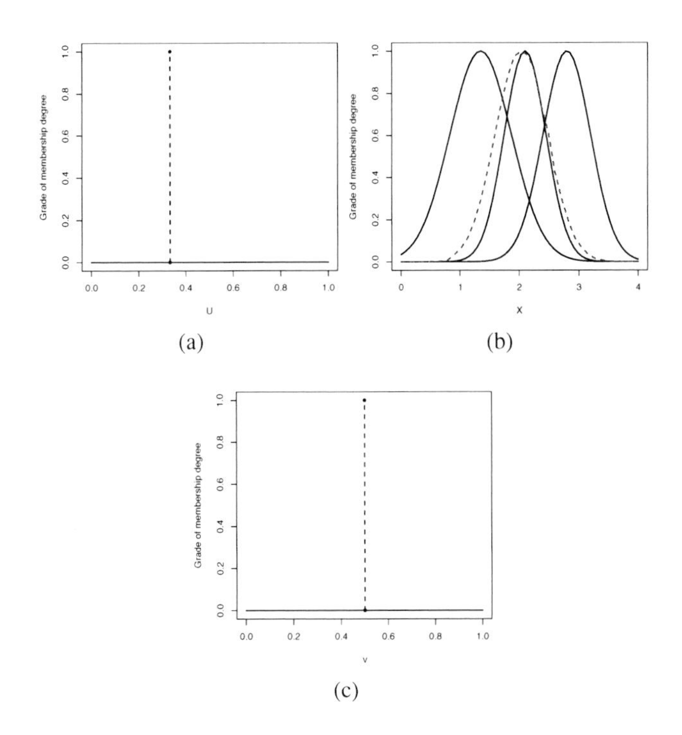

Fig. 6 Mean operator: (a)-Three same singleton fuzzy sets as linguistic weights: $W_i = 1/3$ $(i = 1,2,3)$; (b)-solid lines represent fuzzy sets to be aggregated, dashed line represents aggregation result; (c)- *joinness* of mean operator.

Mean-like type-1 OWA operators can be obtained by selecting the linguistic weights appropriately. For example, Figure 7a shows three identical linguistic weights in the forms of triangular type-1 fuzzy numbers whose cores locate at 1/3 as follows,

$$\mu_{W_i}(u) = max\{0, min\{3u, 2-3u\}\} \tag{13}$$

The type-1 OWA operator with these three linguistic weights behaves as the mean-like operation on type-1 fuzzy sets, which is corroborated by its *joinness* as illustrated in Figure 7c. Its *joinness* clearly indicates that this mean-like type-1 OWA operator neither has the tendency to the join operation nor the meet operation. The result of aggregating three type-1 fuzzy sets by this mean-like OWA operator is shown in Figure 7b.

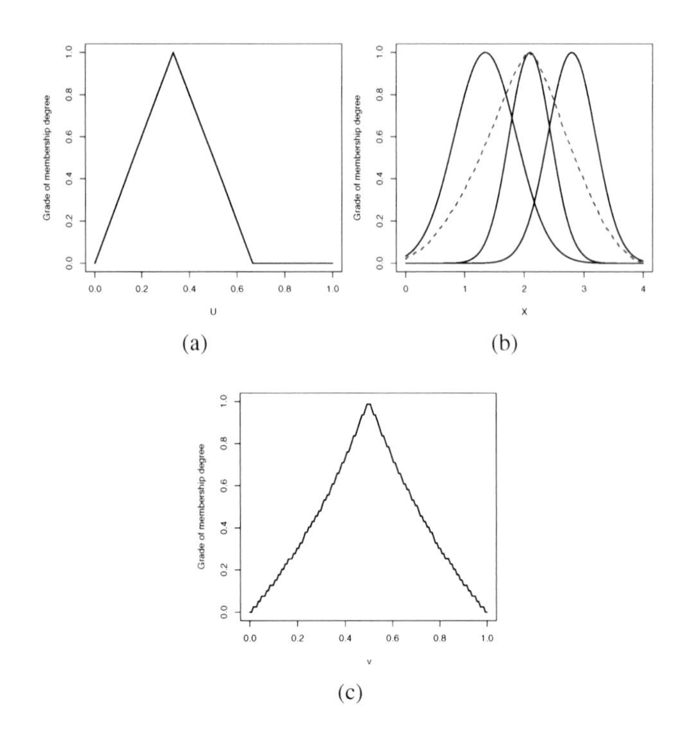

Fig. 7 Mean like operator: (a)-three same type-1 fuzzy sets as linguistic weights with cores locating at 1/3: $W_i(i = 1,2,3)$; (b)- solid lines represent aggregated fuzzy sets, dashed line represents aggregation result; (c)- *joinness* of this mean like operator.

4 Type-2 OWA Operators for Aggregating Type-2 Fuzzy Sets

In the above sections, we have discussed the type-1 OWA operators and some related issues. In this section, we define a termed *type-2 OWA operator* to aggregate type-2 fuzzy sets via an OWA mechanism.

4.1 Definition

Let $\widetilde{F}(X) = \{\widetilde{A} | \widetilde{A} \text{ is type-2 fuzzy set on X}\}$. Based on Zadeh's Extension Principle, in the following we extend Yager's OWA operator and define the type-2 OWA operator for aggregating type-2 fuzzy sets.

Definition 3: *Given n linguistic weights* $\left\{\widetilde{W}_i\right\}_{i=1}^{n}$ *in the form of type-2 fuzzy sets defined on the domain of discourse* $U = [0,\ 1]$, *a type-2 OWA operator is a mapping* $\widetilde{\Phi}$,

$$\begin{aligned}\widetilde{\Phi} \ : \ \widetilde{F}(X) \times \cdots \times \widetilde{F}(X) \ &\rightarrow \ \widetilde{F}(X) \\ \left(\widetilde{A}_1, \cdots, \tilde{A}_n\right) \ &\mapsto \ \widetilde{G}\end{aligned}$$

that is associated with $\left\{\widetilde{W}_i\right\}_{i=1}^{n}$ *to aggregate the type-2 fuzzy sets* $\left\{\widetilde{A}_i\right\}_{i=1}^{n} \subset \widetilde{F}(X)$. *Each slice of the aggregating result,* $\widetilde{G}$, *is defined as*

$$G_x = \bigsqcup_{\substack{\sum_{i=1}^{n} \bar{w}_i a_{\sigma(i)} = x \\ w_i \in U, a_i \in X}} W_{1,\, w_1} \otimes \cdots \otimes W_{n,\, w_n} \otimes A_{1,\, a_1} \otimes \cdots \otimes A_{n,\, a_n} \tag{14}$$

in which $W_{i,\, w_i} \triangleq \mu_{\widetilde{W}_i}(w_i, \cdot)$ *and* $A_{i,\, a_i} \triangleq \mu_{\widetilde{A}_i}(a_i, \cdot)$ *are type-1 fuzzy sets,* $\bar{w}_i = w_i / \sum_{i=1}^{n} w_i$; $\sigma : \{1, \cdots, n\} \rightarrow \{1, \cdots, n\}$ *is a permutation function such that* $a_{\sigma(i)} \geq a_{\sigma(i+1)}$, $\forall i = 1, \cdots, n-1$, *i.e.,* $a_{\sigma(i)}$ *is the ith largest element in the set* $\{a_1, \cdots, a_n\}$; $\sqcup$ *is the join operator defined in (5), whereas* $\otimes$ *is a t-norm operator that applies to type-1 fuzzy sets, for example,* $A_{i,\, a_i}$ *and* $A_{j,\, a_j}$, *as follows:*

$$\mu_{A_{i,\, a_i} \otimes A_{j,\, a_j}}(r) = \sup_{\substack{s \otimes t = r \\ s \in J_{a_i}, t \in J_{a_j}}} \mu_{\widetilde{A}_i}(a_i, s) * \mu_{\widetilde{A}_j}(a_j, t) \tag{15}$$

where $*$ *is a t-norm operator for crisp numbers and can be different from* $\otimes$. *Similar operations are performed on* $W_{i,\, w_i} \otimes W_{j,\, w_j}$ *and* $W_{i,\, w_i} \otimes A_{j,\, a_j}$.

It can be seen that the aggregation result of type-2 fuzzy sets by the type-2 OWA (14), $\widetilde{G} = \widetilde{\Phi}\left(\widetilde{A}_1, \cdots, \tilde{A}_n\right)$, is a type-2 fuzzy set. However, type-2 OWA operations on general type-2 fuzzy sets are computationally intensive. Fortunately, if the linguistic weights and aggregated objects are interval type-2 fuzzy sets (IT2FSs) , type-2 OWA operations can be greatly simplified.

It can be proved that if the linguistic weights $\left\{\widetilde{W}_i\right\}_{i=1}^{n}$ and aggregated objects $\left\{\widetilde{A}_i\right\}_{i=1}^{n}$ are IT2FSs, then the type-2 OWA aggregating result $\widetilde{G} = \widetilde{\Phi}\left(\widetilde{A}_1, \cdots, \tilde{A}_n\right)$ is an IT2FS. So in the IT2FS-oriented type-2 OWA aggregation we only need to calculate the the footprint of uncertainty (FOU) of $\Phi\left(\widetilde{A}_1, \cdots, \tilde{A}_n\right)$, i.e. $FOU(\widetilde{G}) = \bigcup_{x \in X} J_x$. Hence given a point x, we need to calculate the primary membership grade J_x of $\widetilde{G}$.

It can be seen from (15) that for the IT2FSs $\widetilde{A}_i$ and $\widetilde{A}_j$, the domain of $A_{i,\, a_i} \otimes A_{j,\, a_j}$ is

$$J_{a_i a_j} = J_{a_i} \otimes J_{a_j} \triangleq \left\{ s \otimes t \,\middle|\, s \in J_{a_i}, t \in J_{a_j} \right\} \tag{16}$$

then, for the IT2FSs $\widetilde{W}_1, \cdots, \widetilde{W}_n, \widetilde{A}_1, \cdots \widetilde{A}_n$, the domain of $W_{1,w_1} \otimes \cdots \otimes W_{n,w_n} \otimes A_{1,a_1} \otimes \cdots \otimes A_{n,a_n}$ is

$$J_{w_1 \cdots w_n a_1 \cdots a_n} = J_{w_1} \otimes \cdots \otimes J_{w_n} \otimes J_{a_1} \otimes \cdots \otimes J_{a_n} \tag{17}$$

Then we have:

$$J_x = \bigvee_{\substack{\sum_{i=1}^{n} \bar{w}_i a_{\sigma(i)} = x \\ w_i \in U, a_i \in X}} J_{w_1 \cdots w_n a_1 \cdots a_n} \tag{18}$$

which is called IT2FS-oriented type-2 OWA operator.

Without loss of generality, for aggregating interval type-2 fuzzy sets, let $J_{w_i} = \left[g^l_{i,w_i}, g^r_{i,w_i}\right]$, $J_{a_i} = \left[g^l_{i,a_i}, g^r_{i,a_i}\right]$, and $J_{w_1 \cdots w_n a_1 \cdots a_n} = \left[J^l_{w_1 \cdots w_n a_1 \cdots a_n}, J^r_{w_1 \cdots w_n a_1 \cdots a_n}\right]$. Then for t-norm operator $\otimes = min$ or *product* [22],

$$J^l_{w_1 \cdots w_n a_1 \cdots a_n} = g^l_{1,w_1} \otimes \cdots \otimes g^l_{n,w_n} \otimes g^l_{1,a_1} \otimes \cdots \otimes g^l_{n,a_n} \tag{19}$$

$$J^r_{w_1 \cdots w_n a_1 \cdots a_n} = g^r_{1,w_1} \otimes \cdots \otimes g^r_{n,w_n} \otimes g^r_{1,a_1} \otimes \cdots \otimes g^r_{n,a_n} \tag{20}$$

That is to say, the left and right end points of the interval $J_{w_1 \cdots w_n a_1 \cdots a_n}$ only depends on the left and right end points of the aggregated intervals separately. In fact, the left and right end points of the maximum of intervals also only depend on the left and right end points of the individual intervals, i.e., $[b^l_1, b^r_1] \vee \cdots \vee [b^l_n, b^r_n] = [\vee b^l_i, \vee b^r_i]$. Hence, we can calculate the left end point and right end point of $J_x = [J^l_x, J^r_x]$ as follows respectively,

$$J^l_x = \bigvee_{\substack{\sum_{i=1}^{n} \bar{w}_i a_{\sigma(i)} = x \\ w_i \in U, a_i \in X}} J^l_{w_1 \cdots w_n a_1 \cdots a_n} \tag{21}$$

and

$$J^r_x = \bigvee_{\substack{\sum_{i=1}^{n} \bar{w}_i a_{\sigma(i)} = x \\ w_i \in U, a_i \in X}} J^r_{w_1 \cdots w_n a_1 \cdots a_n} \tag{22}$$

4.2 A Procedure for Performing IT2FSs-Oriented Type-2 OWA Operations

Given the linguistic weights $\left\{\widetilde{W}_i\right\}_{i=1}^{n} \subset \widetilde{F}(U)$, as usual, the domains of X and U need to be discretised during calculation in order for the associated IT2FSs-oriented type-2 OWA operator to aggregate IT2FSs $\left\{\widetilde{A}_i\right\}_{i=1}^{n} \subset \widetilde{F}(X)$ on a computer. Let the discretised domains be $\hat{X} = \{\hat{x_1}, \cdots, \hat{x_p}\}$ and $\hat{U} = \{\hat{u_1}, \cdots, \hat{u_k}\}$, which are partitions of the spaces X and U respectively. We have noticed that the type-2 OWA operations in terms of (21) and (22) suffer from the *over-partition* problem as the type-1 OWA operator does due to another partitioning of X produced by $\sum_{k=1}^{n} \bar{w}_i a_{\sigma(i)}$ with all the combinations $(w_1, \cdots, w_n, a_1, \cdots, a_n)$ of weighting points in $\hat{U}$ and aggregating points in $\hat{X}$, i.e,

$$\overline{X} = \{\bar{x}_j\} = \left\{ \sum_{k=1}^{n} \bar{w}_i a_{\sigma(i)} \middle| w_i \in \hat{U}, a_i \in \hat{X}, i = 1, \cdots, n \right\} \tag{23}$$

For example, Figure 8a illustrates one set with poor understandability generated by an IT2FSs-oriented type-2 OWA operator on the $\overline{X}$, the *over-partition* version of discretised X. The following procedure is proposed to perform IT2FSs-oriented type-2 OWA operation while resolving the *over-partition* problem.

Step 1: Initialisation

1. Given the linguistic weights $\left\{\widetilde{W}_i\right\}_{i=1}^{n} \subseteq \widetilde{F}(U)$ in the form of IT2FSs for aggregating IT2FS objects $\left\{\widetilde{A}_i\right\}_{i=1}^{n} \subseteq \widetilde{F}(X)$.
2. Given the discretised domains of linguistic weights, $\hat{U}$, and that of aggregated objects, $\hat{X}$.
3. Let the initial $\overline{G} = (\overline{X}, \mu_{\overline{G}})$, where $\overline{X} = \emptyset, J_0^l = 0, J_0^r = 0$.

Step 2: Calculate $\overline{G}$.

1. Select $w_1 \in \hat{U}, \cdots, w_n \in \hat{U},\ a_1 \in \hat{X}, \cdots, a_n \in \hat{X}$,
2. Normalise $(w_1, \cdots, w_n)$ as $\bar{w}_i = w_i / \sum_{i=1}^{n} w_i$
3. Perform Yager's OWA operation:

$$\bar{y} = \phi_{\bar{w}_1, \cdots, \bar{w}_n}(a_1, \cdots, a_n)$$

4. Calculate $J^l_{w_1 \cdots w_n a_1 \cdots a_n}$ and $J^r_{w_1 \cdots w_n a_1 \cdots a_n}$:

$$J^l_{w_1 \cdots w_n a_1 \cdots a_n} = g^l_{1,w_1} \otimes \cdots \otimes g^l_{n,w_n} \otimes g^l_{1,a_1} \otimes \cdots \otimes g^l_{n,a_n}$$

$$J^r_{w_1 \cdots w_n a_1 \cdots a_n} = g^r_{1,w_1} \otimes \cdots \otimes g^r_{n,w_n} \otimes g^r_{1,a_1} \otimes \cdots \otimes g^r_{n,a_n}$$

5. If there exists $\bar{x} \in \overline{X} : \bar{x} = \bar{y}$, then update the potential primary membership grade $J_{\bar{x}}$:

$$J^l_{\bar{x}} \leftarrow max\left(J^l_{\bar{x}}, J^l_{w_1 \cdots w_n a_1 \cdots a_n}\right)$$

and

$$J^r_{\bar{x}} \leftarrow max\left(J^r_{\bar{x}}, J^r_{w_1 \cdots w_n a_1 \cdots a_n}\right)$$

Otherwise, $\bar{y}$ is added to $\overline{X}$, and the primary membership grade at $\bar{y}$: $J_{\bar{y}} \triangleq J_{w_1 \cdots w_n a_1 \cdots a_n}$.

6. Go to ***Step 2***-1, and continue until all the weight vectors and aggregating points are selected.

   Step 3: Induce the IT2FS G on $\hat{X}$:

$$J^l_{\hat{x}} = \bigvee_{\bar{x}_j \in \Theta_{\hat{x}}} J^l_{\hat{x}_j}$$

and

$$J^r_{\hat{x}} = \bigvee_{\bar{x}_j \in \Theta_{\hat{x}}} J^r_{\hat{x}_j}$$

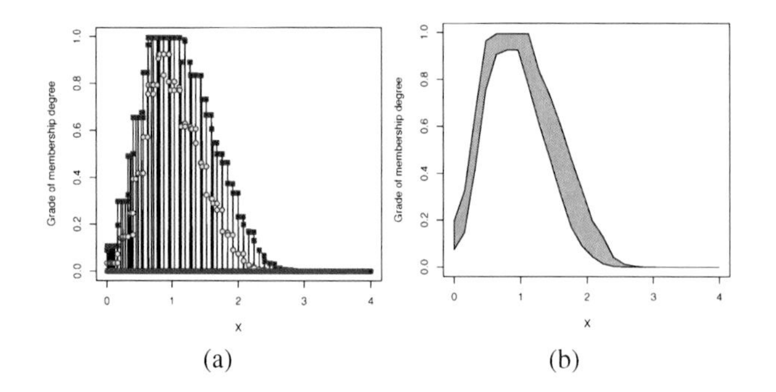

Fig. 8 (a)-A set generated on the *over-partition* version of discretised X; (b)-The associated set generated on the discretised X.

Figure 8b illustrates the result generated on the discretised X of the *over-partition* in Figure 8a.

Example 2. Three linguistic weights $\widetilde{W}_1, \widetilde{W}_2$, and $\widetilde{W}_3$ in the form of IT2FSs as shown in Figure 9 are used to define a type-2 OWA operator with *min* t-norm, $\Phi_{\widetilde{W}_1\widetilde{W}_2\widetilde{W}_3}$. This IT2FSs-oriented type-2 OWA operator is then used to aggregate three interval type-2 fuzzy sets as depicted in Figure 10a, while the Figure 10b shows the corresponding aggregation result.

Example 3. In this example, the three same linguistic weights with different order as shown in Figure 11a are used to define a type-2 OWA operator with *min* t-norm, $\Phi_{\widetilde{W}_1\widetilde{W}_2\widetilde{W}_3}$. This IT2FSs-oriented type-2 OWA operator is also used to aggregate three interval type-2 fuzzy sets as depicted in Figure 10a, while the Figure 11b shows the corresponding aggregation result.

It can be seen from Examples 2 and 3 that the aggregation results obtained via type-2 OWA operators are consistent with the compensative property of Yager's OWA operator [1]: Yager's OWA operators can vary from the "min" (i.e, most-left aggregated object on the domain X) to "max" (i.e, most-right aggregated object on the domain X) aggregation.

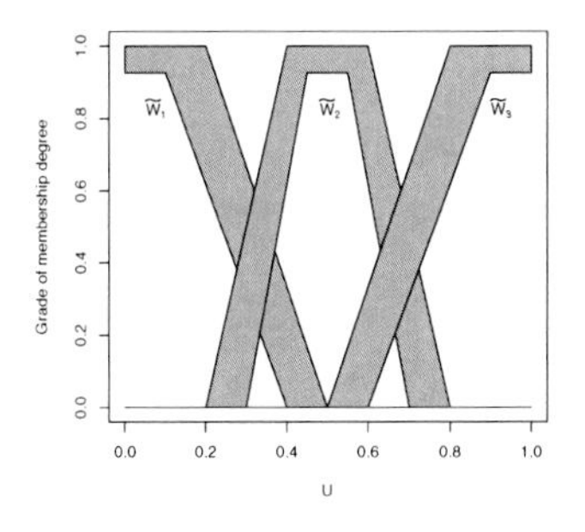

Fig. 9 Three linguistic weights defining a type-2 OWA operator

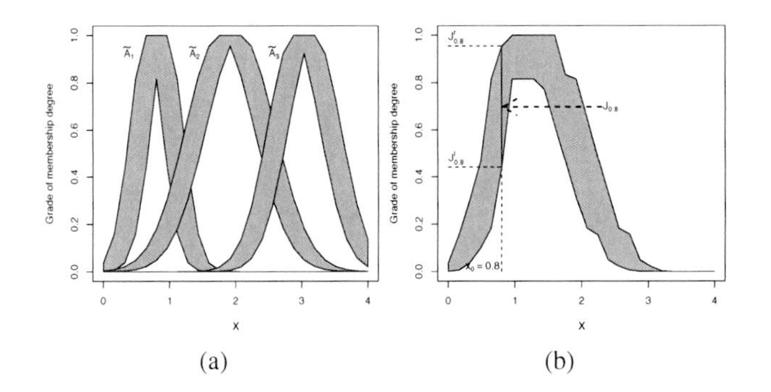

Fig. 10 (a)-Three interval type-2 fuzzy sets to be aggregated by type-2 OWA operators; (b)-Aggregating result by the type-2 OWA operator with linguistic weights in Figure 9

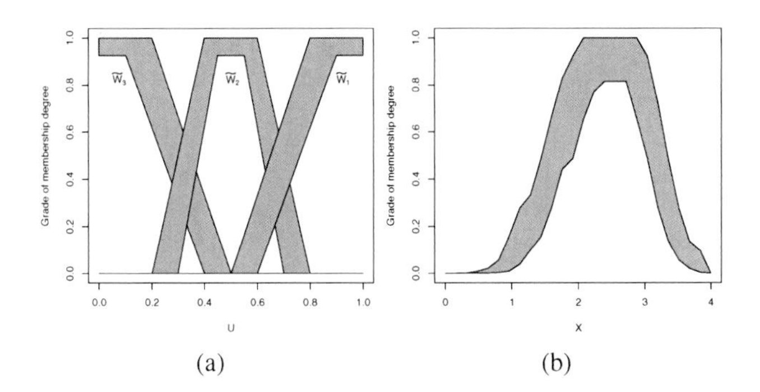

Fig. 11 (a)-Three linguistic weights defining a type-2 OWA operator; (b)-Aggregating result by the type-2 OWA operator with linguistic weights in Figure 11a

5 Conclusion

The type-1 OWA operator is capable of aggregating type-1 fuzzy sets by taking into account the whole membership function of the sets in the aggregation process. In this chapter, some special cases of type-1 OWA operators are addressed, the joinness and the meetness of type-1 OWA operators are proposed as a way of linguistically expressing the degrees of the aggregation being like a join operation and meet operation respectively. Moreover, the type-2 OWA operators are defined to aggregate uncertain information modelled by type-2 fuzzy sets via OWA mechanism.

Some interesting new issues arise including the possibility of applying the type-1 OWAs to merge similar fuzzy sets for improving fuzzy model interpretability/transparency and parsimony [27, 28, 29]. etc. In particular, the type-1 OWA aggregation operators may have great potential in being applied to multi-expert decision making and multi-criteria decision making.

Acknowledgements. This work has been supported by the EPSRC Research Grant EP/C542215/1 and EP/C542207/1.

References

1. Yager, R.R.: On ordered weighted averaging aggregation operators in multi-criteria decision making. IEEE Trans. on Systems, Man and Cybernetics 18, 183–190 (1988)
2. Dubois, D., Prade, H.: A review of fuzzy set aggregation connectives. Information Sciences 36, 85–121 (1985)
3. Fodor, J., Roubens, M.: Fuzzy Preference Modelling and Multicriteria Decision Support. Kluwer, Dordrecht (1994)
4. McClean, S., Scotney, B., Shapcott, M.: Aggregation of imprecise and uncertain information in databases. IEEE Transactions on Knowledge and Data Engineering 13(6), 902–912 (2001)
5. Chen, A.L.P., Chiu, J.-S., Tseng, F.S.C.: Evaluating aggregate operations over imprecise data. IEEE Transactions on Knowledge and Data Engineering 8(2), 273–284 (1996)
6. Dubois, D., Prade, H.: On the use of aggregation operations in information fusion processes. Fuzzy Sets and Systems 142, 143–161 (2004)
7. Xu, Z.S., Da, Q.L.: An overview of operators for aggregating information. International Journal of Intelligent Systems 18, 953–969 (2003)
8. Chiclana, F., Herrera-Viedma, E., Herrera, F., Alonso, S.: Induced ordered weighted geometric operators and their use in the aggregation of multiplicative preference relations. International Journal of Intelligent Systems 19, 233–255 (2004)
9. Chiclana, F., Herrera-Viedma, E., Herrera, F., Alonso, S.: Some induced ordered weighted averaging operators and their use for solving group decision-making problems based on fuzzy preference relations. European Journal of Operational Research 182, 383–399 (2007)
10. Dong, W.M., Wong, F.S.: Fuzzy weighted averages and implementation of the extension principle. Fuzzy Sets and Systems 21, 183–199 (1987); weights, Fuzzy Sets and Systems 94, 157–169 (1998)
11. Guh, Y.-Y., Hon, C.-C., Lee, E.S.: Fuzzy weighted average: the linear programming approach via Charnes and Cooper's rule. Fuzzy Sets and Systems 117, 157–160 (2001)

12. Herrera, F., Herrera-Viedma, E.: Aggregation operators for linguistic weighted information. IEEE Trans. on Systems, Man and Cybernetics-part A 27, 646–656 (1997)
13. Herrera, F., Herrera-Viedma, E., Chiclana, F.: A study of the origin and uses of the ordered weighted geometric operator in multicriteria decision making. International Journal of Intelligent Systems 18, 689–707 (2003)
14. Zimmermann, H.-J.: Fuzzy Sets, Decision Making and Expert Systems. Kluwer Academic Publishers, Boston (1987)
15. Zimmermann, H.-J., Zysno, P.: Latent connectives in human decision making. Fuzzy Sets and Systems 4, 37–51 (1980)
16. Ahn, B.S.: The OWA aggregation with uncertain descriptions on weights and input arguments. IEEE Trans. Fuzzy Systems 15(6), 1130–1134 (2007)
17. Mitchell, H.B., Schaefer, P.A.: On ordering fuzzy numbers. Internaltional Journal of Intelligent Systems 15, 981–993 (2000)
18. Meyer, P., Roubens, M.: On the use of the Choquet integral with fuzzy numbers in multiple criteria decision support. Fuzzy Sets and Systems 157(7), 927–938 (2006)
19. Wang, Z., Yang, R., Heng, P.-A., Leung, K.-S.: Real-valued Choquet integrals with fuzzy-valued integrands. Fuzzy Sets and Systems 157(2), 256–269 (2006)
20. John, R.I., Innocent, P.: Modelling uncertainty in clinical diagnosis using fuzzy logic. IEEE Trans. on Systems, Man and Cybernetics-part B 35, 1340–1350 (2005)
21. Kao, C., Liu, S.-T.: Fractional programming approach to fuzzy weighted average. Fuzzy Sets and Systems 120, 435–444 (2001)
22. Mendel, J.M.: Uncertain Rule-based Fuzzy Logic Systems: Introduction and New Directions. Prentice Hall PTR, Englewood Cliffs (2001)
23. Mizumoto, M., Tanaka, K.: Some Properties of fuzzy sets of type 2. Information and Control 31, 312–340 (1976)
24. Zadeh, L.A.: Fuzzy sets. Information and Control 8, 338–353 (1965)
25. Zadeh, L.A.: The concept of a linguistic variable and its application to approximate reasoning-1. Information Sciences 8, 199–249 (1975)
26. Zhou, S.-M., Chiclana, F., John, R.I., Garibaldi, J.M.: Type-1 OWA operators for aggregating uncertain information with uncertain weights induced by type-2 linguistic quantifiers. Fuzzy Sets and Systems 159, 3281–3296 (2008)
27. Zhou, S.-M., Gan, J.Q.: Constructing accurate and parsimonious fuzzy models with distinguishable fuzzy sets based on an entropy measure. Fuzzy Sets and Systems 157, 1057–1074 (2006)
28. Zhou, S.-M., Gan, J.Q.: Constructing parsimonious fuzzy classifiers based on L2-SVM in high-dimensional space with automatic model selection and fuzzy rule ranking. IEEE Trans. on Fuzzy Systems 15, 398–409 (2007)
29. Zhou, S.-M., Gan, J.Q.: Extracting Takagi-Sugeno fuzzy rules with interpretable submodels via regularization of linguistic modifiers. IEEE Trans. on Knowledge and Data Engineering (2009) (in press)

A Majority Guided Aggregation Operator in Group Decision Making

Gabriella Pasi and Ronald R. Yager

Abstract. This chapter is about majority modelling in the context of group (multi-expert) decision making, to the aim of defining a decision strategy which takes into account the individual opinions of the decision makers. The concept of majority plays in this context a key role: what is often needed is an overall opinion which synthesizes the opinions of the *majority* of the experts. The reduction of the individual experts' opinions into a representative value (which we call the *majority opinion*) is usually performed through an aggregation process. In this chapter we describe two distinct approaches to the definition and consequent computation of a majority opinion within fuzzy set theory, where majority can be expressed by a linguistic quantifier (such as *most*). We first consider the case where linguistic quantifiers are associated with aggregation operators; in this case a majority opinion is computed by aggregating the individual opinions. To model this semantics of linguistic quantifiers the Induced Ordered Weighted Averaging operators (IOWA) are used with a modified definition of their weighting vector. We then consider a second case where the concept of majority is modelled as a vague concept. Based on this interpretation a formalization of a fuzzy majority opinion as a fuzzy subset is described.

1 Introduction

In group decision making (multi-expert decision making) a set of experts are involved in a decision process concerning the evaluation of a set of alternatives.

Gabriella Pasi
Università degli Studi di Milano Bicocca
Viale Sarca 336, 20131 Milano, Italy
e-mail: pasi@disco.unimib.it

Ronald R. Yager
IonaCollege
New Rochelle, NY10801, USA
e-mail: Ryager@Iona.edu

R.R. Yager et al. (Eds.): Recent Developments in the OWA Operators, STUDFUZZ 265, pp. 111–133.
springerlink.com

The first step of this decision process is constituted by the individual evaluations of the decision makers (experts): each expert rates each alternative on the basis of an adopted evaluation scheme [4]. At the end of this step each alternative has an associated performance judgment (evaluation or opinion) on a predefined scale (either numeric or linguistic). The second step of a multi-expert decision process consists in determining for each alternative a consensual judgment, which synthesizes the experts' individual opinions. The consensual judgment is representative of a collective evaluation and is usually computed by means of an aggregation of the individual experts' opinions. Usually also a consensus degree is computed for each alternative, based on a comparaison of the decision makers' opinions. In the case of unanimous consensus, the evaluation process ends with the selection of the best alternative(s). As in real situations humans rarely come to an unanimous agreement, in the literature some fuzzy approaches to evaluate a majority guided aggregation have been proposed. In these approaches full consensus (degree=1) is not necessarily the result of unanimous agreement, but it can be obtained even in case of agreement among a majority of the decision makers [2,6,7,8,9].

This chapter considers the problem of constructing a *majority opinion*, intended as the collective evaluation of a majority of the experts involved in the decision problem. A majority opinion is then intended as the consensual judgment of a given alternative by a majority of the decision makers who have similar opinions. Formally we consider n agents who have expressed individual judgments (opinions) on an alternative. In this chapter we only consider the case of a single alternative, because in the case of multiple alternatives the process of construction of a majority opinion can be independently applied to each alternative. So we have n judgments (opinions) $a_1, \ldots, a_n$ which have to be reduced to an overall majority opinion (related to the considered alternative).

In fuzzy approaches to multi-agent decision making the concept of majority is usually modeled by means of linguistic quantifiers such as *at least 80%* and *most*. In this context linguistic quantifiers are used to indicate a fusion strategy to guide the process of aggregating the experts' opinions. In fuzzy set theory a linguistic quantifier is formally defined as a fuzzy subset of a numeric domain (either non negative real numbers or the unit interval), the membership function of which describes the compatibility of a given absolute or percentage quantity to the concept expressed by the linguistic quantifier. The notion of quantifier guided aggregation has been formally defined by means of Ordered Weighted Averaging operators [18,22] and by means of the concept of fuzzy integrals [5]. In this chapter we consider only the use of OWA operators in group decision making. An example of linguistic expression which employs a quantifier guided aggregation is the following: *Q experts are satisfied by solution a*, where Q denotes a linguistic quantifier, for example *most*, which expresses a majority. To evaluate the satisfaction of this proposition the experts' opinions are aggregated by the OWA aggregation operator which captures the semantics of the concept expressed by the quantifier Q. To associate a linguistic quantifier Q with an OWA aggregation operator, an approach was suggested in [21,22] which makes use of the definition of the linguistic quantifier Q as a fuzzy subset.

In this chapter two distinct approaches to the definition of a *majority opinion* in the context of multi-experts decision making are described. These approaches were introduced in [13].

In the first approach an Induced Ordered Weighted Averaging (IOWA) operator is used to obtain a scalar value for a majority opinion. The second approach is based upon the calculation of the concept of the majority opinion as an imprecise value. Under this second interpretation we propose a formalization of the idea of a fuzzy majority as a fuzzy subset. As we shall see this approach provides in addition to a value for a majority opinion an indication of the strength of that value as the majority opinion.

The main goal of both approaches is to obtain a value which can truly be considered as the opinion of a majority of the experts, that is, a value that is similar for any large group of people. Both methods require we have both information about the similarity between the experts' opinions, and some information about what quantity constitutes the idea of a majority.

The chapter is structured as follows: in section 2 the problem of the definition of aggregation operators with a semantics of majority is presented, in section 3 the first approach to construct Induced Ordered Weighted Averaging operators for linguistic quantifiers with a semantics of majority is explained. In sections 4, 5 and 6 a formal definition of the concept of fuzzy majority opinion is described.

2 The Semantics of OWA Operators in an Aggregation Guided by "Majority" Linguistic Quantifiers

To model the majority concept we consider monotonic non decreasing linguistic quantifiers, such as *most* and *at least 80%*. In particular we are interested in the use of such linguistic quantifiers in guiding an aggregation process aimed at computing the "*majority opinion*" related to a given alternative. The majority opinion is a value which synthesizes the majority of values to be aggregated in a multi-expert decision process. As explained in [18] OWA operators can be constructed based on the fuzzy definition of a linguistic quantifier; by applying this procedure, in the literature related to fuzzy group decision making the concept of majority has been usually modelled. However in this section we will show that by applying this procedure the resulting aggregated value may not be representative of the majority of values. In section 3 we will then introduce a new approach, based on the use of IOWA operators, which better captures an opinion shared by a majority of the experts involved in the decision process.

Objectively, the OWA operator is an aggregation operator taking a collection of argument values and returning a single value. The weights of the OWA weighting vector determine the behaviour of the aggregation operator. These weights have the effect of emphasizing or demphasizing different components in the aggregation. Subjectively, there are a number of different semantics that can be associated with an OWA operator and which determine strategies of construction of its weighting vector.

One semantics that can be associated with the OWA operators is as a generalization of the idea of an averaging or summarizing operator. Here for example $w_i = 1/n$ for all i gives us the simple average.

Another semantics that has been associated with the OWA is a generalization of the logical quantifiers *there exists* and *for all*. The weighting vector with $w_1=1$ and $w_j=0$ for $j \neq i$ corresponds to the quantifier "*there exists*" and the one with $w_n=1$ and $w_j=0$ for $j \neq n$ corresponds to the quantifiers "*for all*". In this framework the arguments are seen as truth values or degrees of satisfaction.

If Q is a quantifier then the OWA aggregation provides a value which can be seen as the truth or fulfilment value of the statement "*Q the elements being aggregated are satisfied*". As outlined in [18], the weights of the weighting vector of an OWA operator are interpreted as the increase in satisfaction in having i+1 criteria "fully" satisfied with respect to having "fully" satisfied i criteria. It is this semantics that has been most often used in applications of OWA operators to multi-criteria and multi-expert decision making. We try to clarify with an example. Let us suppose we define the weighting vector of the aggregation operator associated with the linguistic quantifier *at least 80%,* in the case in which we have to aggregate 5 values. Let us suppose that the 5 values to be aggregated are the evaluations scores of 5 experts related to a given alternative. A crisp definition of this aggregation operator can be: $W_{\text{at least } 80\%} = [0\ 0\ 0\ 1\ 0]$ as full satisfaction can be achieved if the 80% of the experts have judged the alternative in a positive way (non null evaluation) . Let us suppose that the evaluations scores to be aggregated are [1 1 1 0.1 0]. By applying the *at least 80%* aggregation operator to these values we obtain the aggregated value 0.1. However, we would obtain the same overall score when aggregating the values [0.1 0.1 0.1 0.1 0]. These results highlight the fact that the linguistic quantifier is intended to guide the aggregation with a semantics like "It is true that at least 80% of the criteria are *fully* satisfied". In both previous examples, the aggregated value corresponds to the 4th decreasing value, which corresponds to the evaluation of the 80% of the values, independently on the degrees of satisfaction of the previous values, which are greater than or equal to the fourth value. So in this case there is no compensantion between the values to be aggregated.

The semantics of the aggregation guided by the quantifier *at least 80%* modeled with the previous definition of the weighting vector is then not aimed at producing a synthesis of the most similar values in the quantity specified by the quantifier. This latter semantics can be particularly useful in the context of group decision making. While with the original semantics of the weighting vector of an OWA operator the aggregated value is like a degree of satisfaction (truth) of the proposition "Q of the values are fully satisfied", an operator with the semantics of calculating a majority opinion should produce a value which is representative of the 80% *of the most similar values.* In the first example above this "representative" value could be around the value 0.70, because the 80% of the most similar values is around this value. In the second example the aggregated value should truly be 0.1. In other words what we want to obtain is an aggregation of the most similar opinions held by a quantity of decision makers specified by the linguistic quantifier Q. This situation appears to bring us closer in spirit to interpretation of the OWA operator as averaging operator of a specified quantity rather then as a generalized quantifier. In fact what we want is *an average of "most of the similar values"*.

We now present an example which better shows the two distinct semantics of a quantifier guided aggregation; in this example we define the weighting vector of an OWA operator associated with the linguistic quantifier *most.* A simple interpretation of this linguistic quantifier can be that I desire that *at least 70%* of the criteria are satisfied (in other words at least 70% of the values to be aggregated have to be greater than 0). This quantifier has an and-like semantics. If we consider 6 elements to be aggregated a possible weighting vector associated with *most* can be [0 0 0 0.7 0.2 0.1], which means that the concept of majority corresponds to having at least 4 elements satisfied. Let us now suppose to have the following values to be aggregated: (1 1 1 0.5 0 0). The result of the aggregation under the previous weighting vector produces the value 0.35; this aggregated value does not characterize the value of the majority of the most similar values, which is intuitively a value closer to 1. As outlined before the main reason for this result is that the OWA aggregation produces a value which reflects the satisfaction of the proposition "*most of the criteria have to be satisfied*" instead of "*the satisfaction value of most of the criteria*".

It is important to notice that the semantics of the linguistic quantifier is strongly affected by the non-linear component of the aggregation operator (i.e. by the way in which the arguments are reordered). In the example before with the usual construction of the weighting vector the obtained low aggregated value is due to the fact that the values to be aggregated are in decreasing order.

This kind of semantics does not naturally model the meaning of the concept of majority as typically used in group decision making applications. When we use linguistic quantifiers to express our intent in aggregating the opinions of the group of decision makers, the implicit aim in stating *most* is to remind the fact that we want an evaluation that correspond to a majority of the experts holding a similar opinion, where by majority we intend *most.* This means that we need an aggregation operator that takes an average like aggregation of a majority of values that are similar. To this aim what should be effectively aggregated are the most similar values. The concept of similarity plays here a crucial role. In the next section the approach which has been defined in [13] to model this aggregation behaviour is explained.

3 Using IOWA Operators to Compute a Majority Opinion

To produce an aggregation with a majority semantics, in [13] we have proposed to use the IOWA operators with an inducing ordering variable which is based on a proximity metric over the elements to be aggregated. The basic idea is that the most similar values must have close positions in the induced ordering in order to appropriately be aggregated. We also suggested a new strategy for constructing the weighting vector so as to better model the new "majority-based" semantics of the aggregation.

Before introducing the new method, in subsection 3.1 below the definition of IOWA operators is synthesized. In subsection 3.2 the new quantifier guided aggregation procedure is explained.

3.1 Induced Ordered Weighted Averaging Operators

In [23] Yager and Filev introduced an extension of the OWA called the Induced Ordered Weighted Averaging (IOWA) Operators. Here we have a collection of values (a_1, ..., a_n) to be aggregated, and an associate weighting vector W. However, associated with each of the argument values, the a_j, is another value v_j called the order inducing value. Let v-index be an ordering function such that v-index(i) is the index of the ith largest v_i. Using the induced OWA we calculate:

$$\text{I-F}(a_1, \ldots, a_n) = \sum_{i=1}^{n} w_i \, a_{v-index(i)}$$

Here then while we are still aggregating the a_j values our ordering is done with respect to the v_j value. Here then we have pairs (a_j,v_j), a_j being the argument value and v_j being the order inducing value. If W is a vector of weights and B_V is a vector whose values are the argument values ordered by the order inducing values:

$$\text{I-F}(a_1, a_2, \ldots, a_n) = W^T B_V.$$

As a simple example consider the case when we want to aggregate the following four pairs (a_j,v_j): (0.7,0.4), (0.9,0.3), (0.2,1) and (0.6,0.7). In this case the inducing values are v_1=0.4, v_2=0.3, v_3=1, and v_4=0.7, which are ordered 1>0.7>0.4>0.3; hence v-index is such that v-index(1)=3, v-index(2)=4, v-index(3)=1, v-index(4)=2.

If our weighting vector has the following values: w_1=0.1, w_2=0.2, w_3=0.3, w_4=0.4, we obtain the following aggregated value:

$$\text{I-F}(a_1, a_2, a_3, a_4) = (0.1)(0.2)+(0.2)(0.6)+(0.3)(0.7)+(0.4)(0.9)= 0.71.$$

3.2 Computation of a Majority Opinion in Group Decision Making

What we want is to aggregate a set of n values expressing the evaluations of a given alternative by n experts in order to produce a value which synthesizes the opinion of the majority of the experts. To this aim our intent is to take the most similar values in the quantity specified by the quantifier and apply to them an averaging operator. What we need is a computation of the similarities between the opinion values. The values of the inducing variable of the IOWA operator are obtained by means of a function of the similarities between pairs of the opinion values. We define such a function using a Support function, as defined in [24]. A support function Sup is a binary function which computes a value Sup(a,b) which expresses the support from *b* for *a*; the more similar, close are two values the more they support each other. A simple example of support function is the following:

$$\sup(a_i, a_j) = \begin{cases} 1 & \text{if } |a_i - a_j| < \alpha \\ 0 & \text{otherwise} \end{cases} \quad (1)$$

If we consider a set of values to be aggregated, and we want to order them in increasing order of support we compute for each value the sum of its support values with respect to all the others values to be aggregated. Then for each expert's opinion we sum all the supports it has in order to obtain its overall support. These overall supports for an expert's opinion are used as the values of the order inducing variable.

Concerning the alpha value, its setting is fundamental in the aggregation behaviour, as it determines the maximum admissible "distance" among two values (to be aggregated) in order to consider them "similar".This value roughly determines the number of groups of similar values among those to be aggregated. Its setting strongly depends on the granularity and semantics of the values to be aggregated. Let us clarify by an example. Let us suppose to express opinions (i.e. the values to be aggregated) in the range [0,1] (notice that this is not mandatory). Let us also suppose that the various experts may express opinions on a scale 1 to ten; for example professors' marks evaluating a student's examination may be expressed in tenth. By representing these marks in the interval [0,1] we obtain ten possible evaluations values: 0.1, 0.2, 0.3 ..... 1 (this means that values, say 0.45, 0.52 will never be used to score a student's examination). In this setting, when two values among the above ten can be considered similar? In other words, when professors have the feeling to evaluate a given student in a similar way? A possible (reasonable) choice is to set to 0.1 the maximum admissible distance value between two values, in order to consider them similar. This means that two distinct evaluations (still in tenth) equal to 8 and 9 can be considered similar, while 8 and 10 can not (of course we could be more tolerant by setting, say, the alpha value to 0.2). Now let us consider another possible reference set for expressing evaluation scores. In this context we are admitted to evaluate an examination by just four scores: 1, 2, 3, 4. If we express these values in the [0,1] interval we obtian four possible evalution scores, i.e. 0.25, 0.5, 0.75 and 1. It is evident that in this last case the granulaity of possible evaluation judgments is lower that the one of the first example (four possible evaluation score against ten). An appropriate alpha value in this case cannot be lower than 0.25. So a possible way to select the alpha value is to reason about granularity of opinion's evaluations, and about the admissible absolute distance between the two values in order to consider them "similar".

We now present an example of an application of a support function with α=0.4. Let us suppose we have the following values to aggregate:

$$a_1 = 0.9 \ a_2 = 0.7 \ a_3 = 0.6 \ a_4 = 0.1 \ a_5 = 0$$

On the basis of the previous Support function, we compute the values of the supports for each pairs of values:

$$\begin{array}{llll} Sup(a_1,a_2)=1 & Sup(a_1,a_3)=1 & Sup(a_1,a_4)=0 & Sup(a_1,a_5)=0 \\ Sup(a_2,a_1)=1 & Sup(a_2,a_3)=1 & Sup(a_2,a_4)=0 & Sup(a_2,a_5)=0 \\ Sup(a_3,a_1)=1 & Sup(a_3,a_2)=1 & Sup(a_3,a_4)=0 & Sup(a_3,a_5)=0 \\ Sup(a_4,a_1)=0 & Sup(a_4,a_2)=0 & Sup(a_4,a_3)=0 & Sup(a_4,a_5)=1 \\ Sup(a_5,a_1)=0 & Sup(a_5,a_2)=0 & Sup(a_5,a_3)=0 & Sup(a_5,a_4)=1 \end{array}$$

The overall support for each a_i is obtained by adding the support values for a_i; we denote this value by s_i:

$$s_1=2 \quad s_2=2 \quad s_3=2 \quad s_4=1 \quad s_5=1$$

We can see that we have two main "clusters" of similar values. In fact the use of the adopted support function induces a clustering of the arguments which can be controlled by the choice of the threshold parameter α in function (1). In the example we obtain two clusters with some ties of the overall support values. If we want to "solve" the ties we can impose a "stricter" condition by setting α=0.3; in this way we obtain:

$$s_1=1 \quad s_2=2 \quad s_3=1 \quad s_4=1 \quad s_5=1$$

This result, combined with the previous one allows to order the elements to be aggregated in the following increasing order of similarity:

induced similarity order: I = [0 0.1 0.6 0.9 0.7]

So we see that the use of an appropriate Support function allows us to induce an ordering based on proximity.

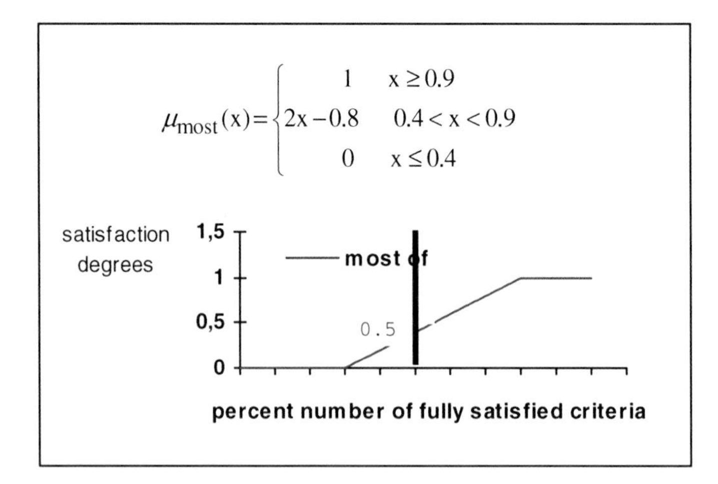

Fig. 1 A possible definition of the linguistic quantifier *most*

Let us suppose that we want now to obtain a majority-based aggregation of the previous values. The selected IOWA operator should then correspond to the linguistic quantifier *most*. We first consider the definition of *most* proposed in Figure 1; in this case the linguistic quantifier is defined by means of a non-decreasing function. Starting from this definition we construct the weighting vector of 5 elements (as 5 are the elements of the example above); it is W = [0 0 0.4 0.4 0.2]. Let us notice that the fifth element of this vector is smaller than the fourth element. By aggregating the vector I above we obtain: I * W = 0.76. We note that this value is a much better representative of the majority of the values to be aggregated, than the value which we would obtain with the usual OWA operator associated with *most* (with the elements in decreasing order of their value): B = [0.9 0.7 0.6 0.1 0], and B*W = 0.28.

However, as outlined before, what can be noticed in the considered W vector is that the last weight (on the right hand side of the vector W) is smaller than the previous value; this is coherent with the interpretation of the weights as increase in satisfaction in having i+1 with respect to having i criteria satisfied. However, in an aggregation with the semantics of majority what would be expected is that the weights of the weighting vector are non decreasing; in fact as in the induced order of the arguments the top value is the most "supported" one from the all the other values (the most representative) it should be more emphasized than the others, or at least not less emphasized. For this reason, and in order to obtain a value which better represents a majority of the aggregated elements, we propose a new strategy for the construction of the weighting vector. This strategy has the aim of emphasizing in the aggregation the most supported values; in other words the values which appear on the right hand side of the vector of values to be aggregated have more influence in the aggregation. In the following we suggest a procedure for the construction of the weighting vector which produces a weighting vector with non decreasing weights. First let us consider the overall support (similarity) values computed for the n values to be aggregated: $s_1, s_2, \ldots, s_n$. In order to compute the non decreasing weights of the weighting vector, we first define the values $t_1, t_2, \ldots, t_n$ based on a modification of the $s_1, s_2, \ldots, s_n$ values: $t_i = s_i + 1$. In this way the similarity of the value a_i with itself (similarity value equal to 1) is also included in the definition of the overall support for a_i. The t_i values are in increasing order, that is t_1 is the smallest value among the t_i.

On the basis of the t_j values, the weights of the weighting vector are computed as follows:

$$w_i = Q(t_i/n)/\sum_{i=1..n} Q(t_i/n)$$

The value $Q(t_i/n)$ denotes the degree to which a given member of the considered set of values represents the majority.

Based on this formula we define the weighting vector of the OWA operator associated with the quantifier *most* presented in Figure 1; if we want to aggregate five elements we obtain: W_1=[0 0 0.333 0.333 0.333].

Let us now aggregate with this weighting vector the induced ordered elements in the example illustrated before I = [0 0.1 0.6 0.9 0.7]; we obtain W_1*I = 0.733.

Although this value is smaller than the value obtained with the weighting vector W above, it is closer to 0.7 that is the most representative value among the values to be aggregated.

Let us now aggregate the six elements $a_1=1$, $a_2=1$, $a_3=1$ $a_4=0.5$, $a_5=0$, $a_6=0$. First of all we have to induce their similarity based ordering. We perform this step by adopting the simple support function defined above. If we set a value $\alpha = 0.4$ we obtain $s_1 = 2$, $s_2 = 2$, $s_3 = 2$, $s_4 = 0$, $s_5 = 1$, $s_6 = 1$. As it can be noticed, in this case the fact that 0.5 has an overall support equal to 0 is due to the fact that it is a value equidistant from both 0 and 1, which are the other values to be aggregated. With this choice of the parameter α, the induced ordering of the values is then I = [0.5 0 0 1 1 1]. We now compute the weights of the weighting vector of six elements, based on the computation procedure shown above: W_I = [0 0 0 0.33 0.33 0.33]. Let us now aggregate the vector I: W_I*I = 0.33 + 0.33 + 0.33 = 1. In this case if we aggregate the vector I with the weighting vector W shown in Table 1 we obtain the lower aggregated value 0.82, which could be considered as worst reflecting the majority of the considered elements. We finally notice that the result 0.35 produced by the classical definition of the OWA operator (based on the definition of the linguistic quantifier *most* presented in Figure 1) applied to the same values [1 1 1 0.5 0 0] is very far from an interpretation based on a the majority–oriented aggregation.

4 The Concept of Fuzzy Majority Opinion

In the previous section the approach proposed in [13] to compute a value (called the majority opinion) synthesizing the majority of a collection of values has been described . In the following an approach based upon the idea of a *fuzzy majority opinion* (also proposed in [13]) is presented. Under this interpretation the majority opinion is no longer represented as a value, but as a fuzzy subset. As we shall see this provides in addition to a value for the majority opinion also an indication of the strength of that value as a representative of the majority opinion.

In the following we shall let A = {a_1 , ..., a_n } be a set of values which constitute the opinions of a group of people. The definition of a fuzzy majority opinion requires that we have information about the similarity between the values provided. It requires also some information about what quantity constitutes the idea of a majority. We assume the availability of a relation on the space from which the values to be aggregated are drawn indicating how similar two values are. In particular we assume a relationship Sim on the domain of A such that for any a_i and a_j Sim[a_i, a_j]$\in$[0,1] and that Sim satisfies the properties Sim(a_i, a_i) = 1 and Sim(a_i, a_j) = Sim(a_j, a_i). We note that the relationship Sim is not a formal similarity relationship as introduced by Zadeh [25], it lacks transitivity, but a proximity relationship [11]. However for linguistic and intuitive convenience we shall refer to Sim(x,y) as indicating the degree of similarity between x and y. Another tool we need is a formal definition of the quantity we consider to constitute a majority. The concept of a majority is user and context dependent idea, however there are certain features common to any definition. We assume a user provided definition of a majority in terms of a fuzzy subset Q on the unit interval. In particular Q: [0,1] → [0,1] such that Q(0) = 0, Q(1) = 1 and Q(x) ≥ Q(y) if x >y. The monotonicity of Q implies that if Q(y) = 1 then for all x > y we have Q(x) = 1. We shall call the point x*, the smallest value for which Q(x) = 1, the Point of Realization, POR.

The concept of majority always has some POR, typically POR < 1. As we shall subsequently see often it will be useful to have definitions of Q that are strictly monotonic, if x > y then Q(x) > Q(y). This strict monotonicity requires Q(x) < 1 if x < 1 and hence POR = 1. This type of strictly monotonic definition of Q allows us to always be able to distinguish between sets of different cardinalities. One solution to this conflict between desiring strict monotonicity and POR>0 is to use a concept of "effective point of realization" EPOR. The quantifier displayed below illustrates this idea.

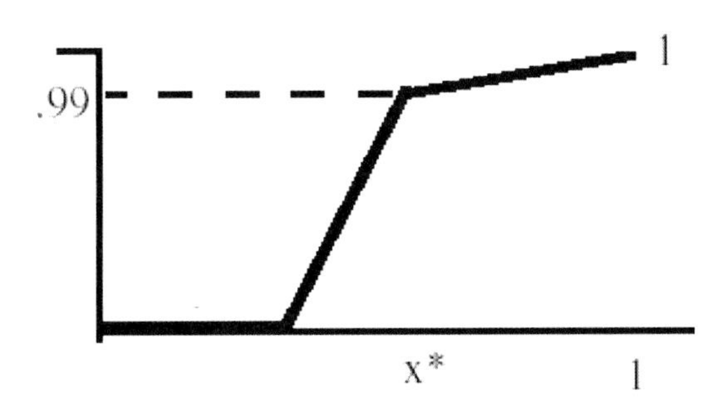

Fig. 2 Implementing an EPOR

Here we allow our definition to be such that x* is our EPOR and we consider 0.99 to effectively denote complete satisfaction. However for x from x* to 1 we use a straight line such that $Q(x) = 1 - \frac{0.1(1-x)}{1-x^*}$. Based on these ideas we introduce a concept of a majority opinion that will as a fuzzy subset that can be interpreted as a possibility distribution on the numeric majority opinions [3].

Let E be a crisp subset of A; the first step is determine the degree to which this is a subset holding a majority opinion. A subset E holds a majority opinion if all the elements in E are similar, and the cardinality of E satisfies our idea of being a majority of elements from A. We shall refer to a subset of values holding a majority opinion, as a gang. Thus a gang is a subset of E that contains a majority of elements having similar values. Let Majop(E) indicate the degree to which the elements in E, constitute a majority opinion, and are a majority of elements from A with similar values. We define

$$\mathrm{Majop}(E) = Q\left(\frac{|E|}{n}\right) \wedge \operatorname*{Min}_{a_i, a_j \in E}[\mathrm{Sim}(a_i, a_j)]$$

For simplicity we shall denote $\text{Sim}(E) = \underset{a_i, a_j \in E}{\text{Min}} [\text{Sim}(a_i, a_j)]$ and hence

$$\text{Majop}(E) = Q\left(\frac{|E|}{n}\right) \wedge \text{Sim}(E)$$

We now express the *opinion* of the elements in E as

$$\text{Op}(E) = \text{Ave}(E) = \frac{\sum_{a_i \in E} a_i}{|E|}$$

It is the average value of the elements in E.

Using both concepts Op(E) and Majop(E) we can define a fuzzy subset F indicating the majority opinion of the set of elements in A

$$F = \text{Majority Opinion} = \bigcup_{E \subseteq A} \left\{ \frac{\text{Majop}(E)}{\text{Op}(E)} \right\}$$

So for each subset E , the value Majop(E) indicates the degree to which the quantity OP(E) represents a majority opinion.

With F the fuzzy subset corresponding to the fuzzy majority opinion we see the $\text{Max}_E[\text{Majop}(E)]$, the maximal membership grade in F, indicates the degree to which there exists a majority opinion.

Here we shall provide an example to illustrate the construction of a fuzzy majority opinion.

Example: We assume that our values are drawn from a scale of 0 to 10. We assume the following simple similarity relation:

$$\text{Sim}(x, y) = 1 \qquad \text{if } |x - y| \leq 2$$

$$\text{Sim}(x, y) = \frac{1}{2}(4 - x) \qquad \text{if } 2 \leq |x - y| \leq 4$$

$$\text{Sim}(x, y) = 0 \qquad \text{if } |x - y| > 4$$

We assume the situation in which our concept majority, Q, is defined as shown in figure 3.

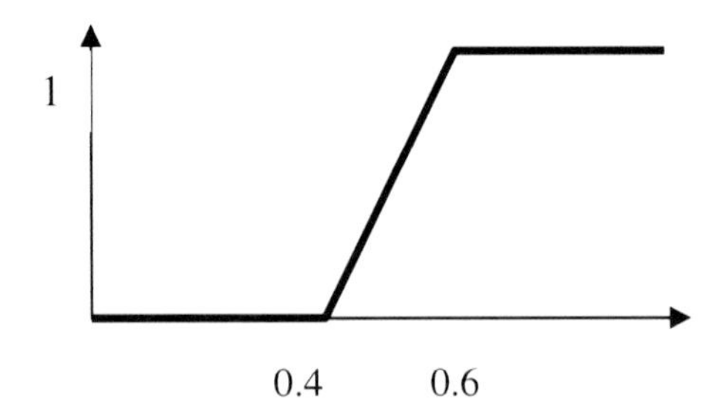

Fig. 3 Definition of the quantity majority

Thus

$$Q(x) = 0 \qquad \text{if } x \leq 0.4$$
$$Q(x) = 5(x - 0.4) \qquad 0.4 < x \leq 0.6$$
$$Q(x) = 1 \qquad x \geq 0.6$$

I. Let us consider the case where A = {1, 4, 5, 5, 6, 9}. Since n = 6 we have 2^6 possible subsets. However any subset having 2 or less elements has $Q(\frac{|E|}{6}) = 0$. In addition any subset having elements with a distance between any two of its members of four or more has Sim(E) = 0. Thus the following are the only subsets for which Majop(E) ≠ 0

$$E_1 = \{4, 5, 5, 6\}$$
$$E_2 = \{4, 5, 5\}$$
$$E_3 = \{4, 5, 6\}$$
$$E_4 = \{5, 5, 6\}$$

E	**AVE(E)**	**Q(E/N)**	**Sim(E)**	**Majop(E)**
E_1	5	1	1	1
E_2	4.7	0.5	1	0.5
E_3	5	0.5	1	0.5
E_4	5.3	0.5	1	0.5

Thus in this case our fuzzy majority F is $F = (\frac{0.5}{4.7}, \frac{1}{5}, \frac{0.5}{5.3})$, which we can see can be expressed as *around 5*. In fact, in describing the subset F use can be made of the connection between fuzzy subsets and natural languages to allow, if possible, the expression of F as a linguistic term.

II. Let us consider the case where A ={1, 1, 4.5, 6.5, 10, 10}

Here again if we eliminate subsets with two or less elements and those which have elements at a distance of four from each other we only two subsets:

$$E_1 = \{1, 1, 4.5\}$$
$$E_2 = \{6.5, 10, 10\}$$

In this

E	**AVE(E)**	**Q(E/N)**	**Sim(E)**	**Majop(E)**
E_1	2.16	0.5	0.25	0.25
E_2	8.833	0.5	0.25	0.25

In this case we get as our majority opinion F = $\{\frac{0.25}{2.16}, \frac{0.25}{8.833}\}$. Here we see very little support for any fuzzy majority opinion. As matter of fact $\text{Max}_x F(x) = 0.25$, there is no gang, no subset of A constituting a majority of people with similar opinions.

We note that at a formal level F is a fuzzy subset of the real line such that

$$F(r) = \underset{\text{all } E \in A \text{ s.t.} Ave(E)=r}{\text{Max}} (Q(\frac{|E|}{n}) \wedge Sim(|E|)$$

5 The Uniqueness of the Majority Opinion

With the use of the fuzzy majority we get a fuzzy subset F where for each value Op(E) we have Majop(E) indicating the extent to which Op(E) can be considered as a majority opinion. As we noted $\text{Max}_E[\text{Majop}(E)]$ indicates the degree to which there exists at least one value that can be considered as a majority opinion. The possibility exists that there exists multiple majority opinions.

In the following we present a measure called the clarity or uniqueness of the majority opinion, aimed at calculating the degree to which there seems to be some unique value that is a majority opinion; this unique value can be a cluster of similar values.

In the following the subset F corresponding to the majority opinion is represented as consisting of a collection of pairs, (u_i, r_i) where u_i = Majop(E) and r_i = Op(E). We shall refer to u_i as the strength of the pair and r_i as the value of the pair. Since F is obtained by using subsets where Majop(E) > 0 here we assume $u_i > 0$. We let q denote the number of pairs.

As a first step in the computation of this uniqueness measure, the pairs (u_i, r_i) are ordered by these u_i values in descending order. Tied values can be arbitrarily adjudicated. Using this ordering, let index(j) be the index of the jth largest of the u_i. Thus $(u_{index(j)}, r_{index(j)})$ is the pair having the jth largest degree of membership in F.

Using this we now introduce the idea of Unique(F) as the degree to which F has a unique majority opinion. This concept is closely related to the idea of specificity introduced by Yager in [17,20]. We recall specificity tries to measure the degree to which a fuzzy subset has one and only element. We also recall that specificity can be denied if a subset has too many members or no members, a low value for largest membership grade. A feature distinguishing the proposed measure of uniqueness from that of specificity relates to the assumption of an underling similarity relationship in the case of uniqueness. Essentially here we want to consider two members of F that are similar as corresponding to element when calculating the degree of uniqueness. We note that in [19] Yager looked at closely related issues in investigating measures of specificity in the face of similarity relations. We now turn to the formal definition of Unique(F). We first introduce a related concept called OverShadowed which we denote OS. We define OS as a mapping on the ordered pairs {1, 2,...,q} into the unit interval such that

$$OS(1) = 0$$
$$OS(j) = \text{Max}_{i=1 \text{ to } j-1}[(Sim(r_{index(j)}, r_{index(i)})) \wedge (1 - OS(i))] \qquad \text{for } j = 2 \text{ to } q$$

We see that OS(j) essentially measures the degree to which $r_{index(j)}$, the value corresponding to the element with j largest membership grade in F, is similar to some other value which has a higher membership grade in F and which has not been overshadowed. Using this we express our measure of uniqueness.

$$\text{Unique}(F) = u_{index(1)} - \text{Max}_{j=2 \text{ to } q}[u_{index(j)} \wedge (1 - OS(j))]$$

Let us apply this concept to our preceding example.

Example:

Case I: Here A = {1, 4, 5, 5, 6, 9} and we got F = $(\frac{0.5}{4.7}, \frac{1}{5}, \frac{0.5}{5.3})$. From this we get three pairs <(0.5, 4.7), (1,5), (0.5, 5.3)>. Using this we get

J	$u_{index(j)}$	$r_{index(j)}$	OS(j)	1 - OS(j)	$u_{index(1)} \wedge (1 - OS(r_j))$
1	1	5	0	1	0
2	0.5	5.3	1	0	0
3	0.5	4.7	1	0	0

Thus in this case we get Unique(F) = 1. There is one unique majority opinion

Case II: Here A = {1,1, 4.5, 4.5} and we got F = $(\frac{0.25}{2.10}, \frac{0.25}{8.833})$. From this we get two pairs <(0.25, 2.16), (0.25, 8.833)>. Using this we get

j	$u_{index(j)}$	$r_{index(j)}$	OS(j)	1 - OS(j)	$u_{index(1)} \wedge (1 - OS(r_j))$
1	0.25	2.16	-	-	-
2	0.25	8.833	0	1	0.25

Thus here Unique(F) = 0.25 - 0.25 = 0. Here we see that it doesn't exist any clear majority opinion.

We note that Unique(F) ≤ $u_{index(1)}$ thus if the strongest element in F is small then Unique(F) will be small, we will have no unique majority opinion. On the other hand if $u_{index(1)}$ is large this doesn't assure us a unique majority opinion as we may have multiple diverse majority opinions.

As previously defined Unique(F) indicates the degree to which F contains a single gang, a single majority opinion. In particular the measure Unique(F) does provide much understanding in the situation when it exists, was this caused by having no majority opinion or by having multiple majority opinions. We introduce a related concept which helps provide some understanding. Let us first define g_j for j = 1 to q such that

$$g_1 = u_{index(1)}$$
$$g_j = u_{index(j)} \wedge (1 - OS(j)) \text{ for } j > 1$$

We now define h_i as the ith largest of g_j. We can now use h_i to indicate the degree to which there exists *at least i distinct majority opinions.* Since $g_1 \geq g_j$ for all j, then $h_1 = g_1$ and hence g_1 is the degree to which there exists at least one majority opinion.

In some sense our idea of fuzzy majority is related to, although more general than, the concept of the mode of the set of observations A. We recall the mode indicates the value in A occurring the most times. Let us see these relationship here between F and the mode. Instead of considering one value we consider collections of similar values. Furthermore we assign a degree of modeness of one to the subset with the most equal elements. With F we use the concept Q to determine how satisfactory is a subset of a given cardinality.

The relationship Sim indicates our idea of what scores are considered as compatible. Let us look at some special cases of Sim. First consider the case of a strong condition for compatibility here Sim(x, x) = 1 and Sim(x, y) = 0 for $x \neq y$. In this case Sim(E) = 0 except in the case in which all elements are the same, in which case Sim(E) = 1. Furthermore if all elements in E are equal to a then Op(E) = a. In this case of similarity the fuzzy majority opinion takes a very interesting form. Let $A = \{a_1, ..., a_n\}$ be our data and let $D = \{d_1, ..., d_t\}$ be the set of distinct values in A, here of course $t \leq n$. Furthermore let n_j be the number of elements in A having the value d_j. Using this we get

$$F = \bigcup_{j=1}^{t} \{\frac{Q(\frac{n_j}{n})}{d_j}\} = \bigcup_{j=1}^{t} \{\frac{u_j}{d_j}\}$$

Here we clearly see the connection with the mode. Furthermore since Q is monotonic no element in A will have a higher membership grade in F then the one with the biggest count in A. Actually we can formally obtain the mode by a using a special definition for Q. In the preceding we defined Q as a pointwise function of its argument, Q(x) just depends upon x. Here we must define Q not in a pointwise fashion. In particular we define $F = \bigcup_{j=1}^{t} \{\frac{Q(n_j, N)}{d_j}\}$ thus u_j = $Q(n_j, N)$. Here $N = \{n_j, j = 1 \text{ to } t\}$, it is the set of all arguments. We now define $Q(n_j, N)$ as follows

$$\text{If } n_j \geq \text{Max}(N) \text{ then } Q(n_j, N) = 1$$
$$\text{If } n_j < \text{Max}(N) \text{ then } Q(n_j, N) = 0$$

Using this definition for Q and the preceding definition for Sim we get F = Mode.

Let us now consider the other extreme for Sim where we assume all the values are compatible, Sim(x, y) = 1. In this case Sim(E) = 1 for all E. In this case Majop(E) = $Q(\frac{|E|}{n})$. Thus

$$F = \bigcup_{E \subset A} \{\frac{Q(\frac{|E|}{n})}{Ave(E)}\}.$$

In this case since Sim(x, y) = 1 for all x and y, then Sim(Ave(E_1), Ave(E_2)) = 1 for all E_1 and E_2.

Further since $Q(\frac{|A|}{n}) = 1$ then we easily show OS(j) = 1 for all j > 1. Thus in this case

$\text{Unique}(F) = u_{index(1)} - \text{Max}_{j=2 \text{ to } q} [u_{index(j)} \wedge (1 - OS(j))] = u_{index(1)} = 1$

Thus in this case there appears one unique majority opinion, Ave(A) the average of the observations.

We further note that if we define F where we define Q using the rule base above, F(E) = Q(|E|, N) where N = {|B| for $B \subseteq A$} = {k = 1 to |A|} then $F = \bigcup_{E \subset A} \{\frac{Q(|E|, N)}{Ave(E)}\}$. In this case with Sim(x, y) = 1 we get F is the average, F = Ave(E).

6 Ordinal Environment

In the preceding sections we considered the situation in which the values to be aggregated were assumed to be numbers. In [13] the problem has been also considered of calculating the majority opinion in the case in which the individual opinions are assumed only to have an ordered nature. We let $S = \{s_1, s_2, ..., s_m\}$ be an ordinal scale such that $s_i > s_j$ if $i > j$. We assume that the opinions to be aggregated $A = \{a_1, a_2, ..., a_n\}$ are drawn from S. A prototypical situation of this kind is the case in which opinions are expressed using linguistic terms such as *good*, *very good*, *perfect.*

In order to develop a method for obtaining a majority opinion we need to provide some information about our idea of what is a majority as well as information about the similarity of the objects in S. We need an ordinal scale for expressing this information; while this scale can be the scale S we need not require it be the same scale. An ordinal scale $T = (t_1, t_2, ..., t_p\}$ is assumed with $t_i > t_j$ if $i > j$; the assumption that T is not necessarily the same as S is less restrictive than assuming them to be the same. A negation can be provided on this scale as $Neg(t_j) = t_{p+1-j}$. It must be pointed out that the assumption that this is an ordinal scale does not preclude us from using a numeric scale such as the unit interval. It must be also pointed out that while the information about the definition of similarity and the concept of majority do not have to be on the same scale as the opinions being aggregated the information about the definition of similarity and the concept of majority do have to be on the same scale.

Here then we assume the availability of a relationship Sim on S such for any s_i, $s_j \in S$ we have $Sim(s_i, s_j) \in T$. Here we assume that $Sim(s_i, s_j) = Sim(s_j, s_i)$ and $Sim(s_j, s_j) = t_p$. In addition we assume the availability of a definition of majority, Q such that $Q:[0,1] \rightarrow T$ where $Q(0) = t_1$, $Q(1) = t_p$ and $Q(x) \geq Q(y)$ if $x>y$. Using these tools a concept of a majority opinion as a fuzzy subset has been defined in a manner analogous to the preceding. Here again the majority opinion is defined as a fuzzy subset F such that

$$F = \bigcup_{E \subseteq A} \{\frac{Majop(E)}{Op(E)}\}$$

Again for any subset E of A we define

$$Majop(E) = Q(\frac{|E|}{n}) \wedge Sim(E)$$

Where $Sim(E) = \underset{a_i, a_j \in E}{Min} [Sim(a_i, a_j)]$. Here Majop determines the degree to which the subset E constitutes a majority of values that are compatible, similar. Since the elements of E are chosen from an ordinal scale S we can indicate $E^* = \underset{a_i \in E}{Max}[a_i]$ and $E_* = \underset{a_i \in E}{Min}[a_i]$. Since Sim(E) is the minimum similarity between any two elements in E, it is the similarity between the two most distant elements in E with respect to the scale S. From this we see that $Sim(E) = Sim(E^*, E_*)$.

The term Op(E) is the aggregated opinion of the elements in the subset E. In the preceding we used the average of the elements in E for OP(E), however here the ordered nature of the elements precludes our using this operation. In this case to calculate the aggregated opinion of the elements E, the median of E is used, thus Op(E) = Med(E). We recall that the median of E is obtained by ordering the elements in E and then taking the middle element. Without loss of generality assume the ordered elements in E are $b_1 \geq b_2 \geq b_{|E|}$. If E is odd then we can obtain the middle element, $Med(E) = b_{\frac{|E|+1}{2}}$. If |E| is even then the Med(E) is not unique, it is between $b_{\frac{|E|}{2}}$ and $b_{\frac{|E|}{2}+1}$. One protocol is to take one of these values as the median. For example we can take the bigger thus if |E| is even $Med(E) = b_{\frac{|E|}{2}+1}$.

Using the median we get as our majority opinion the fuzzy subset

$$F = \bigcup_{E \subseteq A} \{\frac{Majop(E)}{Med(E)}\}$$

We shall now turn to some pragmatic issues related to this ordinal environment. First we note that occurrence of a scale S on which we obtain the individual opinions is generally a natural phenomenon and is not difficult to formulate. It is assumed often of a linguistic nature and as noted it may involve terms like *very poo*r, *goo*d, *very good*. The requirement that the measure of similarity between the observed elements and the definition of the concept Q be on the same scale is required because of the need to perform the operator) $Q(\frac{|E|}{n}) \wedge Sim(E)$.

This requirement can be somewhat relaxed in a pragmatic spirit. In particular if in determining Sim(E) we can be satisfied in only establishing whether the elements in E are compatible with each other or not we greatly reduce the information required with respect to the similarity. That is here we need only assign a value 1 or 0 (True or False) to Sim(E). Furthermore since Sim(E) = $Sim(E^* , E_*)$ all we need to determine is if the boundary elements in E are compatible or not. At a formal level the use of this binary type of measurement of similarity can be seen as defining the relationship Sim on a sub scale of T, the subset scale being $\{t_l, t_p\}$. In particular if the elements in E are compatible Sim(E) = t_p and if they are not compatible Sim(E) = t_l. Thus in the case in which Sim(E) = t_p we get Majop(E) = $Q(\frac{|E|}{n})$ and if elements in E are not compatible, Sim(E) = t_l then Majop(E) = t_l. It will be convenient to refer to these special elements in T as 0 and 1.

Using this more simplified measurement of the similarity we get

$$F = \bigcup_{Sim(E)=t_p} \{\frac{Majop(E)}{Med(E)}\}$$

Let us look more carefully at the collection of subsets of A with compatible elements, those with Sim(E) = t_p. We first note that if $E \subseteq E'$ then if the elements in E are not compatible then those in E' are not compatible. On the other hand if the elements in E' are compatible then all those in E are compatible.

We now consider the idea of maximal compatibility set in this environment. Without loss of generality, assume the elements in A have been indexed such that $a_1 \leq a_2 \leq .. a_i \leq a_{j...} \leq a_n$ if i < j. Let us start with a_1 and find the largest a_j such that $Sim(a_1 , a_j) = t_p = 1$ it is the farthest element in A still compatible with a_1. Let us denote $E_1 = \{a_j$ such that $j \geq 1$ and $Sim(a_1, a_j) = 1\}$ all the elements in E_1 are compatible and all the subsets of E_1 are made of compatible elements. More generally for any $a_i \in A$ let us denote $E_i = \{a_j \mid j \geq i, Sim(a_j, a_i) = 1\}$.

Consider a compatible set E_i and let G and H be any subsets of E_i then since all elements in G and H are compatible and since $Med(G) \in G$ and $Med(H) \in H$ then $Sim(Med(G), Med(H)) = 1$. In particular $Sim(Med(G), Med(E_i) = 1$. Thus the similarity between the compatible set E_i and any of its subset in one.

Let us denote C_A as the collection of compatible subsets of A. We see that

$$C_A = \bigcup_{i=1}^{n} 2^{E_i}$$

It is the union of the power sets of all of the E_i. We can further refine this definition. For any compatibility set E_i let us denote e_i^* as the maximal element, it is the element farthest from a_i, the seed of E_i. Let E_i and E_j be two compatible sets where $j > i$, $a_j > a_i$. We first note that $e_j^* \geq e_i^*$ that is the largest element in E_j must be at least as large as e_i^*. Further we note that if $e_j^* = e_i^*$ then $E_j \subseteq E_i$. We shall call E_i a maximal compatible set if $e_j^* > e_i^*$ for all $i < j$.

We shall denote the collection of indices corresponding to maximal some sets as E. We see that we can express $C_A = \bigcup_{i \in E} 2^{E_i}$.

Thus using this binary similarity relationship we have obtained $F = \bigcup_{E \subset C_A} \{\frac{Q(\frac{|E|}{n})}{Med(E)}\}$ where $C_A = \bigcup_{i \in E} 2^{E_i}$ and E is the collection of maximal compatible sets.

Let us now consider the definition of Q. First we recall that Q is monotonic. We can consider Q to be of the form shown as figure 4.

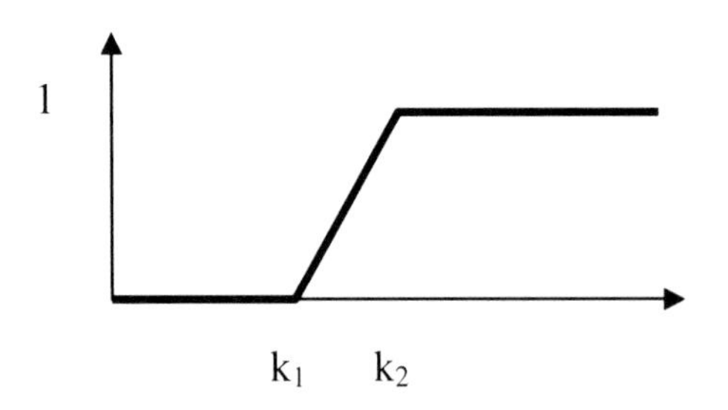

Fig. 4 Basic form for Q

Thus there is some quantity of elements k_1 below which the degree of majority is zero and some quantity of elements k_2 for which we assume complete concept of majority. Using this we easily define Q in a natural manner

$$Q(|E|) = 0 \qquad |E| = k_1$$

$$Q(|E|) = \frac{|E| - k_1}{k_2 - k_1} \qquad k_1 < |E| < k_2$$

$$Q(|E|) = 1 \qquad |E| \geq k_2$$

Furthermore we note that if $B \subset E_i$ then $Q(|B|) \leq Q(|E_i|)$ this implies that for any maximal compatible set $Q(|E_i|$ is at least as large as $Q(|B|)$ for its subsets. Furthermore since if $B \subset E_i$ then $Sim(Med(E_i), Med(B)) = 1$ we can effectively represent the majority opinion as

$$F = \bigcup_{i \in E} \{ \frac{Q(E_i)}{Med(E_i)} \}$$

Thus all we need do is to find all the maximal compatibility sets, the degree to which each constitutes a majority of elements and obtain their respective medians.

7 Conclusions

In this chapter a research work has been reported, which proposed a new modeling of majority opinion in the context of group decision making [13]. The proposed method relies on the use of IOWA. The motivation of this work is based on the fact that one of the main problems related to group decision making is to synthesize an overall opinion shared by the majority of the experts. This requires an aggregation of the individual opinions into an overall value reflecting the concept of majority. Within fuzzy set theory the concept of majority has been modeled by means of fuzzy quantifiers defined as fuzzy subsets of the unit interval. In this chapter two possible definitions of a majority opinion related to a linguistic quantifier over a considered set of values have been presented. The first proposal is aimed at constructing an aggregation operator the semantics of which reflects the concept of majority. The operator described in this chapter when applied to aggregate a considered set of values (opinions) produces an aggregated value which is representative of the majority of the values to be aggregated (called the majority opinion). A modeling of the concept of majority opinion as a fuzzy subset has also been described in this chapter. Connected with this latter formalization is the concept of uniqueness of majority opinion.

References

1. Barwise, J., Cooper, R.: Generalized Quantifiers in Natural Language. Linguistics and Philosophy 4, 159–220 (1981)
2. Bordogna, G., Fedrizzi, M., Pasi, G.: A linguistic modeling of consensus in Group Decision Making based on OWA operators. IEEE Trans. on System Man and Cybernetics 27(1) (1997)

3. Dubois, D., Prade, H.: Possibility Theory: an Approach to Computerized Processing of Uncertainty. Plenum Press, New York (1988)
4. Fishburn, P.C.: A comparative analysis of Group Decision Methods. Behavioral Science 16(6), 538–544 (1971)
5. Grabisch, M.: Fuzzy integrals in multicriteria decision making. Fuzzy Sets and Systems 69, 279–298 (1995)
6. Herrera, F., Herrera-Viedma, E., Verdegay, J.L.: Direct Approach Processes in Group Decision Making Using Linguistic OWA Operators. Fuzzy Sets and Systems 79, 175–176 (1996)
7. Herrera, F., Herrera-Viedma, E., Verdegay, J.L.: A Linguistic Decision Process in Group Decision Making. Group Decision and Negotiation 5, 165–176 (1996)
8. Kacprzyk, J., Fedrizzi, M., Nurmi, H.: Fuzzy Logic with Linguistic Quantifiers in Group Decision Making and Consensus Formation. In: Yager, R.R., Zadeh, L.A. (eds.) An Introduction to Fuzzy Logic Applications in Intelligent Systems, pp. 263–280. Kluwer, Dordrecht (1992)
9. Kacprzyk, J., Nurmi, H., Fedrizzi, M. (eds.): Consensus under fuzziness. Kluwer Academic Publishers, Dordrecht (1997)
10. Keenan, E.L., Westerstal, D.: Generalized quantifiers in Linguistics and Logics. In: van Benthem, J., ter Meulen, A. (eds.) Handbook of Logic and Language, pp. 837–893. North–Holland, Amsterdam (1997)
11. Kaufmann, A.: Introduction to the Theory of Fuzzy Subsets, vol. 1. Academic Press, New York (1975)
12. Lindstrom, P.: First Order Predicate Logic with Generalized Quantifiers. Theoria 35, 186–195 (1966)
13. Pasi, G., Yager, R.R.: Modeling the concept of majority opinion in group decision making. Information Sciences 176(4), 390–414 (2006)
14. Peláez, J.I., Doña, J.M., Gómez-Ruiz, J.A.: Analysis of OWA operators in decision making for modelling the majority concept. Applied Mathematics and Computation 186(2,15), 1263–1275 (2007)
15. Peláez, J.I., Doña, J.M.: A majority model in group decision making using QMA-OWA operator. International Journal of Intelligent Systems 21(2), 193–208
16. Peláez, J.I., Doña, J.M.: Majority additiveordered weighting averaging: A new neat ordered weighting averaging operators based on the majority process. International Journal of Intelligent Systems 18, 469–481 (2003)
17. Yager, R.R.: Entropy and specificity in a mathematical theory of evidence. Int. J. of General Systems 9, 249–260 (1983)
18. Yager, R.R.: On Ordered Weighted Averaging aggregation Operators in Multi Criteria Decision Making. IEEE Trans. on Systems, Man and Cybernetics 18(1), 183–190 (1988)
19. Yager, R.R.: Similarity based specificity measures. International Journal of General Systems 19, 91–106 (1991)
20. Yager, R.R.: On the specificity of a possibility distribution. Fuzzy Sets and Systems 50, 279–292 (1992)
21. Yager, R.R.: Interpreting Linguistically Quantified Propositions. International Journal of Intelligent Systems 9, 541–569 (1994)
22. Yager, R.R.: Quantifier Guided Aggregation using OWA operators. International Journal of Intelligent Systems 11, 49–73 (1996)

23. Yager, R.R., Filev, D.P.: Induced ordered weighted averaging operators. IEEE Transaction on Systems, Man and Cybernetics 29, 141–150 (1999)
24. Yager, R.R.: The power average operator. IEEE Transaction on Systems, Man and Cybernetics Part A 31, 724–730 (2001)
25. Zadeh, L.A.: Similarity relations and fuzzy orderings. Information Sciences 3, 177–200 (1971)
26. Zadeh, L.A.: A computational Approach to Fuzzy Quantifiers in Natural Languages. Computing and Mathematics with Applications 9, 149–184 (1983)

Generating OWA Weights from Individual Assessments

José Luis García-Lapresta, Bonifacio Llamazares, and Teresa Peña

Abstract. In this contribution we propose a method for generating OWA weighting vectors from the individual assessments on a set of alternatives in such a way that these weights minimize the disagreement among individual assessments and the outcome provided by the OWA operator. For measuring that disagreement we have aggregated distances between individual and collective assessments by using a metric and an aggregation function. We have paid attention to Manhattan and Chebyshev metrics and arithmetic mean and maximum as aggregation functions. In this setting, we have proven that medians and the mid-range are the solutions for some cases. When a general solution is not available, we have provided some mathematical programs for solving the problem.

1 Introduction

In 1988 Yager [10] introduced OWA operators as a tool for aggregating numerical values in multi-criteria decision making. An OWA operator is similar to a weighted mean, but with the values of the variables previously ordered in a decreasing way. Thus, contrary to the weighted means, the weights are not associated with concrete variables and, therefore, they are anonymous. Moreover, they satisfy other interesting properties, such as monotonicity, unanimity, continuity and compensativeness, i.e., the value of an OWA operator is always located between the minimum and the maximum values of the variables.

Initially, the weights of an OWA operator may be fixed taking into account the importance we want to give to the assessments. So, the outcome of an OWA operator may be the maximum, the minimum, the average or a median of the individual assessments, among a large number of possibilities.

José Luis García-Lapresta · Bonifacio Llamazares · Teresa Peña
PRESAD Research Group, Dept. of Applied Economics,
University of Valladolid, Spain
e-mail: {lapresta,boni,maitepe}@eco.uva.es

R.R. Yager et al. (Eds.): Recent Developments in the OWA Operators, STUDFUZZ 265, pp. 135–147.
springerlink.com

It is important to note that the determination of the weights of OWA operators is a relevant issue since the origins of the theory of OWA operators. In this way, Yager [10] proposes to use linguistic quantifiers for generating the OWA weights; O'Hagan [7] generates the OWA weights by maximizing their entropy whenever a degree of orness has been fixed; Filev and Yager [4] consider an exponential smoothing approach for generating the OWA weights. After these seminal papers, a large variety of techniques have been proposed in the literature (see, for instance, Wang and Parkan [8] and Xu [9]).

In our proposal, we do not fix the OWA weighting vector, but we generate an OWA operator for each profile of individual assessments, just one that minimizes the disagreement (or equivalently, maximizes the consensus) in the group with respect to the outcome provided by the OWA operator. More concretely, once the agents opinions are known, we first calculate the distances among individual assessments on the alternatives and the collective assessments generated by an arbitrary OWA operator. Secondly, we use an aggregation function for obtaining a representative measure of disagreement from the individual assessments to the collective one. By solving a mathematical program, we obtain the weighting vector(s) that maximize(s) the consensus among individual and collective opinions.

Within the general framework we have chosen the arithmetic mean and the maximum as aggregation functions. Thus the general mathematical programs falls into the *minisum* and the *minimax* procedures, respectively (see, for instance, Brams, Kilgour and Sanver [2]).

The paper is organized as follows. Section 2 is devoted to introduce notation and some basic notions. Section 3 contains our proposal for generating an OWA operator for each profile of individual assessments and some results for the minisum and minimax procedures with two specific metrics (Manhattan and Chebyshev). Section 4 includes an illustrative example. Finally, Section 5 contains some concluding remarks.

2 Preliminaries

An *aggregation function* is a continuous mapping $A : [0,1]^m \longrightarrow [0,1]$ that satisfies the following conditions:

1. *Monotonicity*: $A(x_1,\ldots,x_m) \leq A(y_1,\ldots,y_m)$ for all $(x_1,\ldots,x_m),(y_1,\ldots,y_m) \in [0,1]^m$ such that $x_i \leq y_i$ for every $i \in \{i,\ldots,m\}$.
2. *Unanimity* (or *idempotency*): $A(x,\ldots,x) = x$ for every $x \in [0,1]$.

It is easy to see that every aggregation function is *compensative*, i.e.,

$$\min\{x_1,\ldots,x_m\} \leq A(x_1,\ldots,x_m) \leq \max\{x_1,\ldots,x_m\},$$

for every $(x_1,\ldots,x_m) \in [0,1]^m$.

On aggregation functions, see Fodor and Roubens [5], Grabisch, Orlovski and Yager [6], Calvo, Kolesárova, Komorníková and Mesiar [3] and Beliakov, Pradera and Calvo [1], among others.

2.1 OWA Operators

OWA operators are anonymous aggregation functions that are defined by weighting vectors in the following way.

Given a weighting vector $\mathbf{w} = (w_1, \ldots, w_m) \in [0,1]^m$ such that $\sum_{i=1}^m w_i = 1$, the *OWA operator associated with* $\mathbf{w}$ is the mapping $F_{\mathbf{w}} : [0,1]^m \longrightarrow [0,1]$ defined by

$$F_{\mathbf{w}}(x_1, \ldots, x_m) = \sum_{i=1}^{m} w_i \cdot y_i$$

where y_i is the i-th greatest number of $\{x_1, \ldots, x_m\}$.

The set of weighting vectors will be denoted by

$$\mathscr{W} = \left\{ \mathbf{w} \in [0,1]^m \;\middle|\; \sum_{i=1}^{m} w_i = 1 \right\}.$$

In some cases it is interesting to consider OWA operators associated with weighting vectors satisfying specific requirements. Some of the most used in the literature are the following:

1. Weighting vectors with a fixed *orness* (or *attitudinal character*) $\alpha \in (0,1)$ (see Yager [10]):

$$\mathscr{W}^1_\alpha = \left\{ \mathbf{w} \in \mathscr{W} \;\middle|\; \frac{1}{m-1} \sum_{i=1}^{m} (m-i) w_i = \alpha \right\}.$$

2. Symmetric weights:

$$\mathscr{W}^2 = \left\{ \mathbf{w} \in \mathscr{W} \mid w_i = w_{m+1-i} \quad \forall i \in \left\{1, \ldots, \left[\tfrac{m}{2}\right]\right\} \right\}.$$

3. Centered weights (after Yager [11]):

$$\mathscr{W}^3 = \left\{ \mathbf{w} \in \mathscr{W}^2 \mid w_1 \leq w_2 \leq \cdots \leq w_{\left[\frac{m+1}{2}\right]} \right\}.$$

4. Trimmed weights:

$$\mathscr{W}^4_1 = \{ \mathbf{w} \in \mathscr{W} \mid w_1 = w_m = 0 \}.$$
$$\mathscr{W}^4_2 = \{ \mathbf{w} \in \mathscr{W} \mid w_1 = w_2 = w_{m-1} = w_m = 0 \}.$$
$$\cdots\cdots\cdots\cdots\cdots\cdots\cdots\cdots\cdots\cdots\cdots\cdots$$
$$\mathscr{W}^4_{\left[\frac{m-1}{2}\right]} = \left\{ \mathbf{w} \in \mathscr{W} \mid w_1 = \cdots = w_{\left[\frac{m-1}{2}\right]} = w_{\left[\frac{m}{2}\right]+2} = \cdots = w_m = 0 \right\}.$$

It is worth noting that $\mathscr{W}^3 \subseteq \mathscr{W}^2 \subseteq \mathscr{W}^1_{0.5} \subseteq \mathscr{W}$ and $\mathscr{W}^4_{\left[\frac{m-1}{2}\right]} \subseteq \cdots \subseteq \mathscr{W}^4_2 \subseteq \mathscr{W}^4_1$.

Some well-known aggregation functions are specific cases of OWA operators. For instance:

1. The *maximum*, given by the weighting vector $(1,0,\ldots,0)$.
2. The *minimum*, given by the weighting vector $(0,\ldots,0,1)$.
3. The *arithmetic mean*, given by the weighting vector $\left(\frac{1}{m},\ldots,\frac{1}{m}\right)$.
4. The *mid-range*, given by the weighting vector $\widehat{\boldsymbol{w}} = (0.5,0,\ldots,0,0.5)$.
5. *Medians*, given by the following weighting vectors $\widetilde{\boldsymbol{w}}$:

   a. If m is odd

$$\widetilde{w}_i = \begin{cases} 1, \text{ if } i = \frac{m+1}{2}, \\ 0, \text{ otherwise.} \end{cases}$$

   b. If m is even

$$\widetilde{w}_i = \begin{cases} \theta, & \text{if } i = \frac{m}{2}, \\ 1-\theta, & \text{if } i = \frac{m}{2}+1, \\ 0, & \text{otherwise,} \end{cases}$$

   for some $\theta \in [0,1]$.

   Notice that $\widetilde{\boldsymbol{w}} \in \mathscr{W}^4_{\left[\frac{m-1}{2}\right]}$.

2.2 Collective Assessments

Consider a set of agents (experts or voters) $V = \{1,\ldots,m\}$ $(m \geq 2)$ who show their opinions on a set of alternatives $A = \{a_1,\ldots,a_n\}$ $(n \geq 2)$ through numbers in the interval $[0,1]$.

A *profile* is a $m \times n$ matrix

$$P = \begin{pmatrix} a_1^1 & \cdots & a_j^1 & \cdots & a_n^1 \\ \cdots & \cdots & \cdots & \cdots & \cdots \\ a_1^i & \cdots & a_j^i & \cdots & a_n^i \\ \cdots & \cdots & \cdots & \cdots & \cdots \\ a_1^m & \cdots & a_j^m & \cdots & a_n^m \end{pmatrix}$$

where $a_j^i \in [0,1]$ is the assessment that agent i assigns to alternative a_j. The set of profiles is denoted by $\mathscr{P}$.

In order to aggregate individual assessments, we consider an OWA operator $F_{\boldsymbol{w}}$ associated with a weighting vector $\boldsymbol{w} \in \mathscr{W}^*$, where $\mathscr{W}^*$ may be $\mathscr{W}$ or any of the subsets of weighting vectors mentioned in the previous subsection. Taking into account the j-th column of P, $(a_j^1,\ldots,a_j^m)$, that includes individual opinions on the alternative a_j, we generate the collective assessment on that alternative through $F_{\boldsymbol{w}}$

$$v_j(\boldsymbol{w}) = F_{\boldsymbol{w}}\left(a_j^1,\ldots,a_j^m\right).$$

With $\boldsymbol{v}(\boldsymbol{w}) = (v_1(\boldsymbol{w}), \ldots, v_n(\boldsymbol{w}))$ we denote the vector that contains the collective assessments on the alternatives of A generated by the OWA operator $F_{\boldsymbol{w}}$. On the other hand, the i-th row of P includes the assessments of individual $i \in \{1, \ldots, m\}$ on the alternatives of A and will be denoted by $\boldsymbol{a}^i = \left(a_1^i, \ldots, a_n^i\right)$.

3 A Model for Generating OWA Weights

In our proposal, we do not fix the OWA weighting vector, but we generate an OWA operator for each profile of individual assessments, just one that maximizes the consensus (or, equivalently, minimizes the disagreement) in the group with respect to the outcome provided by the OWA operator. For this, it is necessary to fix two ingredients:

- A metric $d : \mathbb{R}^n \times \mathbb{R}^n \longrightarrow \mathbb{R}$.
- An aggregation function $A : [0,1]^m \longrightarrow [0,1]$.

By means of the metric d, we calculate the distances among individual assessments on the alternatives and the collective assessments generated by an OWA operator associated with a weighting vector belonging to $\mathscr{W}^*$. On the other hand, we use an aggregation function for obtaining a representative measure of disagreement from the individual assessments to the collective one. Thus, given a profile $P \in \mathscr{P}$, we propose to find weighting vector(s) $\boldsymbol{w} \in \mathscr{W}^*$ being solution(s) of the following mathematical program:

$$\begin{aligned} &\min \quad A\Big(d(\boldsymbol{a}^1, \boldsymbol{v}(\boldsymbol{w})), \ldots, d(\boldsymbol{a}^m, \boldsymbol{v}(\boldsymbol{w}))\Big) \\ &\text{s. t.}: \ \boldsymbol{w} \in \mathscr{W}^* \end{aligned} \tag{1}$$

Notice that from continuity of A and compactness of $\mathscr{W}^*$, the existence of solution(s) in (1) is always guaranteed.

Among the large variety of metrics and aggregation functions that we may use in (1), we present with more detail those cases where Manhattan and Chebyshev metrics are used, and the aggregation functions are the arithmetic mean and the maximum.

The *Manhattan metric* is defined by

$$d_1\big((x_1, \ldots, x_n), (y_1, \ldots, y_n)\big) = \sum_{i=1}^{n} |x_i - y_i|.$$

The *Chebyshev metric* is defined by

$$d_\infty\big((x_1, \ldots, x_n), (y_1, \ldots, y_n)\big) = \max\{|x_1 - y_1|, \ldots, |x_n - y_n|\}.$$

3.1 Minisum Outcomes

If we consider the arithmetic mean as aggregation function, then (1) is equivalent to find the weighting vectors that minimize the sum of distances between the individual assessments and the collective assessments generated by the OWA operator associated with those weighting vectors. In other words, (1) becomes

$$\begin{aligned} \min \quad & \sum_{i=1}^{m} d\Big((a_1^i,\ldots,a_n^i),(v_1(\boldsymbol{w}),\ldots,v_n(\boldsymbol{w}))\Big) \\ \text{s. t.}: \quad & \boldsymbol{w} \in \mathscr{W}^* \end{aligned} \tag{2}$$

1. If we use the Manhattan metric, then (2) is transformed in the following mathematical program:

$$\begin{aligned} \min \quad & \sum_{i=1}^{m} \Big(|a_1^i - v_1(\boldsymbol{w})| + \cdots + |a_n^i - v_n(\boldsymbol{w})|\Big) \\ \text{s. t.}: \quad & \boldsymbol{w} \in \mathscr{W}^* \end{aligned} \tag{3}$$

2. If we use the Chebyshev metric, then (2) is now transformed in the following mathematical program:

$$\begin{aligned} \min \quad & \sum_{i=1}^{m} \max\Big\{ |a_1^i - v_1(\boldsymbol{w})|,\ldots,|a_n^i - v_n(\boldsymbol{w})| \Big\} \\ \text{s. t.}: \quad & \boldsymbol{w} \in \mathscr{W}^* \end{aligned} \tag{4}$$

In the first case, i.e., in Problem (3), it is possible to give the analytical solution of the problem when $\mathscr{W}^* = \mathscr{W}$.

Proposition 1. *For $\mathscr{W}^* = \mathscr{W}$, the solutions of Problem (3) are the medians.*

Proof. Taking into account $\mathscr{W}^* = \mathscr{W}$ in Problem (3), we obtain the following mathematical program:

$$\min_{\boldsymbol{w}\in\mathscr{W}} \sum_{i=1}^{m}\sum_{j=1}^{n} |a_j^i - v_j(\boldsymbol{w})| = \min_{\boldsymbol{w}\in\mathscr{W}} \sum_{j=1}^{n}\sum_{i=1}^{m} |a_j^i - v_j(\boldsymbol{w})|. \tag{5}$$

Moreover, the following inequality is satisfied:

$$\min_{\boldsymbol{w}\in\mathscr{W}} \sum_{j=1}^{n}\sum_{i=1}^{m} |a_j^i - v_j(\boldsymbol{w})| \geq \sum_{j=1}^{n} \min_{\boldsymbol{w}\in\mathscr{W}} \sum_{i=1}^{m} |a_j^i - v_j(\boldsymbol{w})|.$$

On the other hand, it is known that given $x_1,\ldots,x_m \in \mathbb{R}$, medians are the solutions of the following problem:

$$\min_{x\in\mathbb{R}} \sum_{i=1}^{m} |x_i - x|.$$

Therefore, they are also the solution of

$$\min_{\boldsymbol{w}\in\mathscr{W}} \sum_{i=1}^{m} |a_j^i - v_j(\boldsymbol{w})|$$

for every $j \in \{1,\ldots,n\}$. Consequently,

$$\sum_{j=1}^{n} \min_{\boldsymbol{w}\in\mathscr{W}} \sum_{i=1}^{m} |a_j^i - v_j(\boldsymbol{w})| = \min_{\boldsymbol{w}\in\mathscr{W}} \sum_{j=1}^{n} \sum_{i=1}^{m} |a_j^i - v_j(\boldsymbol{w})|$$

and medians are the solutions of Problem (5). □

Since medians belong to $\mathscr{W}_i^4$, $i \in \left\{1,\ldots,\left[\frac{m-1}{2}\right]\right\}$, they are also the solutions of Problem (3) when $\mathscr{W}^* = \mathscr{W}_i^4$. Moreover, the median for $\theta = 0.5$ also belongs to $\mathscr{W}^2$ and $\mathscr{W}^3$. Therefore, the solution of Problem (3) when $\mathscr{W}^* = \mathscr{W}^2$ or $\mathscr{W}^* = \mathscr{W}^3$ is the median with $\theta = 0.5$.

In relation to $\mathscr{W}_\alpha^1$, we are going to calculate the orness of medians. We distinguish two cases:

1. If m is odd:

$$orness(\widetilde{\boldsymbol{w}}) = \frac{1}{m-1} \sum_{i=1}^{m} (m-i)\widetilde{w}_i = \frac{1}{m-1} \frac{m-1}{2} = 0.5.$$

2. If m is even:

$$\begin{aligned} orness(\widetilde{\boldsymbol{w}}) &= \frac{1}{m-1} \sum_{i=1}^{m} (m-i)\widetilde{w}_i = \frac{1}{m-1} \left(\frac{m}{2}\theta + \left(\frac{m}{2} - 1\right)(1-\theta)\right) \\ &= \frac{1}{m-1} \left(\frac{m}{2} - 1 + \theta\right). \end{aligned}$$

In the first case, if $\alpha = 0.5$, then the median belongs to $\mathscr{W}_{0.5}^1$ and is the solution of Problem (3) when $\mathscr{W}^* = \mathscr{W}_{0.5}^1$.

In the second case, the minimum and the maximum orness are reached when $\theta = 0$ and $\theta = 1$, respectively. Therefore, if m is even, then

$$\frac{m-2}{2(m-1)} \le orness(\widetilde{\boldsymbol{w}}) \le \frac{m}{2(m-1)}.$$

Consequently, for these values of α, the median, with $\theta = (m-1)\alpha + 1 - \frac{m}{2}$, belongs to $\mathscr{W}_\alpha^1$ and is the solution of Problem (3) when $\mathscr{W}^* = \mathscr{W}_\alpha^1$.

When m is odd and $\alpha \neq 0.5$ or when m is even and $\alpha \notin \left[\frac{m-2}{2(m-1)}, \frac{m}{2(m-1)}\right]$, it is possible to find a numerical solution of Problem (3). For this, we can replace this problem by the following equivalent smooth linear problem that uses some auxiliary variables:

$$
\begin{aligned}
\min\ & \sum_{i=1}^{m}(\lambda_1^i+\cdots+\lambda_n^i) \\
\text{s.t.}:\ & -\lambda_1^i \le v_1(\boldsymbol{w})-a_1^i \le \lambda_1^i, && i=1,\ldots,m \\
& \cdots\cdots\cdots\cdots\cdots\cdots\cdots \\
& -\lambda_n^i \le v_n(\boldsymbol{w})-a_n^i \le \lambda_n^i, && i=1,\ldots,m \\
& \lambda_j^i \ge 0, && i=1,\ldots,m,\ j=1,\ldots,n \\
& \boldsymbol{w}\in \mathscr{W}_\alpha^1
\end{aligned}
\tag{6}
$$

In a similar way, when Problem (4) is considered, i.e., when the Chebyshev metric is used, the solution of the problem can be obtained by solving the following equivalent smooth linear problem that uses some auxiliary variables:

$$
\begin{aligned}
\min\ & \lambda_1+\cdots+\lambda_m \\
\text{s.t.}:\ & -\lambda_1 \le v_j(\boldsymbol{w})-a_j^1 \le \lambda_1, && j=1,\ldots,n \\
& \cdots\cdots\cdots\cdots\cdots\cdots\cdots \\
& -\lambda_m \le v_j(\boldsymbol{w})-a_j^m \le \lambda_m, && j=1,\ldots,n \\
& \lambda_i \ge 0, && i=1,\ldots,m \\
& \boldsymbol{w}\in \mathscr{W}^*
\end{aligned}
\tag{7}
$$

3.2 Minimax Outcomes

We now consider the maximum as aggregation function. In this case, (1) is equivalent to find the weighting vectors that minimize the maximum distance between the individual assessments and the collective assessments generated by the OWA operator associated with those weighting vectors. In other words, (1) becomes

$$
\begin{aligned}
\min\ & \max_{i=1,\ldots,m}\left\{d\Big((a_1^i,\ldots,a_n^i),(v_1(\boldsymbol{w}),\ldots,v_n(\boldsymbol{w}))\Big)\right\} \\
\text{s. t.}:\ & \boldsymbol{w}\in \mathscr{W}^*
\end{aligned}
\tag{8}
$$

1. If we use the Manhattan metric, then (8) is transformed in the following mathematical program:

$$
\begin{aligned}
\min\ & \max_{i=1,\ldots,m}\left\{|a_1^i-v_1(\boldsymbol{w})|+\cdots+|a_n^i-v_n(\boldsymbol{w})|\right\} \\
\text{s. t.}:\ & \boldsymbol{w}\in \mathscr{W}^*
\end{aligned}
\tag{9}
$$

2. If we use the Chebyshev metric, then (8) is transformed in the following mathematical program:

$$\begin{aligned}&\min \max_{i=1,\dots,m} \max\left\{ |a_1^i - v_1(\mathbf{w})|,\dots,|a_n^i - v_n(\mathbf{w})| \right\}\\&\text{s. t.}: \ \mathbf{w} \in \mathscr{W}^*\end{aligned} \tag{10}$$

Problem (9) is equivalent to solve the following smooth linear problem that uses some auxiliary variables:

$$\begin{aligned}
&\min \ \delta \\
&\text{s.t.}: \ \lambda_1^i + \cdots + \lambda_n^i \le \delta, && i=1,\dots,m \\
&\quad -\lambda_1^i \le v_1(\mathbf{w}) - a_1^i \le \lambda_1^i, && i=1,\dots,m \\
&\quad \dots\dots\dots\dots\dots\dots\dots\dots \\
&\quad -\lambda_n^i \le v_n(\mathbf{w}) - a_n^i \le \lambda_n^i, && i=1,\dots,m \\
&\quad \lambda_j^i \ge 0, && i=1,\dots,m,\ j=1,\dots,n \\
&\quad \mathbf{w} \in \mathscr{W}^*
\end{aligned} \tag{11}$$

In the following proposition, we show that the solution of Problem (10) is the average between the maximum and the minimum, i.e., the mid-range, whenever $\mathscr{W}^* = \mathscr{W}$.

Proposition 2. *For $\mathscr{W}^* = \mathscr{W}$, the solution of Problem (10) is the mid-range.*

Proof. When $\mathscr{W}^* = \mathscr{W}$ in Problem (10), we obtain the following mathematical program:

$$\min_{\mathbf{w}\in\mathscr{W}} \max_{i=1,\dots,m} \max_{j=1,\dots,n} \left\{ |a_j^i - v_j(\mathbf{w})| \right\} = \min_{\mathbf{w}\in\mathscr{W}} \max_{j=1,\dots,n} \max_{i=1,\dots,m} \left\{ |a_j^i - v_j(\mathbf{w})| \right\}. \tag{12}$$

Moreover, the following inequality is satisfied:

$$\min_{\mathbf{w}\in\mathscr{W}} \max_{j=1,\dots,n} \max_{i=1,\dots,m} \left\{ |a_j^i - v_j(\mathbf{w})| \right\} \ge \max_{j=1,\dots,n} \min_{\mathbf{w}\in\mathscr{W}} \max_{i=1,\dots,m} \left\{ |a_j^i - v_j(\mathbf{w})| \right\}.$$

On the other hand, it is clear that for every $j \in \{1,\dots,n\}$, the weighting vector $\widehat{\mathbf{w}}$ is the solution of the following problem:

$$\min_{\mathbf{w}\in\mathscr{W}} \max_{i=1,\dots,m} \left\{ |a_j^i - v_j(\mathbf{w})| \right\}.$$

Therefore,

$$\max_{j=1,\dots,n} \min_{\mathbf{w}\in\mathscr{W}} \max_{i=1,\dots,m} \left\{ |a_j^i - v_j(\mathbf{w})| \right\} = \min_{\mathbf{w}\in\mathscr{W}} \max_{j=1,\dots,n} \max_{i=1,\dots,m} \left\{ |a_j^i - v_j(\mathbf{w})| \right\}$$

and $\widehat{\mathbf{w}}$ is also the solution of Problem (12). □

Since $\widehat{\mathbf{w}}$ belongs to $\mathscr{W}_{0.5}^1$ and $\mathscr{W}^2$, it is also the solution of Problem (10) for $\mathscr{W}^* = \mathscr{W}_{0.5}^1$ and $\mathscr{W}^* = \mathscr{W}^2$. On the other hand, when $\mathscr{W}^*$ is any of the remaining subsets, the solution of Problem (10) can be obtained by solving the following equivalent linear problem:

$$\begin{aligned}
&\min \ \delta \\
&\text{s.t.}: \ -\delta \leq v_1(\boldsymbol{w}) - a_1^i \leq \delta, \qquad i = 1,\ldots,m, \\
&\qquad \cdots\cdots\cdots\cdots\cdots\cdots \\
&\qquad -\delta \leq v_n(\boldsymbol{w}) - a_n^i \leq \delta, \qquad i = 1,\ldots,m, \\
&\qquad \boldsymbol{w} \in \mathscr{W}^*
\end{aligned} \tag{13}$$

4 An Illustrative Example

In this section we provide an example to show the solutions obtained when the previous models are used. Suppose that the opinions of four experts on three alternatives are given by the following matrix:

$$P = \begin{pmatrix} 0.7 & 0.6 & 0.1 \\ 0 & 0.5 & 0.8 \\ 0.6 & 0.1 & 1 \\ 0.6 & 0.7 & 0 \end{pmatrix}$$

From the individuals' preferences shown in this matrix, the collective assessment on each alternative generated by an OWA operator is given by

$$\begin{aligned}
v_1(\boldsymbol{w}) &= 0.7w_1 + 0.6w_2 + 0.6w_3, \\
v_2(\boldsymbol{w}) &= 0.7w_1 + 0.6w_2 + 0.5w_3 + 0.1w_4, \\
v_3(\boldsymbol{w}) &= w_1 + 0.8w_2 + 0.1w_3.
\end{aligned}$$

Next, using the models proposed in the previous section, we seek the weighting vector (belonging to some of the weighting vector sets given in Subsection 2.1) that minimizes the disagreement among individual assessments on the alternatives and the collective assessments. It is worth noting that when the set of weighting vector is $\mathscr{W}_\alpha^1$, we show the solutions for $\alpha = 0.75$, $\alpha = 0.5$ and $\alpha = 0.25$. On the other hand, when the analytical solution is unknown, we have used LINGO software to solve the corresponding linear programming problems.

4.1 Minisum Outcomes

When the arithmetic mean and the Manhattan metric are used, the solutions obtained are shown in Table 1. It is worth noting that we have solved Problem (6) to obtain the solution when the set of weighting vectors is $\mathscr{W}_{0.75}^1$ or $\mathscr{W}_{0.25}^1$. For the remaining sets of weighting vectors, the solution is already known (see Subsection 3.1).

Table 1 Solutions for the arithmetic mean and the Manhattan metric

	w_1	w_2	w_3	w_4
$\mathscr{W}$	0	θ	$1-\theta$	0
$\mathscr{W}^1_{0.75}$	0.5	0.3334	0.0833	0.0833
$\mathscr{W}^1_{0.5}$	0	0.5	0.5	0
$\mathscr{W}^1_{0.25}$	0	0	0.75	0.25
$\mathscr{W}^2$	0	0.5	0.5	0
$\mathscr{W}^3$	0	0.5	0.5	0
$\mathscr{W}^4_1$	0	θ	$1-\theta$	0

When the arithmetic mean and the Chebyshev metric are used, the solutions can be obtained solving Problem (7) for the different sets of weighting vectors considered in this contribution. Table 2 summarizes these solutions.

Table 2 Solutions for the arithmetic mean and the Chebyshev metric

	w_1	w_2	w_3	w_4
$\mathscr{W}$	0.4706	0	0	0.5294
$\mathscr{W}^1_{0.75}$	0.75	0	0	0.25
$\mathscr{W}^1_{0.5}$	0.5	0	0	0.5
$\mathscr{W}^1_{0.25}$	0	0.375	0	0.625
$\mathscr{W}^2$	0.5	0	0	0.5
$\mathscr{W}^3$	0.25	0.25	0.25	0.25
$\mathscr{W}^4_1$	0	0.1429	0.8571	0

4.2 *Minimax Outcomes*

When the arithmetic mean and the Manhattan metric are used, the solutions can be obtained by solving Problem (11) for the different sets of weighting vectors considered in this contribution. These solutions are given in Table 3.

When the Manhattan metric is replaced by the Chebyshev metric, the solution is known if the set of weighting vector is $\mathscr{W}$, $\mathscr{W}^1_{0.5}$ or $\mathscr{W}^2$. In other cases, the solutions can be obtained by solving Problem (13). Table 4 shows these solutions.

Table 3 Solutions for the maximum and the Manhattan metric

	w_1	w_2	w_3	w_4
$\mathcal{W}$	0.6667	0	0.2222	0.1111
$\mathcal{W}^1_{0.75}$	0.673	0	0.2307	0.0963
$\mathcal{W}^1_{0.5}$	0	0.7222	0.0556	0.2222
$\mathcal{W}^1_{0.25}$	0	0.375	0	0.625
$\mathcal{W}^2$	0	0.5	0.5	0
$\mathcal{W}^3$	0	0.5	0.5	0
$\mathcal{W}^4_1$	0	0.8334	0.1666	0

Table 4 Solutions for the maximum and the Chebyshev metric

	w_1	w_2	w_3	w_4
$\mathcal{W}$	0.5	0	0	0.5
$\mathcal{W}^1_{0.75}$	0.625	0	0.375	0
$\mathcal{W}^1_{0.5}$	0.5	0	0	0.5
$\mathcal{W}^1_{0.25}$	0	0.375	0	0.625
$\mathcal{W}^2$	0.5	0	0	0.5
$\mathcal{W}^3$	0.25	0.25	0.25	0.25
$\mathcal{W}^4_1$	0	0.7143	0.2857	0

5 Concluding Remarks

In this contribution we have proposed a general method for generating weighting vectors of OWA operators from the individual assessments in such a way that the obtained OWA operator maximizes the consensus among the agents. This endogenous procedure is based on a metric and an aggregation function that provide a measure of the disagreement among the agents with respect to the outcome provided by the OWA operator in each case. We have paid special attention to Manhattan and Chebyshev metrics and the arithmetic mean and maximum as aggregation functions. In these cases, the outcomes may be obtained by solving some mathematical linear programs. Moreover, in some specific situations we have obtained that medians and mid-range are the solutions of the problems.

It is worth noting that the solutions obtained in the above mentioned models might not be unique. Even more, depending the weighting vector we choose, the outcomes provided by the corresponding OWA operator could be different. In this way, it would be necessary to provide an appropriate procedure for choosing a single weighting vector among the set of multiple solutions.

Acknowledgements. This work is partially supported by the Junta de Castilla y León (Consejería de Educación y Cultura, Projects VA092A08 and VA002B08), the Spanish Ministry of Education and Science (Project SEJ2006-04267/ECON), and ERDF.

References

1. Beliakov, G., Pradera, A., Calvo, T.: Aggregation Functions: A Guide for Practicioners. Springer, Heidelberg (2007)
2. Brams, S.J., Kilgour, M., Sanver, M.R.: A minimax procedure for electing committees. Public Choice 132, 401–420 (2007)
3. Calvo, T., Kolesárova, A.: Komorníková, M., Mesiar, R.: Aggregation operators: Properties, classes and construction methods. In: Calvo, T., Mayor, G., Mesiar, R. (eds.) Aggregation Operators: New Trends and Applications, pp. 3–104. Physica-Verlag, Heidelberg (2002)
4. Filev, D., Yager, R.R.: On the issue of obtaining OWA operator weights. Fuzzy Set. Syst. 94, 157–169 (1998)
5. Fodor, J., Roubens, M.: Fuzzy Preference Modelling and Multicriteria Decision Support. Kluwer Academic Publishers, Dordrecht (1994)
6. Grabisch, M., Orlovski, S.A., Yager, R.R.: Fuzzy aggregation of numerical preferences. In: Slowinski, R. (ed.) Fuzzy Sets in Decision Analysis, Operations and Statistics, pp. 31–68. Kluwer Academic Publishers, Boston (1998)
7. O'Hagan, M.: Aggregating template rule antecedents in real-time expert systems with fuzzy set logic. In: Proc. 22nd Annual IEEE Asilomar Conference on Signals, Systems and Computers, pp. 681–689. Pacific Grove, California (1988)
8. Wang, Y.M., Parkan, C.: A minimax disparity approach for obtaining OWA operator weights. Inform. Sci. 175, 20–29 (2005)
9. Xu, Z.: An overview of methods for determining OWA weights. Int. J. Intell. Syst. 20, 843–865 (2005)
10. Yager, R.R.: On ordered weighted averaging operators in multicriteria decision making. IEEE Trans. Syst. Man. Cybern. 18, 183–190 (1988)
11. Yager, R.R.: Centered OWA operators. Soft Comput. 11, 631–639 (2007)

The Role of the OWA Operators as a Unification Tool for the Representation of Collective Choice Sets

Janusz Kacprzyk, Hannu Nurmi, and Sławomir Zadrożny

Abstract. We consider various group decision making and voting procedures presented in the perspective of two kinds of aggregation of partial scores related to the individuals' (group's) testimonies with respect to alternatives and individuals. We show that the ordered weighthed averaging (OWA) operators can be viewed as a unique aggregation tool that – via the change of the order of aggregation, type of aggregation, etc. – can be used for a uniform and elegant formalization of basic group decision making, social choice and voting rules under fuzzy and nonfuzzy preference relations and fuzzy and nonfuzzy majority.

1 Introduction

The purpose of this paper is to briefly point out the power of the ordered weighted averaging (OWA) operators in some bacic issues related to the use of fuzzy logic in broadly perceived group/collective decision making and social choice. This paper is implied by two areas of research that resulted, first, in some position papers on the use of fuzzy sets in political science by Nurmi and Kacprzyk [40,41], some relevant papers on the use of fuzzy sets in social and political sciences (cf. Ragin [45]), and a series of papers by Kacprzyk and Zadrożny

Janusz Kacprzyk · Sławomir Zadrożny
Systems Research Institute, Polish Academy of Sciences
ul. Newelska 6,
01-447 Warsaw, Poland
e-mail: {kacprzyk,zadrożny}@ibspan.waw.pl

Hannu Nurmi
Department of Political Science, University of Turku
FI-20014 Turku, Finland
e-mail: hnurmi@utu.fi

R.R. Yager et al. (Eds.): Recent Developments in the OWA Operators, STUDFUZZ 265, pp. 149–166.
springerlink.com

[26,27,28,29] in which some general "protoforms" of group choice functions have been proposed via the OWA operators. In this paper, we will make an attempt at a new exposition, more explicitly political science oriented. In particular, we will try to follow the voting scienario - cf. Nurmi [36].

Voting systems play an important role in political science, and since there are obviously many elements and aspects of voting that are inherently imprecise, then it has been natural to try to use fuzzy sets therein. A natural starting point has been the use of fuzzy preference relations, cf. Blin and Whinston [7]. The earlier contributions assume as the point of departure a collective fuzzy preference relation over candidates, alternatives, options, ... and look for plausible solution sets exemplified by various consensus winners (cf. Nurmi [35], Kacprzyk [20,23,24]). Another direction has been to assume that we just have individual fuzzy preference relations and try to use just them (without or maybe with an aggregation into a collective fuzzy preference relation); various derivations of the core, minimax set, least vulnerable set, etc. have been proposed (cf. Nurmi [35], Kacprzyk [20,23,24]). In this paper we will consider both the above general philosophies.

The study of voting procedures is an extremely relevant problem, both from the point of view of theory and practice (cf. DeGrazia [9], Fiorina and Plott [10], Nurmi [36]). One can say that the theoretical background is provided by social choice theory (cf. Schwartz [49], Sen [50]), and also in this area fuzzy sets have been applied (cf. Nurmi, Fedrizzi and Kacprzyk [37).

The general setting considered here is as follows. We have a non-fuzzy set $S = \{s_1,...,s_M\}$ of decision alternatives (candidates, policies, options, ...). Then the fuzzy binary relation R over S is given by a membership function μ_R: $S^2 \rightarrow [0,1]$. For S of a sufficiently small cardinality, R can be represented as a $M \times M$ matrix such that its entry $r_{ij} \in [0,1]$ denotes the degree (intensity) in which R holds between the s_i and s_j.

In the context considered R is meant as a fuzzy preference relation. Basically, since the first works on fuzzy preference relations (cf. Blin [6], Blin and Whinston [7], Bouyssou [8], Orlovsky [42]), $r_{ij}= 1$ was meant as a definite preference of s_i over s_j, $r_{ij} = 0$ – as a definite preference of x_j over x_i, and $r_{ij}= 0.5$ – as an indifference between the two alternatives.

Notice that by R we mean now a social (group) preference relation that may be obtained in various ways, notably by some aggregation of individual fuzzy preference relations (cf. Billot [5], Blin and Whinston [7]). This view of R as a fuzzy social preference relation will be assumed here.

Now, if we have R, a fuzzy social preference relation, we wish to use it to find plausible - fuzzy or non-fuzzy - choice sets (solutions). Among many solution concepts presented in the literature (cf. Kacprzyk and Nurmi[25]), one can mention:

- the set of α-consensus winners:

$S_\alpha = \{x_i: r_{ij} \geq \alpha, \text{ for all } x_j \in X\}$,

- the set of minimax consensus winners

$S_M = \{x_i: r_i = r'\}$, where $r_i = \min_j r_{ij}$ and $r' = \max_m r_m$

which is always nonempty, and is a straightforward generalization of Kramer's minimax set (cf. Kramer [33]).

- the set of α-Copeland winners

$S_C = \{x_i: s^C_i = \max_j s^C_j\}$

where $s^C_i = \text{card}\{x_j \mid r_{ij} \geq \alpha\}$, and for $\alpha = 0.5$ we obtain a refinement of the classic Copeland rule.

If we take as the point of departure the set of individual fuzzy preference relations, we can proceed in a similar way and obtain, for instance:

- the fuzzy α-core

$X_\alpha = \{x_j$: such that for all $x_i \in X$, $r_{ij} \leq \alpha$ for at least z individuals$\}$

and for α=0.5 and z corresponding to a simple majority, this reduces to the core.

Similar extensions can be defined for other standard solution concepts (cf. Nurmi [35]).

The same applies to the extensions of tournament solutions (cf. Nurmi and Kacprzyk [38]). Since tournaments are complete and asymmetric relations, a natural way of constructing a tournament is to conduct pairwise comparisons of decision alternatives using the majority rule.

Two important solution concepts in non-fuzzy tournament literature are the uncovered set and the Banks set (cf. Banks [3], Nurmi [35]). An alternative s_i covers another alternative s_j if the former defeats (is more preferred) the latter and, moreover, defeats all those alternatives that s_j defeats. A covered alternative will inevitably lose a pairwise majority voting procedure regardless of the order in which the alternatives are brought to pairwise comparisons. Thus, given a profile of individual preferences, an obvious solution concept is the set of uncovered alternatives. However, this set may be too large and among various refinements suggested, the Banks set is relevant. To define the Banks set one begins with an alternative, s_1, and finds out whether another alternative exists that defeats it. If there is no, we finish and assume that s_1 is the end point of the Banks chain which begins at s_1. If, on the other hand, another alternative that defeats $s_{1,}$, s_i, is found, we look for an alternative that defeats both s_1 and s_i. If none exists, then x_i is the end point of the Banks chain beginning at x_1. Otherwise, one looks for an alternative defeating all the preceding ones, and the procedure continues until we eventually find no alternative that defeats all the preceding ones in the chain. We obviously reach an end point of the chain beginning at x_1. Starting from each alternative one necessarily ends up with a chain with an end point. One alternative may, however, give rise to several Banks chains. The Banks set consists of the end points of all Banks chains in the tournament. The main significance of the Banks set is that it coincides with all strategic voting outcomes in binary voting agendas.

Fuzzy analogues of the uncovered set and the Banks set are studied in Nurmi and Kacprzyk [39]. In fact, two covering relations, strong and weak, can be defined, and the set of weakly uncovered alternatives is always a subset of the strongly uncovered ones. Moreover, it can be shown that the set of Copeland winners is necessarily within the set of strongly uncovered alternatives but it is not necessary that the Copeland winners are weakly uncovered ones (cf. Nurmi [35]).

As we can see, there has been much research effort to find reasonable choice rules. The main strategy is to introduce more information about individual preferences than is usually the case in the social choice theory. Now, we will elaborate on this issue by assuming various fuzzy elements of the models, notably the fuzzy majority and fuzzy preference relations.

2 Group Decisions under fuzzy Preferences and a Fuzzy Majority

Suppose that we have a set of $M \geq 2$ options, and a set of $N \geq 2$ *individuals* (experts) $E = \{e_1, \ldots, e_N\}$. An individual fuzzy preference relation in $S \times S$ of individual $e_k \in E$ assigns a value from [0,1] (we do not consider more flexible types in which the value for [0,1] is replaced by, for instance, a linguistic value).

Two lines of reasoning may be followed here to find solutions (cf. Kacprzyk [20,21,22,23,24]):

- a direct approach $\{R_1, ..., R_N\} \rightarrow$ solution
(1)
- an indirect approach $\{R_1, ..., R_N\} \rightarrow R \rightarrow$ solution

that is, in the first case we determine a *solution* just on the basis of individual fuzzy preference relations, and in the second case we form first a social fuzzy preference relation R which is then used to find a solution.

Another basic element underlying group decision making is the concept of a *majority*, i.e. a solution is to be an option (or options) best acceptable by (at least!) some number of the individuals. Some negative result in group decision making are closely related to a (too) strict representation of majority, and a natural attempt is to somehow make that strict concept of majority closer to its real human perception by making it more vague (cf. Intrilligator [17,18], Judin [19], Nurmi [35]).

A natural manifestations of such a ``soft" majority are the so-called linguistic quantifiers as, e.g., most, almost all, much more than a half, etc. that can be dealt with using fuzzy-logic-based calculi of linguistically quantified statements as proposed by Zadeh [58], and – what is more relevant here – by Yager's [54] [cf. Yager and Kacprzyk [57]] ordered weighted averaging (OWA) operators.

These calculi have been applied by the authors to introduce that new concept of a fuzzy majority (represented by a fuzzy linguistic quantifier) into group decision making and consensus formation models (cf. Kacprzyk [20-24]; Kacprzyk and Nurmi [25]; Nurmi and Kacprzyk [38], etc.). See also Kacprzyk, Zadrożny, Fedrizzi and Nurmi [30] for a comprehensive review.

The group decision making boils down to an aggregation process, because individual testimonies are aggregated. On the other hand, it is a choice process, because some options should be chosen.

We will present some unified, joint view of that aggregation and choice process showing that by the OWA operators we can derive a general expression for a multitude of individual and collective choice functions under fuzzy preferences and a fuzzy majority. We will extend our previous papers Kacprzyk and Zadrożny [**Błąd! Nie zdefiniowano zakładki.**,26,2627], use more general analyses from Nurmi and Kacprzyk [39,40], and some solutions given in Zadrożny [60] and Zadrożny and Kacprzyk [62].

As mentioned earlier, we assume that the set of options is $S=\{s_1,\dots,s_M\}$, and the set of individuals (experts) is $E=\{e_1,\dots,e_N\}$. An individual e_k presents his or her preference relation represented for convenience as a matrix:

$$[r_{ij}^k]=[\mu_{R_k}(s_i,s_j)],\forall i,j,k \tag{2}$$

Under a simple assumption (fuzzy tournament), $\mu_R(s_i,s_j)+\mu_R(s_j,s_i)=1,\forall i\neq j$; cf. Nurmi and Kacprzyk [38], a plausible interpretation may be:

$$\mu_R(s_i,s_j)=\begin{cases}1 & \text{definite preference of } s_i \text{ over } s_j\\ c\in(0.5,1) & \text{preference to some extent}\\ 0.5 & \text{indifference}\\ d\in(0,0.5) & \text{preference to some extent}\\ 0 & \text{definite preference of } s_j \text{ over } s_i\end{cases} \tag{3}$$

One can also use some weaker definitions, like weak preference relations, and also a so-called preference structure (cf. Fodor and Roubens [12]) but we will not deal with this issue.

We will consider both the general forms of fuzzy preference relations and their special cases such as fuzzy tournaments (3) and *crisp linear orderings*, i.e. crisp relations R defined on $S\times S$ and possessing the properties of:

- reflexivity: $\forall s_i\in S\ R(s_i,s_i)$
- anti-symmetry: $\forall s_i,s_j\in S\ R(s_i,s_j)\wedge R(s_j,s_i)\rightarrow s_i=s_j$
- transitivity: $\forall s_i,s_j,s_k\in S\ R(s_i,s_j)\wedge R(s_j,s_k)\rightarrow R(s_i,s_k)$

The latter are considered in many classical voting/choice rules. We will show that our general scheme of group choice covers many of these rules when applied to crisp linear orderings.

2.1 *Fuzzy Majority and the OWA Operators*

Fuzzy majority constitutes a natural generalization of the concept of majority in the case of a fuzzy setting. A fuzzy majority was introduced into group decision making under fuzziness by Kacprzyk [21], and then considerably extended in the works of Fedrizzi, Herrera, Herrera-Viedma, Kacprzyk, Nurmi, Verdegay, Zadrożny, etc. (see, e.g., a review by Kacprzyk and Nurmi [25] or Kacprzyk, Zadrożny, Fedrizzi and Nurmi [30], see also Pasi and Yager [44]. It is basically equated with a fuzzy linguistic quantifier exemplified by *most*, *almost all*, etc.

The essence of fuzzy linguistic quantifiers boils down to the looking for truth of a proposition:

"Most objects posses a certain property"

that may be formally expressed as follows:

$$\underset{x \in X}{\mathrm{Q}} \; P(x) \tag{4}$$

where Q denotes a fuzzy linguistic quantifier (in this case "most"), $X = \{x_1, \ldots x_m\}$ is a set of objects, $P(\cdot)$ corresponds to a fuzzy property; $\mathrm{truth}(P(x_i)) = \mu_P(x_i)$.

Zadeh's [58] *calculus of linguistically quantified propositions* deals with a linguistic quantifier represented as a fuzzy set $Q \in \mathrm{F}([0,1])$, where F(A) denotes the family of all fuzzy sets on A. For instance, Q ("most") may be defined by:

$$\mu_Q(y) = \begin{cases} 1 & \text{for} \quad y \geq 0.8 \\ 2y - 0.6 & \text{for} \;\; 0.3 < y < 0.8 \\ 0 & \text{for} \quad y \leq 0.3 \end{cases} \tag{5}$$

The truth of (4) is determined as:

$$\mathrm{truth}(\mathrm{Q}P(X)) = \mu_Q(\sum_{i=1}^{m} \mu_P(x_i) / m) \tag{6}$$

where m=card(X).

The linguistic operator may be treated as a flexible aggregation operator situated somewhere between the AND and OR which in fuzzy logic are usually interpreted as the min and max, respectively.

Another approach to the modeling of fuzzy linguistic quantifiers (and flexible aggregation), pursued here, is by using Yager's OWA (ordered weighted averaging) operators [54] (see also Yager and Kacprzyk's [57] volume). An interesting and powerful tool may here also be the OWmin and/or OWmax operators, the Sugeno and Choquet integrals, etc. Moreover, a different view of fuzzy quantifiers and their related calculi due to Gloeckner [13] may be employed. These tools will not be used here.

An OWA operator O of dimension n may be briefly described as follows:

$$O:\Re^n \to \Re$$

$$W = [w_1,\ldots,w_n], \quad w_i \in [0,1], \quad \sum_{i=1}^{n} w_i = 1$$

$$O(a_1,\ldots,a_n) = \sum_{j=1}^{n} w_j b_j, \quad b_j \text{ is } j\text{-th largest of the } a_i \tag{7}$$

Here, a_i are values to be aggregated and W is a vector of weights defining the operator. Depending on W, the aggregation schemes as the minimum and maximum, median or average can be obtained. In the context of fuzzy (multivalued) logic, the classical general ($\forall$) and existential ($\exists$) quantifiers may be interpreted as the min and max, respectively. We denote:

$$\begin{aligned} O_\forall &: \forall \to W = [0,\ldots,0,1] \\ O_\exists &: \exists \to W = [1,0,\ldots,0] \end{aligned} \tag{8}$$

The following weight vectors define other OWA operators that are useful for our purposes:

$$\begin{aligned} &O_A: \textbf{\textit{average}} && W = [1/n,\ldots,1/n] \\ &O_{most}: \textbf{\textit{most}} && \\ &O_{maj}: \textbf{\textit{classical crisp majority}} && W = [0,\ldots,0,1,0,\ldots,0] \quad w_{(n/2)+1} \; or \; w_{(n+1)/2} = 1 \end{aligned} \tag{9}$$

The O_A yields the usual arithmetic mean but when applied to arguments from $\{0,1\}$, it may be regarded as counting the number of 1's in the argument. The result of aggregation using O_A is also proportional to the sum of its arguments. This will be useful for our purposes.

The OWA operators possess the following properties relevant for our considerations:

- neutrality: $O(a_1,\ldots,a_n) = O(a_{i1},\ldots,a_{in})$ for all $(a_1,\ldots,a_n) \in R^n$,
- monotonicity: $a_i < b_i \Rightarrow O(a_1,\ldots,a_i,\ldots,a_n) \le O(a_1,\ldots,b_i,\ldots,a_n)$,
- idempotence: $O(a,a,\ldots,a) = a$,
- compensativeness: $\min_i a_i \le \cdot O(a_1,\ldots,a_n) \cdot \le \max_i a_i$,

Many extensions of the original concept of an OWA operator have been proposed, and some are discussed in this volume. We will not deal with them as the purpose of our paper is more basic and conceptual.

2.2 Individual Choice under Fuzzy Preferences

The very purpose of the group decision process is ultimately to make a plausible or rational choice according to some rule. Such rules are called *choice rules* (*functions*) in general, and are defined as

$$C(X,R) = X_0 \quad X_0 \subseteq X, \quad \forall X \subseteq S \tag{10}$$

i.e. for each X, it produces its subset meant as the *choice set* of the best options in X with respect to Ri (cf. Aizerman and Aleskerov [1], Arrow [2]).

Particular choice functions refer to some *rationality concepts* exemplified by a classic concept of a choice function based on *greatestness*,[50] i.e.:

$$C_1(X,R) = \{s_i \in X : \forall_{\substack{s_j \in X \\ s_i \neq s_j}} s_i R s_j\} \tag{11}$$

Hence, the set of chosen (preferred) options, $C_1(X,R)$, consists of options greater than all other options in X, in the sense of R. The set of the greatest elements may however easily be empty. Thus, often a similar notion of *non-dominance* is employed implying the following definition of the choice function:

$$C_2(X,R) = \{s_i \in X : \neg \exists_{\substack{s_j \in X \\ s_i \neq s_j}} (s_j R s_i)\} \tag{12}$$

and if R *is* anti-symmetric and complete, C_1 and C_2 are identical.

Among some other perspectives one can mention Aizerman and Aleskerov's [1] view of those choice functions as integrals (cf. Schwartz [49] and Kitainik [32]), and the so-called GETCHA and GOCHA [49] concepts, that form generalizations of C_1 and C_2, respectively. They always produce a non-empty choice set.

In the setting of fuzzy preference relations (3), Nurmi [35] defnes the *consensus winner* and α-*consensus winner* choice sets as follows, respectively:

$$C_3(X,R) = \{s_i \in X : \forall_{i \neq j} r_{ij} > 0.5\} \text{ and } C_4(X,R) = \{s_i \in X : \forall_{i \neq j} r_{ij} \geq \alpha > 0.5\}\,.$$

where $r_{ij} = \mu_R(s_i,s_j)$, cf. Section 2.

These choice functions are a proper generalization of (11): if $r_{ij} \in \{0,1\}$, then it is identical with (11) and, moreover, even for $r_{ij} \in [0,1]$ the choice set is either of a single-element or empty.

Nurmi [35] introduces also a counterpart of (12), the *minimax consensus winner*:

$$C_5(X,R) = \{s_i : i = \arg\min_j \max_k \; r_{kj}\} \tag{13}$$

Notice that these choice functions produce crisp choice sets.

In order to obtain non-empty choice sets for a wider class of fuzzy preference relations, Kacprzyk [21, 22,24,] introduced the concept of a Q-consensus winner defined, informally, as follows:

$$C_6(X,R) = \{s_i \in X : \underset{i \neq j}{Q}\, (r_{ij} > \alpha)\} \tag{14}$$

where Q is a linguistic quantifiers in the sense of Zadeh [58]. The idea is to use a quantifier referring to the concept of a *fuzzy majority*, thus to choose options that are preferred over, e.g., *most* of the other options.

Some other approaches, also intended for other types of fuzzy preference relations, include the work of Świtalski [52], Ovchinnikov [43], Roubens [47], Kitainik [32], Barrett et al. [4], and others [5,8,42].

Barrett et al.'s [4] paper is relevant here as they noticed that most of the known individual choice functions assess the membership of s_i in the choice set via an aggregation of the preferences of all other options against s_i, *i.e.*:

$$\mu_{C(X,R)}(s_i) = \text{AGG}\{r_{ij}\}_{i \neq j} \tag{15}$$

or

$$\mu_{C(X,R)}(s_i) = \text{AGG}\{r_{ji}\}_{i \neq j} \tag{16}$$

where the left hand side denotes the membership degree of a given option in the (fuzzy) choice set and AGG denotes some aggregation operator.

We will consider now some definitions of individual choice functions by using the OWA operator. Obviously, C_1 defined by (11) may be expressed as follows:

$$\mu_{C_1}(s_i) = O_{\forall} R(s_i, s_j) \tag{17}$$

Assuming the traditional definition of neg(x) = 1-x, C_2 may be expressed as: $\mu_{C_2}(s_i) = 1 - O_{\exists} R(s_j, s_i)$.

The C_5 (13) may be expressed as, using the defuzzification via the minimum:

$$\mu_{C_7}(s_i) = O_{\exists} R(s_j, s_i) \tag{18}$$

We obtain therefore a general scheme for individual choice functions ICR:

$$\mu_{ICR}(s_i) = O(a_1,\ldots,a_M) \tag{19}$$

$$a_j = \mu_R(s_i,s_j) \quad \text{or} \tag{20}$$

$$a_j = \mu_R(s_j,s_i) \tag{21}$$

Various choice rules can be obtained by the selection of the OWA operator O in (19), the use of (20) or (21), the defuzzification via the maximum or minimum, etc.

3 Collective Choice under Fuzzy Preferences

A collective choice rule describes how to determine a set of preferred options starting from the set of individual preference relations, that is using the direct approach:

$$\{R_1,\ldots,R_N\} \to 2^N$$

For our next discussion it is only important that the individual preferences should somehow be aggregated so as to produce a set of options satisfying preferences of all involved parties according to some rationality principles, and not if we use the direct or indirect approach.

One of the most popular rules of aggregation is the simple majority rule (known also as the Condorcet rule) – cf. Nurmi [36] or Risse [46], which may be described as:

$$R(s_i,s_j) \Leftrightarrow \text{Card}\{k : R_k(s_i,s_j)\} \geq \text{Card}\{k : R_k(s_j,s_i)\} \tag{22}$$

$$S_0 = \{s_i \in S : \underset{i \neq j}{\forall} R(s_i,s_j)\} \tag{23}$$

where $\text{Card}\{A\}$ denotes cardinality of the set A and S_0 is the set of collectively preferred options.

As its counterpart in the fuzzy case Nurmi [35] proposed:

$$R(s_i,s_j) \Leftrightarrow \text{Card}\{k : R_k(s_i,s_j) > \alpha\} \geq threshold \tag{24}$$

$$S_0 = \{s_i \in S : \neg \underset{j}{\exists} R(s_j,s_i)\} \tag{25}$$

which means that a more flexible concept of majority used is still crisp.

Later, Kacprzyk [21,22,23] interpreted rule (22) – (23) employing the concept of a fuzzy majority equated with a linguistic quantifier and introduced a Q-core that may be informally stated in a slightly modified version, as the $Q1/Q2$-core (cf. Zadrozny [61]) as:

$CC_{Q1,Q2}$: *Set of options, which are for most (Q2) of individuals "better" than most (Q1) of the rest of options from the set S.*

$$CC_{Q1,Q2} \in F(S)$$
$$\mu_{CC_{Q1,Q2}}(s_i) = \underset{s_j}{Q1} \underset{x_k \in X}{Q2} R_k(s_i,s_j) \tag{26}$$

where F(S) denotes a family of all fuzzy sets on S. Then, using Zadeh's fuzzy linguistic quantifiers, we obtain:

$$h_i^j = \frac{1}{N}\sum_{k=1}^{N} r_{ij}^k \quad h_i = \frac{1}{M-1}\sum_{\substack{j=1\\ j\neq i}}^{M} \mu_{Q2}(h_j^k)$$
$$\mu_{CC_{Q1,Q2}}(s_i) = \mu_{Q1}(h_i) \tag{27}$$

Where: h_i^j is the degree to which, according to all individuals, option s_i is better than option s_j; h_i is the degree to which, according to all individuals, s_i is better than most ($Q1$) other options; $\mu_{Q1}(h_i)$ is the degree (sought) to which, in the opinion of most ($Q2$) individuals, s_i is better than most ($Q1$) other options. Thus, the membership function degree for option s_i in the choice set $CC_{Q1,Q2}$ equals the truth value given by (26).

Formula (26) serves as a prototype for our generic collective choice rule to be proposed in the next section. The aggregation scheme implied by (26) is based on: first, by using various linguistic quantifiers we obtain different collective choice rules; second, by changing the order of aggregation we can obtain another family of collective choice rules, including some other well-known rules; third, by replacing the original Zadeh linguistic quantifiers with the OWA operators we obtain a more flexible aggregation scheme.

This implies the transformation of (26) into:

$$\underset{s_j}{Q1}\ \underset{x_k\in X}{Q2}\ R_k(s_i,s_j) \rightarrow O_{most}^j\, O_{most}^k\, R_k(s_i,s_j)$$

with j and k indexing the set of options and individuals, respectively.

Thus, O_{most}^j (O_{most}^k) denotes an OWA operator guided aggregation over all options (individuals) with the underlying weight vector indicated by the lower index, here *most.*

Now, the proposed generic collective choice rule (CCR) may be expressed as:

$$\mu_{CCR}(s_i) = O_1\, O_2\, R_k(s_p,s_q) \tag{28}$$

This generic scheme has a number of "degrees of freedom":

1. the upper indexes of the OWA operators, i.e., if we first aggregate over individuals and then over options or in the reverse order,
2. the weights vectors of both OWA operators,
3. whether (p, q) corresponds to (i, j) or to (j, i)

This leads to the four types of collective choice rules:

Type I $\mu_{CCR}(s_i) = O_1^k O_2^j R_k(s_i, s_j)$

Type II $\mu_{CCR}(s_i) = O_1^j O_2^k R_k(s_i, s_j)$

Type III $\mu_{CCR}(s_i) = O_1^k O_2^j R_k(s_j, s_i)$

Type IV $\mu_{CCR}(s_i) = O_1^j O_2^k R_k(s_j, s_i)$

The rules of type III and IV should be properly understood: they identify the fuzzy sets of options that are collectively *rejected* by the group.

Clearly, to identify relations with classical rules (or concepts), an appropriate defuzzification should be employed, notably via the max or min operation:

- for type I and II rules choose an s_i such that $i = \arg\max_j \mu_{CCR}(s_j)$
- for type III and IV rules choose an s_i such that $i = \arg\min_j \mu_{CCR}(s_j)$

Now we can show some well-known choice rules covered by our generic scheme using some specific OWAs, i.e. their correcponding weight vectors.

First, for the rules that are both of type I and type II:

1. $O_\forall O_\forall$ the "consensus solution" (29)

An option is a member of this collective choice set to a degree to which it dominates *all* options according to the preferences of *all* individuals.

2. $O_{avg} O_{avg}$ the Borda rule (30)

In the collective choice rule proposed by Borda for crisp orderings (called the *Borda count*) (see, e.g., [9]) an option gets *M-1, M-2, M-3*, ..., 0 points for being first, second, third,..., last, respectively, in an individual ordering (*M* is the number of options under consideration, as previously). The points assigned to an option in particular orderings are then summed up which yields the Borda count of the option. The alternative with the largest Borda count is declared the winner.

The following rule may be classified as type III or IV:

3. $O_\exists O_\exists$ - the minimax degree set (Nurmi) [35] (31)

This choice rule introduced by Nurmi is an extension of his minimax consensus winner defined by (13), and it contains such options that are *least dominated* when confronted with *all* other options according to preferences of *all* individuals.

Some examples of type I rules are:

4. $O_{avg}^k O_\forall^j$ - the plurality voting (32)

An option is here chosen if it appears as the first in the largest number of individual crisp linear orderings. In (32) $O_{\forall}^{j}$ requires that an option dominates all other options and O_{avg}^{k} "counts" for how many individuals it occurs. Thus, for crisp linear orderings $O_{\forall}^{j}(r_{i1}^{k},\ldots,r_{iM}^{k})$ returns 1 if s_i is the first in the ordering of individual k and 0 otherwise. Then, O_{avg}^{k} gives the number of 1's obtained by $O_{\forall}^{j}(r_{i1}^{k},\ldots,r_{iM}^{k})$ for the particular individuals k, divided by N – the total number of individuals.

5. $O_{maj}^{k}O_{\forall}^{j}$ - the qualified plurality voting (33)

This extends the above "regular" plurality voting so that an option is chosen if and only if an it is first in the orderings of a qualified majority of individuals.

6. $O_{avg}^{k}O_{maj}^{j}$ - an approval voting-like rule (34)

Now, each individual chooses as many options as he or she likes, and then each option is ranked according to the number of individuals who have chosen it. The (34) is equivalent to the approval voting in the sense that the individual preferences are crisp linear orderings and O_{maj}^{j} models the choice of an individual if he or she has to choose a subset of preferred options – as required by the classical approval voting procedure.

7. $O_{\forall}^{k}O_{maj}^{j}$ - the "consensus+approval voting" rule (35)

This collective choice rule does not have a classic counterpart. It may be viewed as a stronger version of approval voting because it includes an option only if it belongs to the sets of options chosen by *all* individuals.

Some examples of type II rules are:

8. $O_{\forall}^{j}O_{maj}^{k}$ - the simple majority (Condorcet) (36)

This rule yields the same result for crisp linear orderings as the classic Condorcet (majority) rule chooses such an option that dominates all other options in the majority of linear orderings expressed by particular individuals.

9. $O_{\forall}^{j}O_{\exists}^{k}$ - the Pareto optimal options (37)

An option s_i is *undominated in the Pareto sense* if no option s_j exists such that $R^k(s_j,s_i)$ for all k and $P^k(s_j,s_i)$ for at least one k; with P denoting the strict preference. For the linear orderings, the combination of the OWA operators $O^j_\forall O^k_\exists$ produces exactly the same result: for each option s_j there must be an individual k such s_i precedes s_j. in her or his ordering.

10. $O^j_{avg} O^k_{maj}$ - the Copeland rule (38)

The Copeland choice set (see also Introduction) consists of those options that dominate the largest number of other ones in the majority of linear orderings of the individuals. This is a weaker variation of the Condorcet choice rule giving higher chances for a resulting non-empty choice set. In order to cover it with our generic scheme we replace $O_\forall$ in (36) with O_{avg} in (38). Here again O_{avg} makes it possible to "count" the number of options that are dominated by s_i in the linear orderings of the majority of individuals.

An example of a type III rule is:

11. $O^k_{most} O^j_{avg}$ - Kacprzyk's Q-minimax set [21,24] (39)

Kacprzyk's Q-minimax set consists of these options which are dominated by the smallest number of other options in the preferences of most of individuals, and this is exactly what (39) stands for.

And finally, some examples of type IV rules are:

12. $O^j_\exists O^k_{avg}$ - the Kramer's minimax set [35] (40)

Kramer's minimax set (see also Introduction) consists of those options which, when confronted with their toughest competitors, fare best, i.e. are dominated by them in the smallest number of linear orderings of the individuals. To determine the membership of option s_i, O^k_{avg} counts in how many orderings option s_j precedes option s_i and $O^j_\exists$ returns the highest of this numbers.

13. $O^j_\forall O^k_{maj}$ - the Condorcet loser (41)

The Condorcet loser is an option that is dominated by all other options in the majority of individual linear orderings, and (41) properly represents this concept for linear orderings.

Thus, the generic scheme proposed covers some classical rules, especially well-known in the context of voting (cf. Miller [34]. Nurmi [36], Saari [48]). Some of the recovered rules are not collective choice rules *sensu stricto*. For crisp preferences (usually linear orderings), the rules proposed yield the identical results as the classic rules. Moreover, these rules are readily applicable to any other forms of preference relations, notably fuzzy preference relations. Clearly, some of the rules proposed concern rejection of options, and not acceptance as traditionally assumed.

More information on the choice rules presented above as well as their properties can be found in Kacprzyk and Zadrożny [29] and Zadrożny [61], while to a more foundational analysis of voting, we refer the reader to Miller [34], Nurmi [36], Saari [48], etc.. A more comprehensive discussion from the perspective of political science can be found in Nurmi and Kacprzyk [40,41].

4 Concluding Remarks

We have presented the OWA operators as a unique aggregation tool that via the change of the order of aggregation, type of aggregation, etc. can be used for a uniform and elegant formalization of basic group decision making, social choice and voting rules under fuzzy and nonfuzzy preference relations and majority.

Acknowledgments. Support from the Ministry of Science and Higher Education under Grant N N519 404734 is gratefully acknowledged.

References

[1] Aizerman, M., Aleskerov, F.: Theory of Choice. North-Holland, Amsterdam (1995)
[2] Arrow, K.J.: Social Choice and Individual Values, 2nd edn. Wiley, New York (1963)
[3] Banks, J.: Sophisticated Voting Outcomes and Agenda Control. Social Choice and Welfare 1, 295–306 (1985)
[4] Barrett, C.R., Pattanaik, P.K., Salles, M.: On choosing rationally when preferences are fuzzy. Fuzzy Sets and Systems 34, 197–212 (1990)
[5] Billot, A.: Aggregation of preferences: The fuzzy case. Theory and Decision 30, 51–93 (1991)
[6] Blin, J.M.: 'Fuzzy relations in group decision theory. Journal of Cybernetics 4, 17–22 (1974)
[7] Blin, J.M., Whinston, A.P.: Fuzzy sets and social choice. Journal of Cybernetics 4, 17–22 (1973)
[8] Bouyssou, D.: A note on the sum of differences choice function for fuzzy preference relations. Fuzzy Sets and Systems 47, 197–202 (1992)
[9] DeGrazia, A.: Mathematical derivation of an election system. Isis 44, 42–51 (1953)
[10] Fiorina, M., Plott, C.: Committee decisions under majority rule: an experimental study. American Political Science Review 72, 575–595 (1978)
[11] Fodor, J., Marichal, J.-L., Roubens, M.: Characterization of the ordered weighted averaging operators. IEEE Transactions on Fuzzy Systems 3(2), 236–240 (1995)
[12] Fodor, J., Roubens, M.: Fuzzy Preference Modelling and Multicriteria Decision Support. Kluwer Academic Publishers, Dordrecht (1994)
[13] Gloeckner, I.: Fuzzy Quantifiers - A Computational Theory. Springer, Heidelberg (2006)

[14] Herrera, F., Herrera-Viedma, E.: Choice Functions and Mechanisms for Linguistic Preference Relations. European Journal of Operational Research 120, 144–161 (2000)
[15] Herrera, F., Herrera-Viedma, E., Verdegay, J.L.: Direct Approach Processes in Group Decision Making using Linguistic OWA operators. Fuzzy Sets and Systems 79, 175–190 (1996)
[16] Herrera, F., Herrera-Viedma, E., Verdegay, J.L.: Applications of the linguistic OWA operators in group decision making. In: Yager, R.R., Kacprzyk, J. (eds.) The Ordered Weighted Averaging Operators. Theory and Applications, pp. 205–218. Kluwer Academic Publishers, Boston (1997)
[17] Intrilligator, M.D.: A probabilistic model of social choice. Review of Economic Studies 40, 553–560 (1973)
[18] Intrilligator, M.D.: Probabilistic models of choice. Mathematical Social Sciences 2, 157–166 (1982)
[19] Judin, A.B.: Computational Methods of Decision Making (in Russian, Vychislitelnye Metody Priniatya Reshenniy), Nauka, Moscow (1989)
[20] Kacprzyk, J.: Some commonsense solution concepts in group decision making via fuzzy linguistic quantifiers. In: Kacprzyk, J., Yager, R.R. (eds.) Management Decision Support Systems Using Fuzzy Sets and Possibility Theory, pp. 125–135. Verlag TÜV Rheinland, Cologne (1985)
[21] Kacprzyk, J.: Group decision making with a fuzzy majority via linguistic quantifiers. Part I: A consensory - like pooling. Cybernetics and Systems: an Int. Journal 16, 119–129 (1985)
[22] Kacprzyk, J.: Group decision making with a fuzzy majority via linguistic quantifiers. Part II: A competitive - like pooling. Cybernetics and Systems: an Int. Journal 16, 131–144 (1985)
[23] Kacprzyk, J.: Group decision making with a fuzzy linguistic majority. Fuzzy Sets and Systems 18, 105–118 (1986)
[24] Kacprzyk, J.: On some fuzzy cores and soft consensus measures in group decision making. In: Bezdek, J.C. (ed.) The Analysis of Fuzzy Information, vol. 2, pp. 119–130. CRC Press, Boca Raton (1987)
[25] Kacprzyk, J., Nurmi, H.: Group decision making under fuzziness. In: Słowiński, R. (ed.) Fuzzy Sets in Decision Analysis, Operations Research and Statistics, pp. 103–136. Kluwer, Boston (1998)
[26] Kacprzyk, J., Zadrożny, S.: Computing with words in decision making through individual and collective linguistic choice rules. International Journal of Uncertainty, Fuzziness and Knowledge – Based Systems 9, 89–102 (2001)
[27] Kacprzyk, J., Zadrożny, S.: Collective choice rules in group decision making under fuzzy preferences and fuzzy majority: a unified OWA operator based approach. Control and Cybernetics 31, 937–948 (2002)
[28] Kacprzyk, J., Zadrożny, S.: Dealing with imprecise knowledge on preferences and majority in group decision making: towards a unified characterization of individual and collective choice functions. Bulletin of the Polish Academy of Sciences (Tech.) 51, 279–302 (2003)
[29] Kacprzyk, J., Zadrożny, S.: Towards a general and unified characterization of individual and collective choice functions under fuzzy and nonfuzzy preferences and majority via the Ordered Weighted Averaging Operators. International Joornal of Intelligent Systems 24, 4–26 (2009)

[30] Kacprzyk, J., Zadrożny, S., Fedrizzi, M., Nurmi, H.: On group decision making, consensus reaching, voting and voting paradoxes under fuzzy preferences and a fuzzy majority: a survey and a granulation perspective. In: Pedrycz, W., Skowron, A., Kreinovich, V. (eds.) Handbook of Granular Computing, pp. 906–929. Wiley, Chichester (2008)
[31] Kim, J.B.: Fuzzy rational choice functions. Fuzzy Sets and Systems 10, 37–43 (1983)
[32] Kitainik, L.: Fuzzy Decision Procedures with Binary Relations: Towards a Unified Theory. Kluwer Academic Publishers, Boston (1993)
[33] Kramer, G.: A dynamical model of political equilibrium. Journal of Economic Theory 16, 310–334 (1977)
[34] Miller, N.: Committees, Agendas, and Voting. Harwood Academic Publishers, Chur (1995)
[35] Nurmi, H.: Approaches to collective decision making with fuzzy preference relations. Fuzzy Sets and Systems 6, 249–259 (1981)
[36] Nurmi, H.: Comparing Voting Systems, Reidel, Dordrecht (1987)
[37] Nurmi, H., Fedrizzi, M., Kacprzyk, J.: Vague notions in the theory of voting. In: Kacprzyk, J., Fedrizzi, M. (eds.) Multiperson Decision Making Models Using Fuzzy Sets and Possibility Theory, pp. 43–52. Kluwer, Dordrecht (1990)
[38] Nurmi, H., Kacprzyk, J.: On fuzzy tournaments and their solution concepts in group decision making. EJOR 51, 223–232 (1991)
[39] Nurmi, H., Kacprzyk, J.: Social choice under fuzziness: a perspective. In: Fodor, J., De Baets, B., Perny, P. (eds.) Preferences and Decisions under Incomplete Knowledge, pp. 107–130. Springer, Heidelberg (2000)
[40] Nurmi, H., Kacprzyk, J.: Fuzzy sets in political science: an overview. New Mathematics and Natural Computation 3, 281–299 (2007)
[41] Nurmi, H., Kacprzyk, J.: Political representation: perspective from fuzzy systems theory. New Mathematics and Natural Computation 3, 153–163 (2007)
[42] Orlovsky, S.A.: Decision-making with a fuzzy preference relation. Fuzzy Sets and Systems 1, 155–167 (1978)
[43] Ovchinnikov, S.V.: Modelling valued preference relations. In: Kacprzyk, J., Fedrizzi, M. (eds.) Multiperson Decision Making Models using Fuzzy Sets and Possibility Theory, pp. 64–70. Kluwer Academic Publishers, Dordrecht (1990)
[44] Pasi, G., Yager, R.R.: Modeling the concept of majority opinion in group decision making. Information Sciences 176, 390–414 (2006)
[45] Ragin, C.C.: Fuzzy-Set Social Science. The University of Chicago Press, Chicago (2000)
[46] Risse, M.: Why the Count de Borda cannot beat the Marquis de Condorcet. Social Choice and Welfare 25, 95–113 (2005)
[47] Roubens, M.: Some properties of choice functions based on valued binary relations. EJOR 40, 309–321 (1989)
[48] Saari, D.: Decisions and elections. Explaining the unexpected. Cambridge University Press, Cambridge (2000)
[49] Schwartz, T.: The Logic of Collective Choice. Columbia University Press, New York (1986)
[50] Sen, A.K.: Collective Choice and Social Welfare. Oliver & Boyd, Edinburgh (1970)
[51] Skala, H.J.: Arrow's impossibility theorem: some new aspects. In: Gottinger, H.W., Leinfellner, W. (eds.) Decision Theory and Social Ethics, D. Reidel, Dordrecht (1978)

[52] Świtalski, Z.: Choice functions associated with fuzzy preference relations. In: Kacprzyk, J., Roubens, M. (eds.) Non-conventional Preference Relations in Decision Making, Springer, Berlin (1988)

[53] Tanino, T.: On group decision making under fuzzy preferences. In: Kacprzyk, J., Fedrizzi, M. (eds.) Multiperson Decision Making Models using Fuzzy Sets and Possibility Theory, pp. 172–185. Kluwer, Dordrecht (1990)

[54] Yager, R.R.: On ordered weighted averaging aggregation operators in multi-criteria decision making. IEEE Transactions on Systems, Man and Cybernetics 18, 183–190 (1988)

[55] Yager, R.R.: Induced aggregation operators. Fuzzy Sets and Systems 137, 59–69 (2003)

[56] Yager, R.R., Filev, D.P.: Induced ordered weighted averaging operators. IEEE Trans. Systems Man Cybern. 29, 141–150 (1999)

[57] Yager, R.R., Kacprzyk, J. (eds.): The Ordered Weighted Averaging Operators: Theory and Applications. Kluwer, Boston (1997)

[58] Zadeh, L.A.: A computational approach to fuzzy quantifiers in natural languages. Comp. and Maths. with Appls. 9, 149–184 (1983)

[59] Zadeh, L.A., Kacprzyk, J.: Computing with words in information/intelligent systems 1: Foundations, 2. Applications. Physica-Verlag, Heidelberg (1999)

[60] Zadrożny, S.: An approach to the consensus reaching support in fuzzy environment. In: Kacprzyk, J., Nurmi, H., Fedrizzi, M. (eds.) Consensus under Fuzziness, pp. 83–109. Kluwer, Boston (1996)

[61] Zadrożny, S., Kacprzyk, J.: Collective choice rules: a classification using the OWA operators. In: Proc. EUSFLAT-ESTYLF Joint Conference, Palma, Spain, pp. 21–24 (1999)

[62] Zadrożny, S., Kacprzyk, J.: An approach to individual and collective choice under linguistic preferences. In: Proc. Eight International Conference on Information Processing and Management of Uncertainty in Knowledge-based Systemsm, IPMU 2000, Madrid, pp. 462–469 (July 2000)

Applying Linguistic OWA Operators in Consensus Models under Unbalanced Linguistic Information

E. Herrera-Viedma, F.J. Cabrerizo, I.J. Pérez and M.J. Cobo,
S. Alonso, and F. Herrera

Abstract. In Group Decision Making (GDM) the automatic consensus models are guided by different consensus measures which usually are obtained by aggregating similarities observed among experts' opinions. Most GDM problems based on linguistic approaches use symmetrically and uniformly distributed linguistic term sets to express experts' opinions. However, there exist problems whose assessments need to be represented by means of unbalanced linguistic term sets, i.e., using term sets which are not uniformly and symmetrically distributed. The aim of this paper is to present different Linguistic OWA Operators to compute the consensus measures in consensus models for GDM problems with unbalanced fuzzy linguistic information.

1 Introduction

In a classical Group Decision Making (GDM) situation there is a problem to solve, a solution set of possible alternatives, and a group of two or more experts, who express their opinions about this solution set of alternatives. These problems consists of multiple individuals interacting to reach a decision. Each expert may have unique

E. Herrera-Viedma · I.J. Pérez · M.J. Cobo · F. Herrera
Dept. of Computer Science and A.I, University of Granada, 18071 Granada, Spain
e-mail: viedma@decsai.ugr.es, ijperez@decsai.ugr.es
mjcobo@decsai.ugr.es, herrera@decsai.ugr.es

F.J. Cabrerizo
Department of Software Engineering and Computer Systems,
Distance Learning University of Spain, 28040 Madrid, Spain
e-mail: cabrerizo@issi.uned.es

S. Alonso
Dept. of Software Engineering, University of Granada, 18071 Granada, Spain
e-mail: zerjioi@ugr.es

R.R. Yager et al. (Eds.): Recent Developments in the OWA Operators, STUDFUZZ 265, pp. 167–186.
springerlink.com

motivations or goals and may approach the decision process from a different angle, but have a common interest in reaching eventual agreement on selecting the "best" option(s) [5, 14, 40].

Usually, many problems present quantitative aspects which can be assessed by means of precise numerical values [6, 27, 26, 37]. However, some problems present also qualitative aspects that are complex to assess by means of precise and exact values. In these cases, the fuzzy linguistic approach [15, 31, 46, 47, 53, 54, 55] can be used to obtain a better solution. This is the case, for instance, when experts try to evaluate the "comfort" of a car, where linguistic terms like "good", "fair", "poor" are used [38]. Many of these problems use linguistic variables assessed in linguistic term sets whose terms are uniformly and symmetrically distributed, i.e., assuming the same discrimination levels on both sides of mid linguistic term (see Fig. 1). However, there exist problems that need to assess their variables with linguistic term sets that are not uniformly and symmetrically distributed [17, 30]. This type of linguistic term sets are called *unbalanced linguistic term sets* (see Fig. 2).

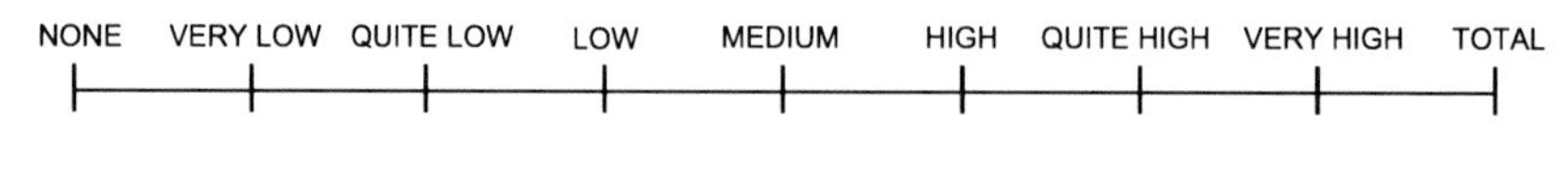

Fig. 1 Example of a linguistic term set of 9 labels

Fig. 2 Example of an unbalanced linguistic term set of 8 labels

To solve GDM problems, the experts are faced by applying two processes before obtaining a final solution [18, 22, 28, 36, 37]: *the consensus process* and *the selection process* (see Fig. 3). The former consists in obtaining the maximum degree of consensus or agreement between the set of experts on the solution set of alternatives. Normally, the consensus process is guided by a human figure called moderator [18, 22, 36], who is a person that does not participate in the discussion but monitors the agreement in each moment of the consensus process and is in charge of supervising and addressing the consensus process toward success, i.e., to achieve the maximum possible agreement and to reduce the number of experts outside of the consensus in each new consensus round. The latter refers to obtaining the solution set of alternatives from the opinions on the alternatives given by the experts. It involves two different steps [23, 41]: aggregation of individual opinions and exploitation of the collective opinion. Clearly, it is preferable that the set of experts achieves a great agreement among their opinions before applying the selection process and, therefore, in this paper we focus on the consensus process.

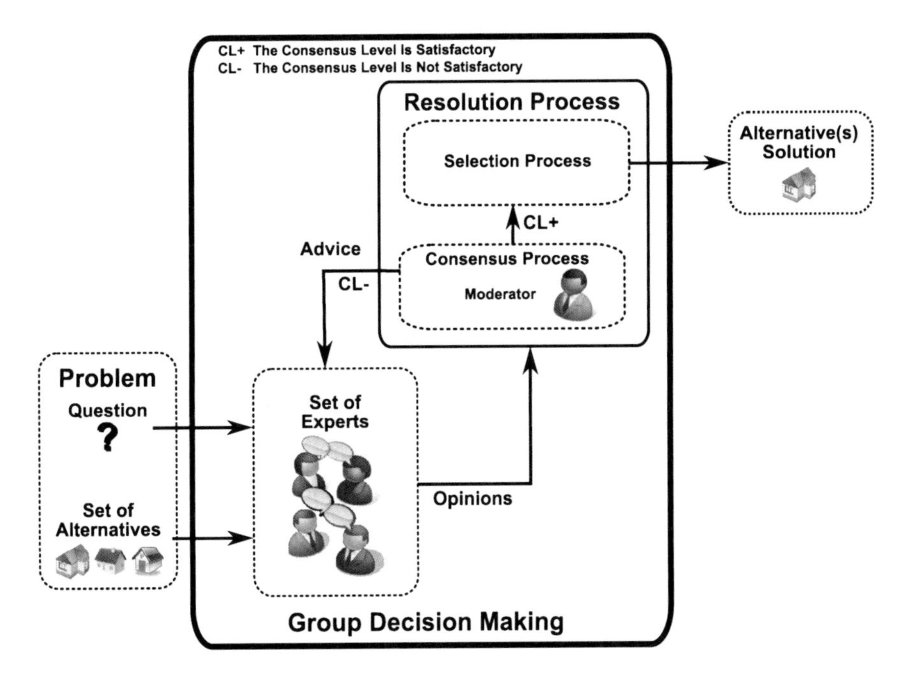

Fig. 3 Resolution process of a GDM problem

A consensus process is defined as a dynamic and iterative group discussion process, coordinated by a moderator helping experts bring their opinions closer. The moderator uses a consensus measure to assess the consensus level existing among experts. If the consensus level is lower than a specified threshold, the moderator would urge experts to discuss their opinions further in an effort to bring them closer. On the contrary, when the consensus level is higher than the threshold, the moderator would apply the selection process in order to obtain the final consensus solution to the GDM problem.

In such a framework, we find different aspects to solve:

1. An important question is how to substitute the actions of the moderator in the group discussion process in order to automatically model the whole consensus process. Some automatic consensus approaches have been proposed in [4, 26, 28, 31, 39].
2. Most of these consensus models use only consensus measures to control and guide the consensus process. However, if a consensus process is seen as a type of persuasion model [12], other criteria could be used to guide consensus reaching processes as, for example, the *cooperation* or *consistency criterion*. Some fuzzy consensus approaches based on both consistency and consensus measures can be found in [11, 13, 21, 26].
3. On the other hand, a natural question in the consensus process is how to measure the closeness among experts' opinions in order to obtain the consensus

measure. To do so, different approaches have been proposed. For instance, several authors have introduced *hard consensus measures* varying between 0 (no consensus or partial consensus) and 1 (full consensus o complete agreement) [1, 2, 43, 44]. However, consensus as a full and unanimous agreement is far from being achieved in real situations, and even if it is, in such a situation, the consensus reaching process could be unacceptably costly. So, in practice, a more realistic approach is to use *softer consensus measures* [33, 34, 35], which assess the consensus degree in a more flexible way, and therefore reflect the large spectrum of possible partial agreements, and guide the consensus process until widespread agreement (not always full) is achieved among experts. The soft consensus measures are based on the concept of coincidence [22], measured by means of similarity criteria defined among experts' opinions.
4. Sometimes, we find problems when it is not possible to compute directly the similarity among opinions because experts provide incomplete preferences [27], or use different elements of preference representation [6], or different expression domains of preferences as multi-granular fuzzy linguistic contexts [25] or unbalanced fuzzy linguistic contexts [17, 30]. In [28, 31, 26], we have presented consensus models dealing with different elements of preference representation, multi-granular linguistic preferences and incomplete preferences, respectively, and, in this paper, we focus on consensus models under unbalanced fuzzy linguistic preferences.
5. Other aspect to study is how to obtain the consensus measures from closeness values measured among experts' opinions. Usually, this is done by aggregating those closeness values by means of adequate aggregation operators. The OWA type operators [48] are very useful to develop such aggregations because they allows us to include different semantics in the aggregation process, as for example, the concept of fuzzy majority [32] or consistency semantics [9].

The aim of this paper is to present some Linguistic OWA operators to compute consensus measures in GDM problems under unbalanced linguistic preferences. As in [18, 21], we assume a consensus model which is guided by two types of consensus measures, consensus degrees and proximity measures. We present two LOWA operators to compute those consensus measures in an unbalanced linguistic context: an unbalanced LOWA operator guided by the concept of fuzzy majority to compete the consensus degrees and an unbalanced Induced LOWA operator guided by the consistency semantics to compute the proximity measures.

In order to do this, the paper is structured as follows. In Section 2, we present some preliminaries. In Section 3 we define a methodology to manage unbalanced fuzzy linguistic information together with the unbalanced LOWA operators. Section 4 presents the application of those unbalanced LOWA operators in a consensus model for GDM problems with unbalanced fuzzy linguistic preferences. Finally, some concluding remarks are pointed out in Section 5.

2 Preliminaries

In this section, we make a review of the 2-tuple fuzzy linguistic representation model[24] and the concept of hierarchical linguistic contexts[25] which are used to define the methodology to manage unbalanced fuzzy linguistic information.

2.1 *The 2-Tuple Fuzzy Linguistic Representation Model*

The 2-tuple fuzzy linguistic representation model was introduced in [24] to carry out processes of computing with words in a precise way when the linguistic term sets are symmetrically and uniformly distributed and to improve several aspects of the ordinal fuzzy linguistic approach [15, 19, 20]. This model is based on the concept of *symbolic translation* and represents the linguistic information by means of a pair of values, (s, α), where s is a linguistic label and α is a numerical value that represents the value of the symbolic translation.

Definition 1. [24]Let $\beta \in [0, g]$ be the result of an aggregation of the indexes of a set of labels assessed in a linguistic term set $S = \{s_0, s_1, \ldots, s_{g-1}, s_g\}$, where g stands for cardinality of S, i.e., the result of a symbolic aggregation operation. Let $i = round(\beta)$ and $\alpha = \beta - i$ be two values, such that, $i \in [0, g]$ and $\alpha \in [-0.5, 0.5)$, then α is called a symbolic translation.

This model defines a set of transformation functions to manage the linguistic information expressed by linguistic 2–tuples.

Definition 2. Let S be a linguistic term set and $\beta \in [0, g]$ a value supporting the result of a symbolic aggregation operation, then the 2–tuple that expresses the equivalent information to β is obtained with the following function:

$$\begin{aligned} \Delta : [0, g] &\longrightarrow S \times [-0.5, 0.5) \\ \Delta(\beta) &= (s_i, \alpha) \\ i &= round(\beta) \\ \alpha &= \beta - i \end{aligned} \tag{1}$$

where "round" is the usual round operation, s_i has the closest index label to β and α is the value of the symbolic translation.

Proposition 1. *Let $S = \{s_0, \ldots, s_g\}$ be a linguistic term set and (s_i, α) be a linguistic 2-tuple. There is always a function Δ^{-1}, such that, from a 2-tuple value it returns its equivalent numerical value $\beta \in [0, g] \subset \mathscr{R}$:*

$$\begin{aligned} \Delta^{-1} : S \times [-0.5, 0.5) &\longrightarrow [0, g] \\ \Delta^{-1}(s_i, \alpha) &= i + \alpha = \beta \end{aligned} \tag{2}$$

Remark 1. We should point out that a linguistic term can be seen as a linguistic 2–tuple by adding to it the value 0 as symbolic translation, $s_i \in S \Longrightarrow (s_i, 0)$.

The 2-tuples linguistic computational model presents different techniques to manage the linguistic information [24]:

1. *A 2-tuple comparison operator:* The comparison of linguistic information represented by 2-tuples is carried out according to an ordinary lexicographic order. Let (s_k, α_1) and (s_l, α_2) be two 2-tuples, with each one representing a counting of information:

   a. if $k < l$ then (s_k, α_1) is smaller than (s_l, α_2).
   b. if $k = l$ then
      i. if $\alpha_1 = \alpha_2$ then (s_k, α_1), (s_l, α_2) represent the same information.
      ii. if $\alpha_1 < \alpha_2$ then (s_k, α_1) is smaller than (s_l, α_2).
      iii. if $\alpha_1 > \alpha_2$ then (s_k, α_1) is bigger than (s_l, α_2).

2. *A 2-tuple negation operator:* It is defined as

$$Neg(s_i, \alpha) = \Delta(g - \Delta^{-1}(s_i, \alpha)). \tag{3}$$

3. *2-tuple aggregation operators:* Using the function Δ and Δ^{-1} any aggregation operator can be easily extended for dealing with linguistic 2-tuples, such as the Linguistic OWA operator [20], the weighted average operator, the OWA operator, etc., (see [24]).

2.2 Hierarchical Linguistic Contexts

In [25] the hierarchical linguistic contexts were introduced to improve the precision of processes of computing with words in multi-granular linguistic contexts [31]. In this work, we use them to manage the unbalanced fuzzy linguistic information.

A *Linguistic Hierarchy* is a set of levels, where each level represents a linguistic term set with different granularity from the remaining levels of the hierarchy. Each level is denoted as $l(t, n(t))$, where t is a number indicating the level of the hierarchy, and $n(t)$ is the cardinality of the linguistic term set of t. Moreover, we assume levels containing linguistic terms whose membership functions are triangular-shaped, uniformly and symmetrically distributed in $[0,1]$, and linguistic term sets having an odd value of granularity where the central label represents the value of *indifference*. A graphical example of a linguistic hierarchy is shown in Fig. 4.

The levels belonging to a linguistic hierarchy are ordered according to their granularity, i.e., for two consecutive levels t and $t+1$, $n(t+1) > n(t)$. Hence, the level $t+1$ could be considered as a refinement of the previous level t. Then, a linguistic hierarchy LH can be defined as the union of all levels t:

$$LH = \bigcup_t l(t, n(t)). \tag{4}$$

Given a LH, we denote as $S^{n(t)}$ the linguistic term set of LH corresponding to the level t of LH characterized by a granularity of uncertainty $n(t)$: $S^{n(t)} =$

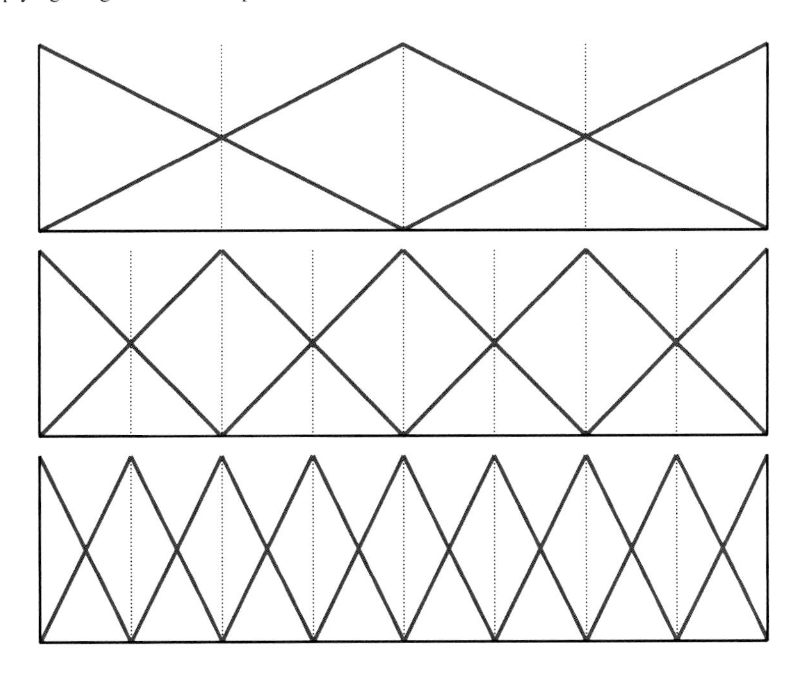

Fig. 4 Linguistic hierarchy of 3, 5 and 9 labels

$\{s_0^{n(t)}, \ldots, s_{n(t)-1}^{n(t)}\}$. Furthermore, the linguistic term set of the level $t+1$ is obtained from its predecessor as:

$$l(t, n(t)) \rightarrow l(t+1, 2 \cdot n(t) - 1). \tag{5}$$

Transformation functions between labels from different levels to make processes of computing with words in multigranular linguistic information contexts without loss of information were defined in [25].

Definition 3. [25] Let $LH = \bigcup_t l(t, n(t))$ be a linguistic hierarchy whose linguistic term sets are denoted as $S^{n(t)} = \{s_0^{n(t)}, \ldots, s_{n(t)-1}^{n(t)}\}$, and let us consider the 2-tuple fuzzy linguistic representation. The transformation function from a linguistic label in level t to a label in level t' is defined as $TF_{t'}^{t} : l(t, n(t)) \longrightarrow l(t', n(t'))$ such that

$$TF_{t'}^{t}(s_i^{n(t)}, \alpha^{n(t)}) = \Delta_{t'} \left(\frac{\Delta_t^{-1}(s_i^{n(t)}, \alpha^{n(t)}) \cdot (n(t') - 1)}{n(t) - 1} \right). \tag{6}$$

3 A Model to Manage Unbalanced Fuzzy Linguistic Information

Following those results presented in [17, 30], a model to manage unbalanced fuzzy linguistic term sets based on the linguistic 2-tuple model is presented. It carries out computational operations of unbalanced fuzzy linguistic information using the 2-tuple computational model and different levels of a LH. This model presents two components:

- A *representation model of unbalanced fuzzy linguistic information.*
- A *computational model of unbalanced fuzzy linguistic information.*

3.1 An Unbalanced Fuzzy Linguistic Representation Model

The procedure to represent unbalanced fuzzy linguistic information defined in [30] works as follows:

1. Find a level t^- of LH to represent the subset of linguistic terms S^L_{un} on the left of the mid linguistic term of unbalanced fuzzy linguistic term set S_{un}. This level of LH should support the distribution of the labels of S^L_{un} on the discourse universe.
2. Find a level t^+ of LH to represent the subset of linguistic terms S^R_{un} on the right of the mid linguistic term of S_{un}.
3. Represent the mid term of S_{un} using the mid terms of the levels t^- and t^+.

The problem appears when there does not exist a level t^- or t^+ in LH to represent S^L_{un} or S^R_{un}, respectively. Then, we propose to overcome this problem by applying the following algorithm, which is defined assuming that there does not exist t^-, as it happens with the unbalanced fuzzy linguistic term set given in Fig. 2:

1. Represent S^L_{un}:
   a. Identify the mid term of S^L_{un}, called S^L_{mid}. To do so, we have to observe the distribution of the labels of S^L_{un} on the discourse universe.
   b. Find a level t^-_2 of the left sets of LH^L to represent the left term subset of S^L_{un}, where LH^L represents the left part of LH.
   c. Find a level t^+_2 of the right sets of LH^L to represent the right term subset of S^L_{un}.
   d. Represent the mid term S^L_{mid} using the levels t^-_2 and t^+_2.
2. Find a level t^+ of LH to represent the subset of linguistic terms S^R_{un}.
3. Represent the mid term of S_{un} using the levels t^+ and t^+_2.

For example, applying this algorithm, the representation of the unbalanced fuzzy linguistic term set $S_{un} = \{N, VL, L, M, H, QH, VH, T\}$ shown in Fig. 2 with the linguistic hierarchy LH shown in Fig. 4 would be as it is shown in Fig. 5. In this example,

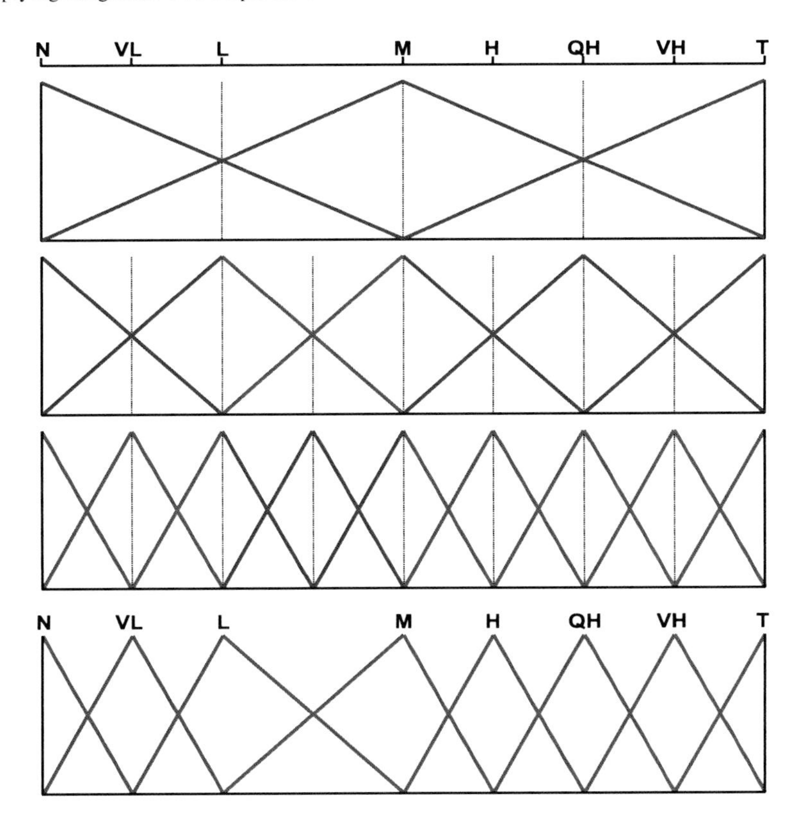

Fig. 5 Representation for an unbalanced term set of 8 labels

- $S^L_{un} = \{N, VL, L\}$,
- $S^L_{mid} = L$,
- $LH^L = \{s_0^{n(1)}\} \bigcup \{s_0^{n(2)}, s_1^{n(2)}\} \bigcup \{s_0^{n(3)}, s_1^{n(3)}, s_2^{n(3)}, s_3^{n(3)}\}$.

Thus, we have that $t_2^- = 3$, $t_2^+ = 2$, the mid label $S^L_{mid} = L$ (due to its position on the discourse universe) is represented using both levels, 3 and 2, and the mid term of S_{un} is represented using the levels 2 and 3.

3.2 An Unbalanced Fuzzy Linguistic Computational Model: Some Unbalanced Linguistic OWA Operators

In any fuzzy linguistic approach we need to define a computational model to manage and aggregate linguistic information. As in [24] we have to define three types of computation operators to deal with unbalanced fuzzy linguistic information: comparison operators, negation operator and aggregation operators. In an unbalanced linguistic context, previously to carry out any computation task of unbalanced fuzzy linguistic information we have to choose a level $t' \in \{t^-, t_2^-, t^+, t_2^+\}$, such that $n(t') = max\{n(t^-), n(t_2^-), n(t^+), n(t_2^+)\}$:

1. *An unbalanced linguistic comparison operator:* The comparison of linguistic information represented by two unbalanced linguistic 2-tuples $(s_k^{n(t)}, \alpha_1)$, $t \in \{t^-, t_2^-, t^+, t_2^+\}$, and $(s_l^{n(t)}, \alpha_2)$, $t \in \{t^-, t_2^-, t^+, t_2^+\}$ is similar to the usual comparison of two 2-tuples but acting on the values $TF_{t'}^t(s_k^{n(t)}, \alpha_1) = (s_v^{n(t')}, \beta_1)$ and $TF_{t'}^t(s_l^{n(t)}, \alpha_2) = (s_w^{n(t')}, \beta_2)$. Then, we have:

   a. if $v < w$ then $(s_v^{n(t')}, \beta_1)$ is smaller than $(s_w^{n(t')}, \beta_2)$.
   b. if $v = w$ then
      i. if $\beta_1 = \beta_2$ then $(s_v^{n(t')}, \beta_1)$, $(s_w^{n(t')}, \beta_2)$ represent the same information.
      ii. if $\beta_1 < \beta_2$ then $(s_v^{n(t')}, \beta_1)$ is smaller than $(s_w^{n(t')}, \beta_2)$.
      iii. if $\beta_1 > \beta_2$ then $(s_v^{n(t')}, \beta_1)$ is bigger than $(s_w^{n(t')}, \beta_2)$.

2. *An unbalanced linguistic 2-tuple negation operator.* Let $(s_k^{n(t)}, \alpha)$, $t \in \{t^-, t_2^-, t^+, t_2^+\}$ be an unbalanced linguistic 2-tuple, then:

$$NEG(s_k^{n(t)}, \alpha) = Neg(TF_{t''}^t(s_k^{n(t)}, \alpha)), \quad (7)$$

   where $t \neq t''$, $t'' \in \{t^-, t_2^-, t^+, t_2^+\}$.
3. *An unbalanced linguistic aggregation operator.* As aforementioned, in order to deal with unbalanced fuzzy linguistic information we have to represent it in a LH. Hence, any unbalanced linguistic aggregation operator must aggregate unbalanced fuzzy linguistic information by means of its representation in a LH. We use the aggregation processes designed in the 2-tuple computational model but acting on the unbalanced linguistic values transformed by means of $TF_{t'}^t$. Then, once a result is obtained, it is transformed to the correspondent level $t \in \{t^-, t_2^-, t^+, t_2^+\}$ by means of $TF_t^{t'}$ for expressing the result in the unbalanced linguistic term set S_{un}. In such a way, we define the following unbalanced linguistic OWA operators: the $LOWA_{un}$ operator which is an extension of the Linguistic Ordered Weighted Averaging operator proposed in [20] and the $ILOWA_{un}$ operator which is a linguistic extension of the Induced OWA operators [9, 50, 51, 52].

   - **Definition 4.** Let $\{(a_1, \alpha_1), \ldots, (a_m, \alpha_m)\}$ be a set of unbalanced linguistic assessments to aggregate, then the $LOWA_{un}$ operator ϕ_{un} is defined as:

$$\phi_{un}\{(a_1, \alpha_1), \ldots, (a_m, \alpha_m)\} = W \cdot B^T = C_{un}^m\{w_k, b_k,\ k = 1, \ldots, m\} =$$

$$w_1 \otimes b_1 \oplus (1 - w_1) \otimes C_{un}^{m-1}\{\beta_h, b_h,\ h = 2, \ldots, m\}$$

     where $b_i = (a_i, \alpha_i) \in (S^{n(t)} \times [-0.5, 0.5))$, $W = [w_1, \ldots, w_m]$, is a weighting vector, such that, $w_i \in [0,1]$ and $\sum_i w_i = 1$, $\beta_h = \frac{w_h}{\sum_2^m w_k}$, $h = \{2, 3 \ldots, m\}$, and B is the associated ordered unbalanced 2-tuple vector. Each element $b_i \in B$ is the i-th largest unbalanced 2-tuple in the collection $\{(a_1, \alpha_1), \ldots, (a_m, \alpha_m)\}$, and C_{un}^{m} is the convex combination operator of m unbalanced 2-tuples. If $w_j = 1$ and $w_i = 0$ with $i \neq j$, $\forall i,j$ the convex combination is defined as: $C_{un}^{m}\{w_i, b_i,\ i = 1,\ldots,m\} = b_j$. And if $m = 2$ then it is defined as:

$$C_{un}^{2}\{w_l, b_l,\ l = 1,2\} = w_1 \otimes b_j \oplus (1 - w_1) \otimes b_i = TF_{t}^{t'}(s_k^{n(t')}, \alpha)$$

where $(s_k^{n(t')}, \alpha) = \Delta(\lambda)$ and $\lambda = \Delta^{-1}(TF_{t'}^{t}(b_i)) + w_1 \cdot (\Delta^{-1}(TF_{t'}^{t}(b_j)) - \Delta^{-1}(TF_{t'}^{t}(b_i)))$, $b_j, b_i \in (S^{n(t)} \times [-0.5, 0.5))$, $(b_j \geq b_i)$, $\lambda \in [0, n(t') - 1]$, $t \in \{t^-, t_2^-, t^+, t_2^+\}$.

In[48] it was defined an expression to obtain W by means of a fuzzy linguistic non-decreasing quantifier Q [56]:

$$w_i = Q(i/m) - Q((i-1)/m),\ \ i = \{1,2,\ldots,m\}. \tag{8}$$

In such a way, it is possible to incorporate in the aggregation process the semantics of the fuzzy majority [32] represented by the quantifier. When the $LOWA_{un}$ operator uses a quantifier Q then it is called ϕ_{un}^{Q}.

- **Definition 5.** Let $\{(a_1, \alpha_1), \ldots, (a_m, \alpha_m)\}$ and $(u_1, \ldots, u_m)$ $u_i \in \mathscr{R}$ be a set of unbalanced linguistic assessments to aggregate and the set of values used to induce the ordering of the unbalanced linguistic assessments, respectively. Then, an $ILOWA_{un}$ operator Φ_{un} is defined as:

$$\Phi_{un}(\langle u_1, p_1\rangle, \ldots, \langle u_m, p_m\rangle) = TF_{t}^{t'}(\sum_{i=1}^{m} w_i \cdot \Delta^{-1}(TF_{t'}^{t} p_{\sigma(i)})), \tag{9}$$

being $p_i = (a_i, \alpha_i)$ and σ a permutation of $\{1, \ldots, m\}$ such that $u_{\sigma(i)} \geq u_{\sigma(i+1)}$, $\forall i = 1, \ldots, m-1$, i.e., $\langle u_{\sigma(i)}, p_{\sigma(i)}\rangle$ is the pair with $u_{\sigma(i)}$ the i-th highest value in the set $\{u_1, \ldots, u_m\}$.

In the above definition, the reordering of the set of values to be aggregated, $\{p_1, \ldots, p_n\}$, is induced by the reordering of the set of values $\{u_1, \ldots, u_n\}$ associated with them, which is based upon their magnitude. Due to this use of the set of values $\{u_1, \ldots, u_n\}$, Yager and Filev called them the values of an order inducing variable[9, 50, 51, 52]. A natural question in the definition of the unbalanced ILOWA operator is how to obtain the associated weighting vector. Following Yager's ideas on quantifier guided aggregation [49], we could compute the weighting vector of an IOWA operator using a linguistic quantifier Q [56] as:

$$w_i = Q\left(\frac{\sum_{k=1}^{i} u_{\sigma(k)}}{T}\right) - Q\left(\frac{\sum_{k=1}^{i-1} u_{\sigma(k)}}{T}\right), \tag{10}$$

being $T = \sum_{k=1}^{n} u_k$ and σ the permutation used to produce the ordering of the values to be aggregated.

4 A Consensus Model for GDM Problems Based on Unbalanced Linguistic OWA Operators

In this section, we present a consensus model defined for GDM problems with unbalanced fuzzy linguistic preference relations providing support to the experts to reach consensus during the process of making a decision. This consensus model presents the following main characteristics:

1. It is designed to guide the consensus process of unbalanced fuzzy linguistic GDM problems.
2. It is based on two consensus criteria: *consensus degrees* and *proximity measures*. The first ones are used to measure the agreement amongst all the experts, while the second ones are used to learn how close the collective and individual expert's preference are. Both consensus criteria are calculated at three different levels: pair of alternatives, alternatives and relation.
3. It uses unbalanced linguistic OWA operators to compute the above consensus criteria.
4. A *feedback mechanism* is defined using the above consensus criteria. It substitutes the moderator's actions, avoiding the possible subjectivity that he/she can introduce, and gives advice to the experts to find out the changes they need to make in their opinions in order to obtain the highest degree of consensus possible.

This consensus model presents three phases:

1. *Computing consensus degrees.*
2. *Controlling the consensus state.*
3. *Feedback mechanism.*

In the following subsections, we describe them in detail.

4.1 Computing Consensus Degrees

A GDM problem based on preference relations is classically defined as a decision situation where there are a set of experts, $E = \{e_1,\ldots,e_m\}$ $(m \geq 2)$, and a finite set of alternatives, $X = \{x_1,\ldots,x_n\}$ $(n \geq 2)$, and each expert e_i provides his/her preferences about X by means of a preference relation, $P_{e_i} \subset X \times X$, where the value $\mu_{P_{e_i}}(x_j,x_k) = p_i^{jk}$ is interpreted as the preference degree of the alternative x_j over x_k for e_i. In this paper, we deal with unbalanced fuzzy linguistic preference relations, i.e., $P_{e_i} = (p_i^{jk}) \in S_{un}$, and therefore, p_i^{jk} represents the preference of alternative x_j over alternative x_k for the expert e_i assessed on an unbalanced fuzzy linguistic term set S_{un}.

Then, consensus degrees are used to measure the current level of consensus in the decision process. As aforementioned, they are given at three different levels: pairs of alternatives, alternatives and relations. To calculate them, some similarity or coincidence function are required to obtain the level of agreement amongst all the experts [18, 22, 31]. Moreover, these similarity functions detect how far each individual expert is from the rest. In such a way, the computation of the consensus degrees is carried out as follows:

1. For each pair of experts (e_i, e_j) $(i = 1, \ldots, m-1,\ j = i+1, \ldots, m)$, an unbalanced linguistic similarity matrix, $SM_{ij} = (sm_{ij}^{lk})$, $sm_{ij}^{lk} \in (S^{n(t)} \times [-0.5, 0.5))$, is defined as

$$sm_{ij}^{lk} = NEG(TF_t^{t'}(\Delta(|\Delta^{-1}(TF_{t'}^{t}(p_i^{lk})) - \Delta^{-1}(TF_{t'}^{t}(p_j^{lk}))|))). \tag{11}$$

   being $p_i^{lk} = (s_v^{n(t)}, \alpha_1)$, $t \in \{t^-, t_2^-, t^+, t_2^+\}$, $p_j^{lk} = (s_w^{n(t)}, \alpha_2)$, $t \in \{t^-, t_2^-, t^+, t_2^+\}$, and $t' \in \{t^-, t_2^-, t^+, t_2^+\}$.
2. An unbalanced linguistic consensus matrix, $CM = (cm^{lk})$, is calculated by aggregating all the similarity matrices using the $LOWA_{un}$ operator ϕ_{un}^{Q} as the aggregation function:

$$cm^{lk} = \phi_{un}^{Q}(sm_{ij}^{lk},\ i = 1, \ldots, m-1,\ j = i+1, \ldots, m). \tag{12}$$

3. Once the consensus matrix, CM, is computed, we proceed to calculate the consensus degrees at the three different levels:

   a. **Level 1.** *Unbalanced linguistic consensus degree on pairs of alternatives.* The consensus degree on a pair of alternatives (x_l, x_k), called cp^{lk}, is defined to measure the consensus degree amongst all the experts on that pair of alternatives. The closer $\frac{\Delta^{-1}(cp^{lk})}{n(t)-1}$ to 1, the greater the agreement amongst all the experts on the pair of alternatives (x_l, x_k). Thus, this measure is used to identify those pairs of alternatives with a poor level of consensus and it coincides is with the element (l, k) of the consensus matrix CM:

$$cp^{lk} = cm^{lk};\ \forall\, l, k = \{1, 2, \ldots, n\} \wedge l \neq k. \tag{13}$$

   b. **Level 2.** *Unbalanced linguistic consensus degree on alternatives.* The consensus degree on an alternative x_l, called ca^l, is defined to measure the consensus degree amongst all the experts on that alternative:

$$ca^l = \phi_{un}^{Q}(cp^{l1}, \ldots, cp^{ln}). \tag{14}$$

   c. **Level 3.** *Unbalanced linguistic consensus degree on the relation.* The consensus degree on the relation, called cr, is defined to measure the global consensus degree amongst all the experts' opinions and is used by the consensus model to control the consensus situation. It is calculated as:

$$cr = \phi_{un}^{Q}(ca^1, \ldots, ca^n). \tag{15}$$

4.2 Controlling the Consensus State

The consensus state control process involves deciding if the feedback mechanism should be applied to provide advice to the experts or if the consensus process should be finished. To do so, a minimum consensus threshold, $\gamma \in [0,1]$, is fixed before applying the consensus model. When the consensus measure, cr, satisfies the minimum consensus threshold, γ, the consensus model finishes and the selection process is applied to obtain the solution. Additionally, the consensus model should avoid situations in which the global consensus measure may not converge to the minimum consensus threshold. To do that, a maximum number of rounds *MaxRounds* should be fixed and compared to the current number of round of the consensus model *NumRound*.

Then, the operation of the consensus state control process is as follows: Firstly, the global consensus measure, cr, is checked against the minimum consensus threshold, γ. If $\frac{\Delta^{-1}(cr)}{n(t)-1} > \gamma$, the consensus process finishes and the selection process is applied. Otherwise, it will check if the maximum number of rounds, *MaxRounds*, has been reached. If so, it finishes and the selection process is applied too, and if not, it activates the feedback mechanism.

4.3 Feedback Mechanism

If the global consensus measure is lower than the minimum consensus threshold then the experts' opinions must be modified. The goal of the feedback mechanism is to provide recommendations to support the experts in changing their opinions. The feedback mechanism uses proximity measures to identify those experts furthest away from the collective opinion. In the following, both the computation of the proximity measures and the production of advice are explained in detail.

4.3.1 Computation of Proximity Measures

These measures evaluate the agreement between the individual experts' opinions and the group opinion. To compute them for each expert, we need to obtain the collective unbalanced fuzzy linguistic preference relation, $P_{e_c} = (p_c^{lk})$, which summarizes preferences given by all the experts. We compute it by means of the aggregation of the set of individual unbalanced fuzzy linguistic preference relations $\{P_{e_1}, \ldots, P_{e_m}\}$ using an $IOWA_{un}$ operator, Φ_{un}^{Q}, which allows to obtain each collective preference degree p_{ik}^{c} according to the most consensual individual preference degrees using the consensus scores of each expert e_h for each pair of alternatives x_i and x_k

$$\{z_{ik}^1, z_{ik}^2, \ldots, z_{ik}^m\}$$

$$z_{ik}^h = \frac{\sum_{l=h+1}^{n}(\Delta^{-1}(sm_{ik}^{hl})/(n(t)-1)) + \sum_{l=1}^{h-1}(\Delta^{-1}(sm_{ik}^{lh})/(n(t)-1))}{n-1}. \quad (16)$$

as the values of the order inducing variable, i.e.,

$$p_{ik}^c = \Phi_{un}^Q(\langle z_{ik}^1, \bar{p}_{ik}^1\rangle, \ldots, \langle z_{ik}^m, \bar{p}_{ik}^m\rangle) = TF_t^{t'}(\Delta(\sum_{h=1}^{m} w_h \cdot \Delta^{-1}(TF_{t'}^t(\bar{p}_{ik}^{\sigma(h)})))), \quad (17)$$

where

- σ is a permutation of $\{1,\ldots,m\}$ such that $z_{ik}^{\sigma(h)} \geq z_{ik}^{\sigma(h+1)}, \forall h = 1,\ldots,m-1$, i.e., $\langle z_{ik}^{\sigma(h)}, \bar{p}_{ik}^{\sigma(h)}\rangle$ is the pair of the linguistic unbalanced 2-tuples with $z_{ik}^{\sigma(h)}$ the h-th highest consensus score in the set $\{z_{ik}^1 \ldots, z_{ik}^m\}$;
- the weighting vector is computed according to the following expression:

$$w_h = Q\left(\frac{\sum_{j=1}^{h} z_{ik}^{\sigma(j)}}{T}\right) - Q\left(\frac{\sum_{j=1}^{h-1} z_{ik}^{\sigma(j)}}{T}\right), \quad (18)$$

with $T = \sum_{j=1}^{m} z_{ik}^j$ and Q a fuzzy linguistic quantifier.

Once P_{e_c} is obtained, we can compute the proximity measures carrying out the following two steps:

1. For each expert, e_i, a proximity matrix, $PM_i = (pm_i^{lk})$, is obtained where

$$pm_i^{lk} = NEG(TF_t^{t'}(\Delta(\left|\Delta^{-1}(TF_{t'}^t(p_i^{lk})) - \Delta^{-1}(TF_{t'}^t(p_c^{lk}))\right|))). \quad (19)$$

being $p_i^{lk} = (s_v^{n(t)}, \alpha_1)$, $t \in \{t^-, t_2^-, t^+, t_2^+\}$, $p_c^{lk} = (s_w^{n(t)}, \alpha_2)$, $t \in \{t^-, t_2^-, t^+, t_2^+\}$, and $t' \in \{t^-, t_2^-, t^+, t_2^+\}$.

2. Computation of proximity measures at three different levels:

   a. **Level 1.** *Unbalanced linguistic proximity measure on pairs of alternatives.* The proximity measure of an expert e_i on a pair of alternatives (x_l, x_k) to the group's one, called pp_i^{lk}, is expressed by the element (l,k) of the proximity matrix PM_i:

$$pp_i^{lk} = pm_i^{lk};\ \forall\, l,k = 1,\ldots,n \wedge l \neq k. \quad (20)$$

   b. **Level 2.** *Unbalanced linguistic proximity measure on alternatives.* The proximity measure of an expert e_i on an alternative x_l to the group's one, called pa_i^l, is calculated as follows:

$$pa_i^l = \phi_{un}^Q(pp_i^{l1}, \ldots, pp_i^{ln}). \quad (21)$$

   c. **Level 3.** *Unbalanced linguistic proximity measure on the relation.* The proximity measure of an expert e_i on his/her unbalanced fuzzy linguistic preference relation to the group's one, called pr_i, is calculated as the average of all proximity measures on the alternatives:

$$pr_i = \phi_{un}^Q(pa_i^1, \ldots, pa_i^n). \quad (22)$$

Then, we can use them to provide advice to the experts to change their opinions and to find out which direction that change has to follow in order to obtain the highest degree of consensus possible.

4.3.2 Production of Advice

The production of advice to achieve a solution with the highest degree of consensus possible is carried out in two steps: *Identification rules* and *Direction rules*.

1. **Identification rules (IR).** We must identify the experts, alternatives and pairs of alternatives that are contributing less to reach a high degree of consensus and, therefore, should participate in the change process.

   a. *Identification rule of experts (IR.1)*. It identifies the set of experts that should receive advice on how to change some of their preference values. This set of experts, called $EXPCH$, that should change their opinions are those whose proximity measure on the relation, pr_i, is lower than the minimum consensus threshold γ. Therefore, the identification rule of experts, IR.1, is the following:

$$EXPCH = \{i \mid (\frac{\Delta^{-1}(pr_i)}{n(t)-1}) < \gamma\} \tag{23}$$

   b. *Identification rule of alternatives (IR.2)*. It identifies the alternatives whose associated assessments should be taken into account by the above experts in the change process of their preferences. This set of alternatives is denoted as ALT_i. The identification rule of alternatives, IR.2, is the following:

$$ALT_i = \{x_l \in X \mid (\frac{\Delta^{-1}(ca^l)}{n(t)-1}) < \gamma \wedge e_i \in EXPCH\} \tag{24}$$

   c. *Identification rule of pairs of alternatives (IR.3)*. It identifies the particular pairs of alternatives (x_l, x_k) whose respective associated assessments p_i^{lk} the expert e_i should change. This set of pairs of alternatives is denoted as $PALT_i$. The identification rule of pairs of alternatives, IR.3, is the following:

$$PALT_i = \{(x_l, x_k) \mid x_l \in ALT \wedge e_i \in EXPCH \wedge (\frac{\Delta^{-1}(pp_i^{lk})}{n(t)-1}) < \gamma\} \tag{25}$$

2. **Direction rules (DR).** We must find out the direction of the change to be recommended in each case, i.e., the direction of change to be applied to the preference assessment p_i^{lk}, with $(x_l, x_k) \in PALT_i$. To do this, we define the following two direction rules.

   a. *DR.1*. If $p_i^{lk} > p_c^{lk}$, the expert e_i should decrease the assessment associated to the pair of alternatives (x_l, x_k), i.e., p_i^{lk}.
   b. *DR.2*. If $p_i^{lk} < p_c^{lk}$, the expert e_i should increase the assessment associated to the pair of alternatives (x_l, x_k), i.e., p_i^{lk}.

Remark 2. These direction rules will not be produced when a decrease or increase are suggested to an assessment represented by the first or last label of the unbalanced linguistic term set, respectively.

5 Concluding Remarks

In this paper we have presented an application of LOWA operators in a consensus model for GDM problems with unbalanced fuzzy linguistic preference relations. We have defined two unbalanced LOWA operators to aggregate unbalanced linguistic information to compute the consensus criteria in order to guide the consensus process. In such a way we can incorporate different semantics in computation of the consensus criteria, as the concept of fuzzy majority or consensus semantics.

Acknowledgements. This paper has been developed with the financing of andalucian excellence project TIC05299, Feder Funds in FUZZYLING and FUZZYLING-II projects (TIN2007-61079 and TIN2010-17876) and PETRI project (PET 2007-0460).

References

1. Bezdek, J., Spillman, B., Spillman, R.: Fuzzy measures of preferences and consensus in group decision making. In: Proc. 1977 IEEE Conf. on Decision and Control, pp. 1303–1309 (1977)
2. Bezdek, J., Spillman, B., Spillman, R.: A fuzzy relation space for group decision theory. Fuzzy Sets and Systems 1, 255–268 (1978)
3. Bilgiç, T.: Interval–valued preference structures. European Journal of Operational Research 105(1), 162–183 (1998)
4. Cabrerizo, F.J., Alonso, S., Herrera-Viedma, E.: A consensus model for group decision making problems with unbalanced fuzzy linguistic information. International Journal of Information Technology & Decision Making 8(1), 109–131 (2009)
5. Chen, S.J., Hwang, C.L.: Fuzzy multiple attributive decision making: Theory and its applications. Springer, Berlín (1992)
6. Chiclana, F., Herrera, F., Herrera-Viedma, E.: Integrating three representation models in fuzzy multipurpose decision making based on fuzzy preference relations. Fuzzy Sets and Systems 97(1), 33–48 (1998)
7. Chiclana, F., Herrera, F., Herrera-Viedma, E.: A note on the internal consistency of various preference representations. Fuzzy Sets and Systems 131(1), 75–78 (2002)
8. Chiclana, F., Herrera, F., Herrera-Viedma, E., Martínez, L.: A note on the reciprocity in the aggregation of fuzzy preference relations using owa operators. Fuzzy Sets and Systems 137(1), 71–83 (2004)
9. Chiclana, F., Herrera-Viedma, E., Herrera, F., Alonso, S.: Some induced ordered weighted averaging operators and their use for solving group decision-making problems based on fuzzy preference relations. European Journal of Operational Research 182(1), 383–399 (2007)
10. Chiclana, F., Herrera-Viedma, E., Alonso, S., Herrera, F.: Cardinal Consistency of Reciprocal Preference Relations: A Characterization of Multiplicative Transitivity. IEEE Transactions on Fuzzy Systems 17(1), 14–23 (2009)

11. Chiclana, F., Mata, F., Martínez, L., Herrera-Viedma, E., Alonso, S.: Integration of a consistency control module within a consensus model. International Journal of Uncertainty, Fuzziness and Knowledge Based Systems 16(1), 35–53 (2008)
12. Cialdini, R.B.: Influence: science and practice, 4th edn. Allyn and Bacon, Boston (2001)
13. Fedrizzi, M., Fedrizzi, M., Marques Pereira, R.A.: On the issue of consistency in dynamical consensual aggregation. In: Bouchon-Meunier, B., Gutierrez-Ros, J., Magdalena, L., Yager, R.R. (eds.) Technologies for Constructing Intelligent Systems 1: Tasks. Studies in Fuzziness and Soft Computing, pp. 129–138. Physica-Verlag, Berlin (2002)
14. Fodor, J., Roubens, M.: Fuzzy preference modelling and multicriteria decision support. Kluwer, Dordrecht (1994)
15. Herrera, F., Herrera–Viedma, E.: Linguistic decision analysis: steps for solving decisions problems under linguistic information. Fuzzy Sets and Systems 115, 67–82 (2000)
16. Herrera, F., Herrera-Viedma, E., Chiclana, F.: Multiperson decision making based on multiplicative preference relations. European Journal of Operational Research 129, 372–385 (2001)
17. Herrera, F., Herrera-Viedma, E., Martínez, L.: A fuzzy linguistic methodology to deal with unbalanced linguistic term sets. IEEE Transactions on Fuzzy Systems 16(2), 354–370 (2008)
18. Herrera, F., Herrera-Viedma, E., Verdegay, J.L.: A model of consensus in group decision making under linguistic assessments. Fuzzy Sets and Systems 78(1), 73–87 (1996)
19. Herrera, F., Herrera-Viedma, E., Verdegay, J.L.: A linguistic decision process in group decision making. Group Decision and Negotiation 5, 165–176 (1996)
20. Herrera, F., Herrera-Viedma, E., Verdegay, J.L.: Direct approach processes in group decision making using linguistic owa operators. Fuzzy Sets and Systems 79, 175–190 (1996)
21. Herrera, F., Herrera-Viedma, E., Verdegay, J.L.: A rational consensus model in group decision making using linguistic assessments. Fuzzy Sets and Systems 88(1), 31–49 (1997)
22. Herrera, F., Herrera-Viedma, E., Verdegay, J.L.: Linguistic measures based on fuzzy coincidence for reaching consensus in group decision making. Int. J. of Approximate Reasoning 16(3-4), 309–334 (1997)
23. Herrera, F., Herrera-Viedma, E., Verdegay, J.L.: Choice processes for non-homogeneous group decision making in linguistic setting. Fuzzy Sets and Systems 94, 287–308 (1998)
24. Herrera, F., Martínez, L.: A 2-tuple fuzzy linguistic representation model for computing with words. IEEE Transactions on Fuzzy Systems 8(6), 746–752 (2000)
25. Herrera, F., Martínez, L.: A model based on linguistic 2-tuples for dealing with multigranularity hierarchical linguistic contexts in multiexpert decision-making. IEEE Transactions on Systems, Man and Cybernetics. Part B: Cybernetics 31(2), 227–234 (2001)
26. Herrera-Viedma, E., Alonso, S., Herrera, F., Chiclana, F.: A consensus model for group decision making with incomplete fuzzy preference relations. IEEE Transactions on Fuzzy Systems 15(5), 863–877 (2007)
27. Herrera-Viedma, E., Chiclana, F., Herrera, F., Alonso, S.: A group decision-making model with incomplete fuzzy preference relations based on additive consistency. IEEE Transactions on Systems, Man and Cybernetics, Part B, Cybernetics 37(1), 176–189 (2007)
28. Herrera-Viedma, E., Herrera, F., Chiclana, F.: A consensus model for multiperson decision making with different preference structures. IEEE Transactions on Systems, Man and Cybernetics-Part A: Systems and Humans 32(3), 394–402 (2002)
29. Herrera-Viedma, E., Herrera, F., Chiclana, F., Luque, M.: Some issues on consistency of fuzzy preference relations. European Journal of Operational Research 154(1), 98–109 (2004)

30. Herrera-Viedma, E., López-Herrera, A.G.: A model of information retrieval system with unbalanced fuzzy linguistic information. International Journal of Intelligent Systems 22(11), 1197–1214 (2007)
31. Herrera-Viedma, E., Martínez, L., Mata, F., Chiclana, F.: A consensus support system model for group decision-making problems with multi-granular linguistic preference relations. IEEE Transaction on Fuzzy Systems 13(5), 644–658 (2005)
32. Kacprzyk, J.: Group decision making with a fuzzy linguistic majority. Fuzzy Sets and Systems 18, 105–118 (1986)
33. Kacprzyk, J.: On some fuzzy cores and soft consensus measures in group decision making. In: Bezdek, J. (ed.) The Analysis of Fuzzy Information, pp. 119–130. CRC Press, Boca Raton (1987)
34. Kacprzyk, J., Fedrizzi, M.: Soft consensus measure for monitoring real consensus reaching processes under fuzzy preferences. Control Cybernet 15, 309–323 (1986)
35. Kacprzyk, J., Fedrizzi, M.: A soft measure of consensus in the setting of partial (fuzzy) preferences. European Journal of Operational Research 34, 316–323 (1988)
36. Kacprzyk, J., Fedrizzi, M., Nurmi, H.: Group decision making and consensus under fuzzy preferences and fuzzy majority. Fuzzy Sets and Systems 49(1), 21–31 (1992)
37. Kacprzyk, J., Nurmi, H., Fedrizzi, M.: Consensus under fuzziness. Kluwer Academic Publishers, Boston (1997)
38. Levrat, E., Voisin, A., Bombardier, S., Bremont, J.: Subjective evaluation of car seat comfort with fuzzy set techniques. International Journal of Intelligent Systems 12(12), 891–913 (1997)
39. Mata, F., Martínez, L., Herrera-Viedma, E.: An adaptive consensus support model for group decision making problems in a multi-granular fuzzy linguistic context. IEEE Transaction on Fuzzy Systems 17(2), 279–290 (2009)
40. Marakas, G.H.: Decision Support Systems in the 21th Century, 2nd edn. Pearson Education, Upper Saddle River (2003)
41. Roubens, M.: Fuzzy sets and decision analysis. Fuzzy Sets and Systems 90(2), 199–206 (1997)
42. Saaty, T.L.: The analytic hierarchy process. MacGraw-Hill, New York (1980)
43. Spillman, B., Bezdek, J., Spillman, R.: Coalition analysis with fuzzy sets. Kybernetes 8, 203–211 (1979)
44. Spillman, B., Spillman, R., Bezdek, J.: A fuzzy analysis of consensus in small groups. In: Wang, P.P., Chang, S.K. (eds.) Fuzzy Automata and Decision Processes, pp. 331–356. North-Holland, Amsterdam (1980)
45. Tanino, T.: Fuzzy preference orderings in group decision making. Fuzzy Sets and Systems 12, 117–131 (1984)
46. Xu, Z.S.: An approach to group decision making based on incomplete linguistic preference relations. International Journal of Information Technology & Decision Making 4(1), 153–160 (2005)
47. Xu, Z.S.: A method for multiple attribute decision making with incomplete weight information in linguistic setting. Knowledge-Based Systems 20(8), 719–725 (2007)
48. Yager, R.R.: On Ordered Weighted Averaging Aggregation Operators in Multicriteria Decision Making. IEEE Transactions on Systems, Man, and Cybernetics 18, 183–190 (1988)
49. Yager, R.R.: Quantifier guided aggregation using OWA operators. International Journal of Intelligent Systems 11(1), 49–73 (1996)
50. Yager, R.R.: Induced aggregation operators. Fuzzy Sets and Systems 137(1), 59–69 (2003)

51. Yager, R.R., Filev, D.P.: Operations for granular computing: mixing words and numbers. In: Proceedings of the FUZZ-IEEE World Congress on Computational Intelligence, Anchorage, pp. 123–128 (1998)
52. Yager, R.R., Filev, D.P.: Induced ordered weighted averaging operators. IEEE Transactions on Systems, Man, and Cybernetics 29(2), 141–150 (1999)
53. Zadeh, L.A.: The concept of a linguistic variable and its applications to approximate reasoning. Part I. Information Sciencies 8(3), 199–249 (1975)
54. Zadeh, L.A.: The concept of a linguistic variable and its applications to approximate reasoning. Part II. Information Sciencies 8(4), 301–357 (1975)
55. Zadeh, L.A.: The concept of a linguistic variable and its applications to approximate reasoning. Part III. Information Sciencies 9(1), 43–80 (1975)
56. Zadeh, L.A.: A computational approach to fuzzy quantifiers in natural languages. Computers and Mathematics with Applications 9(1), 149–184 (1983)

Part II
Applications

Fusion Strategies Based on the OWA Operator in Environmental Applications

G. Bordogna*, M. Boschetti, A. Brivio, P. Carrara,
M. Pagani, and D. Stroppiana

Abstract. Ill-known environmental phenomena are often modeled by means of multisource spatial data fusion. Generally, these fusion strategies have to cope with distinct kinds of uncertainty, related to the ill-defined knowledge of the phenomenon, the lack of classified data, the distinct trust of the information sources, the imprecision of the observed variables. In this chapter we discuss the advantage of modeling multisource spatial data fusion in the environmental field based on the OWA operator, and overview two applications. The first application is aimed at defining an environmental indicator of anomaly at continental scale based on a fusion of partial hints of evidence of anomaly. The second application computes seismic hazard maps based on a consensual fusion strategy defined by an extended OWA operator that accounts for data imprecision, and reliability of the data sources. In particular, the proposed fusion function models a consensual dynamics and is parameterized so as to consider a varying spatial neighborhood of the data to fuse.

1 Introduction

The modeling of ill-known environmental phenomena, when training data are unavailable, is often based on the accumulation of hints of evidence, i.e., contributing factors [24]. To this aim multisource spatial data fusion strategies offer an intuitive paradigm to compute maps of global evidence of the hazard or of the susceptibility of a location to the modeled phenomenon. Spatial data fusion consists of a data integration process that combines spatial data from multiple sources to generate spatial data of "higher quality", carrying information not available from

G. Bordogna · M. Pagani
CNR IDPA, via Pasubio 5, 24044 Dalmine (BG), Italy
e-mail: gloria.bordogna@idpa.cnr.it

M. Boschetti · A. Brivio · P. Carrara · D. Stroppiana
CNR IREA, via Bassini 15, 20133 Milano, Italy

* Corresponding author.

R.R. Yager et al. (Eds.): Recent Developments in the OWA Operators, STUDFUZZ 265, pp. 189–207.
springerlink.com

any individual source [10][37]. "Higher quality" can be intended a better description of a spatial feature, or a better signal, or even the opportunity to take a better decision. This last interpretation is the one that we model in this chapter.

We consider the fusion of spatial data that are independently produced by both software models and human experts, hereafter named sources.

Current GISs are inadequate to support the experts in modeling multisource spatial data fusion affected by uncertainty because flexible decision strategies can hardly be defined by using the available aggregation operators [7][21][27]. The fusion operations that GISs offer are generally based on Boolean logic, basically maps overlay and weighted linear combination [24]. Further, these systems do not represent and manage the imprecision and uncertainty of the data, allowing to associate only precise values with each spatial unit. These are some of the motivations for developing software applications allowing users/experts to define and then execute their personalized flexible fusion strategy on geographic data, possibly affected by imperfection [4][2][11][24][36]. In this context, the implementation of multisource fusion strategies based on the OWA operator had drawn great success [3][24][9][18][25]. OWA operators have been indicated as appropriate tools for spatial data fusion since they are a family of mean-like operators that allow realizing distinct fusion strategies [24][29][42]. These approaches are appealing since they can be defined to model a variety of real situations in a flexible way [6][9][18].

The definition of a spatial data fusion strategy first requires to represent the input data in a common space, and then to define the way in which these data must be combined to generate the output. The first step is necessary when the input data are heterogeneous, as in the case of multisource data, and characterized by either different spatial and temporal resolutions, measurement errors, range of values, or distinct reliability of their sources [36]. These are all causes of imprecision and uncertainty that must be appropriately dealt with when fusing spatial information. The second step represents the core of the process, that is, the stage where we synthesize the spatial information available in order to generate spatial data of higher semantic level. Of particular importance within this context are soft fusion strategies defined by linguistic quantifiers [39][43] associated with the concept of fuzzy majority [19] and implemented by an OWA operator [40][41]. Furthermore, there are situations in which the fusion has the objective of synthesizing the results independently produced by the sources, i.e., the experts or the models, by also taking into account their agreements, i.e., the data values variability within a local or global spatial neighborhood, so as to reduce possible semantic errors. In this case, the fusion operator is a context dependent operator according to the classification given in [2]. Moreover, it can be necessary to take into account the sources trust scores, that is, their reliability or presumed credit. These characteristics are very important in the geographic context where information may come from distinct sources with very distinct reliability and acquisition characteristics.

In this chapter we first discuss the problem of modeling environmental phenomena within distinct computational frameworks. Then we illustrate the application of OWA for modeling environmental indicators at continental scale. Finally, we introduce the consensual OWA and its application to produce seismic hazard maps. The conclusions summarize the main contents of the chapter.

2 Modeling Environmental Phenomena

Environmental phenomena present considerable difficulties for modelers mainly due to the poor knowledge of the subsystems and interrelationships, the difficulties of carrying out experiments, the inability to accurately measure the internal system variables and to perturb the system to observe the effects. For these reasons, indicators can be defined, that take Boolean or gradual values in a range, to help conveying information on the state of the phenomenon by reducing a large quantity of data to its simplest form [34][22].

The choice of a mathematical framework for formalizing a model of an environmental phenomenon strongly depends on the quantity, quality, and type of data available, on the type of knowledge of the phenomenon, and the type of process one wants to define, i.e., a transparent (white box) or opaque (black box) modelling.

First of all let us consider the availability of data from which to derive an indicator: they can be either statistically meaningful, providing good quantity of high quality classified data for training and testing purposes or not. This aspect basically determines the method to choose.

Secondly, data can be characterized by distinct quality, defined in terms of both reliability or trust or credibility of the sources of the data and embedded imperfection of the data [4].

Finally, at high level, the data can be of three distinct types: numeric, ordinal, nominal or categorical.

On the other side, the knowledge of the causes or factors and processes influencing or determining the phenomenon can be ill-defined at distinct levels: there are situations in which heuristic vague rules in the form "*if condition and condition ...and condition then conclusion*" determining the occurrence or hint of the phenomenon can be specified so that the interrelationships between factors are qualitatively made explicit, and situations in which the knowledge of the phenomenon or of the hints carrying to the phenomenon is solely based on the reinforcement of evidence carried by distinct data. The worst situation is when the knowledge is completely missing: in this case the phenomenon can be forecasted provided that large quantity of high quality classified data are available.

The mathematical framework for formalizing the process of integration of multisource spatial data can be based on either a black box or a white box paradigm, according to the fact that it provides an implicit or explicit model of the phenomenon. Neural network approaches falls into the first type while exploratory data analysis, data mining and integration techniques are of second type since they allow to explicit association rules between data values. A rich survey on multicriteria integration approaches can be found in [2][36][24].

Probability theory is the most commonly adopted mathematical framework where, often, input data are combined by applying Bayesian rules [20][17]. The main drawback of this approach is that source data are considered independent, an assumption that is rarely true in the analysis of environmental variables. It also models a very strict integration where all factors must contribute to some degree.

The Dempster-Shafer theory of evidence [12] [31] constrains the data integration based on the Dempster-Shafer rule, that generates errors when the degree of conflict among the single sources of evidence that support each of the considered hypotheses becomes relevant [1]. Moreover, this approach is too rigid to model phenomena flexibly and in a robust way so as to be able to generate hazard or susceptivity maps even when some sources of evidence are missing and the integration criterion is ill-defined.

Neural networks are a black box paradigm for modeling complex processes such as those involved in pattern recognition [38]. Their applicability is limited both by the kind of data that can be managed, numeric type data, and even more importantly, by the need for large, high quality classified data sets to train the network which are scarce in many real applications, and are certainly lacking in many situations of environmental field.

CART techniques [5] are an alternative white box paradigm. They are based on the automatic construction from data analysis of decision trees encoding a distribution of the dependent variable in terms of the predictors (factors). They show several advantages since they provide a simple and easily understandable interpretation; they deal with any type of data, do not need to normalize data, and are robust and perform well. Nevertheless, they need classified data, and often generate complex and artificial models due to the Boolean logic they apply. In fact, the splitting of the tree's branches is performed based on the satisfaction of Boolean selection criteria, and do not provide any means to compensate among partial satisfactions of the criteria.

Fuzzy set theory makes it possible to define white box paradigms by flexibly modeling the expert ill-defined knowledge of the integration strategy by means of fuzzy aggregation operators: these operators can be defined with a severe, compromise, or indulgent behavior, corresponding with the modeling of a risk taken, a risk trade-off, or a risk adverse decision attitude respectively [29][27][42]. These approaches do not need statistically meaningful data since they do not make severe assumption on data distributions, nor need classified data.

For these reasons fuzzy approaches are appealing in many real cases [32][33][35]. Furthermore, the advantage of a white paradigm, once it has been validated, is also to enrich the knowledge on the causes of a phenomenon, and then to allow for a virtuous feedback.

3 Modeling an Environmental Anomaly Indicator by the OWA

Problem Statement

This section summarises the application of the OWA operator to fuse multisource spatial data for defining an environmental Anomaly Indicator (AI) at large scale [8][34].

Monitoring environmental systems at continental scale involves the analysis of complex, multi-disciplinary and large-scale phenomena. Indicators are often used in these cases. However, the definition of an indicator relies on the possibility of describing the system under analysis either through a model or the establishment of a quantitative linkage with a process or response variable of interest.

The task of formalising an environmental model becomes harder over large areas (i.e. the continent or the globe) where no agreement has been reached on the dynamics of the phenomena involved. Moreover, the increasing availability of global datasets, mainly with the contribution of Earth Observations (EOs), poses the problem of their aggregation in synthetic indicators that picture, in a simplified but broader way, the state of the environment.

Only a few experiments have been published on the application of fuzzy set approaches for the integration of different kinds of observations and the results obtained in the development of environmental indicators still rely on expert knowledge in order to derive the membership functions [3][26][30]. However, this seems to be feasible only at local and regional scales: at the continental scale, expertise formalisation becomes controversial thus requiring more robust approaches to knowledge representation.

The objective of a research carried out in the framework of the Observatory for Land cover and Forest change (OLF) of the GeoLand Project (EU 6th Framework Program) was to formalise a synthetic indicator for assessing environmental status at continental scale, reducing the need of disputed models that formalise interrelationships among factors influential on the status of the environment and in condition of no reference information. The study proposed a fuzzy Anomaly Indicator (AI), designed for periodical assessment of the vegetation component of the environment, that is based on a simplified model (reinforcement of evidence) exploiting time series of EO data rather than on the formalisation of complex and debatable models. Flexibility characterises the system at different levels:

- First, the contribution of factors are quantified in an automated way, although manual customization is possible when more knowledge is available.
- Second, the expert can choose the fusion strategy by defining the semantics of the fuzzy majority used for evaluating the AI, and by exploiting OWA operators to implement the correspondent fusion function [40].
- Finally, the system can cope with missing data, because the formalisation of the fusion strategy is done through a relative linguistic quantifier, so that the AI can be computed even when a subset of the contributing factors is available; redundancy is useful when some factors are missing.

Workflow to Build the Anomaly Indicator

Fig. 1 shows the work flow of the computation of the AI. The input spatial data are in raster format with a common spatial resolution and geo-reference system.

The AI is built by aggregating the evidence of anomaly represented by scores that are computed as degrees of satisfaction of soft constraints; these soft constraints (in Fig 1 named factor score membership functions) are defined for a set of contributing factors that are deemed to be influential on the status of the environment.

The AI is therefore computed pixel by pixel through the aggregation of N factor scores (p_k, with k = 1,. . .,N) eventually weighted based on their relative importance degrees (i_k, with k = 1,. . .,N) that reflect the significance of the contributing factors in determining the AI.

The definition of the k-th contributing factor value is derived as the difference (hereafter named Δ_k) between the current observation of the k-th variable and LTA_k , that is the long-term average of the k-th variable (Fig. 1, Step 1).

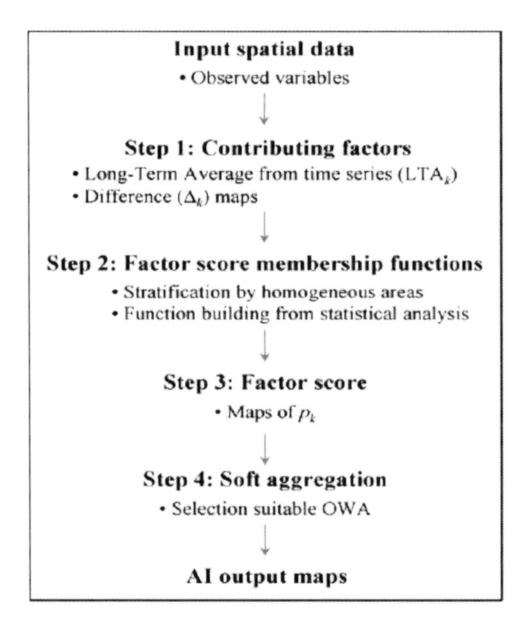

Fig. 1 Flow chart: the steps of the procedure designed to compute the AI

The factor scores, that is the contributions of each factor values Δ_k to the anomaly evaluation, are constrained through factor score membership functions. These functions are formalised, for each factor, from the statistical analysis of the Δ_k historical time series over homogeneous areas identified based on a land cover map (Fig. 1, Step 2). This step relies on stratification, based on the land cover spatial distribution, for identifying areas where ecological conditions can be assumed homogeneous and any deviation from LTA_k can be ascribed to actual changes of the land cover conditions. A membership function is derived from the frequency histogram of the factor's Δ_k for each homogeneous area, thus assuring also a statistically reliable cardinality (all pixels from a land cover class are exploited to build a single membership function). Fig. 2 shows an example for a single factor

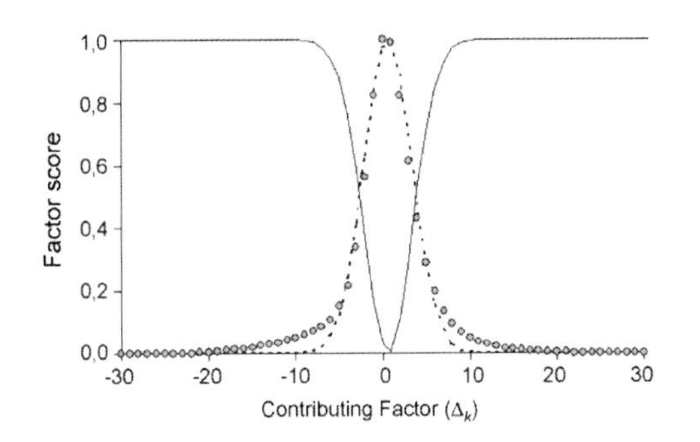

Fig. 2 One factor score membership function. The continuous line identifies the anomaly derived from the frequency histogram of the Δ_k time series (circle markers) fitted by a Gaussian curve (dashed line) for one homogeneous area and one 10-day period.

and land cover class. The shape of the function depends on the frequency distribution of the Δ_k values, for a given period, within the homogeneous area; areas subject to a high variability in both time and space are characterised by wider curves. A wider curve implies more tolerance, i.e. low scores are assigned also to greater deviations from the most frequent values. Expert knowledge, if available, could be used for the definition of shape of the membership functions associated with the soft constraints. However, although intuitive at first, such a customization would require the identification of the factor's contribution to environmental conditions and this is unlikely to be consistent over a continent and throughout the ecosystems. This type of customization would be feasible and more suitable for local studies.

The factor scores p_k (real number in the range [0,1]), are derived by computing the degrees of satisfaction of the membership functions (Fig. 1, Step 3) and they are aggregated using an OWA operator (Fig. 1, Step 4) by taking into account (if it exists) the importance i_k of k-th factor. In the proposed approach the OWA implements a soft linguistic quantifier such as "*most of*", "*at least two*", etc. By varying the quantifier, the OWA aggregation changes as well.

All the methodological steps summarised above are applied to grid spatial data and the AI computation is performed pixel by pixel with a 10-day time step, thus providing periodic output maps.

Evaluating an Environmental Anomaly Indicator in Africa

The approach described above has been applied to obtain monthly AI maps of Africa for the period 1996–2002. The AI was built by using vegetation phenology (start of greenness, season length, and season peak) and rainfall estimates as contributing factors.

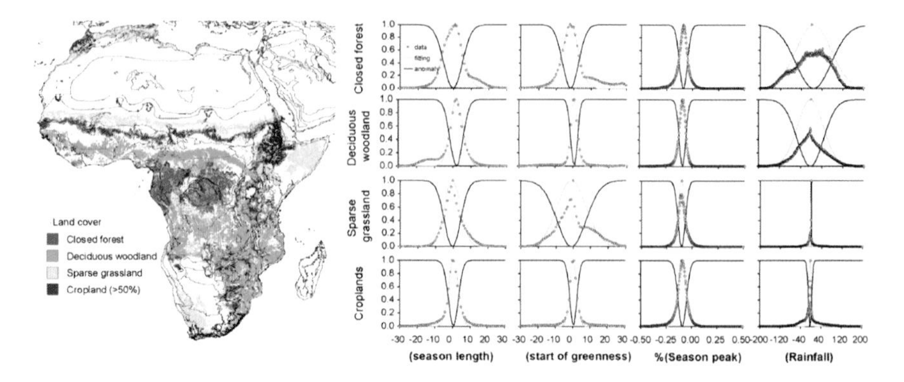

Fig. 3 Examples of soft constraints to evaluate the factor scores. The membership functions are presented only for the four land cover classes shown on the left. The histogram of the Δ_k values for the range 1–10 April in the considered years, derived from the analysis of the historical dataset, is shown by the circle markers, the fitting Gaussian function by the grey continuous line and the membership function by the black continuous line.

Figure 3 illustrates the membership functions produced in the case study for the different factors in some land cover classes. A narrow function implies that even a small departure from the LTA_k can be labelled as highly anomalous. On the contrary, wide functions imply that the class is characterised by high variability that could be due to the frequent occurrence of disturbance. In the cropland areas (last row in the figure) a longer season is likely to have a greater impact (more anomalous) compared to closed forest where on average the season is longer. Note that fitting accuracy varies with both the land cover class and the contributing factor. Fitting accuracy was quantified using the root mean squared error (RMSE) computed for each 10-day period and vegetation class.

The AI maps (Figure 4) were produced by averaging to monthly values 10-day factor score maps of each contributing factor and then by aggregating them. Due to the absence of an expert who could formalise a widely accepted model of factor influence on the AI at the continental scale, in a first test a neutral operator $OWA_{\hat{W}}$ was chosen corresponding to the arithmetic average. The average RMSE values were used to evaluate the importance degrees $\hat{I}$ to be associated with the factors in their aggregation. Then the importance value was assigned based on the accuracy of the interpolation of the histograms: the factor that is characterised by the least accurate interpolation is assigned a lower importance in order to reduce its influence in the AI computation. This way the importance of the factor accounts for the reliability of the data from which the factor was derived. This step was deemed necessary to avoid high anomaly scores of a single factor.

The resulting AI maps show that, in the period 1996–2002, the most anomalous areas of the continent were Southern Africa, the Horn of Africa and the sub-Saharan regions. The temporal profiles extracted over a set of five regions of interest in these key areas pointed out that significant anomalies occurred in 1997–1998 and 2000–2001 when the indicator is characterised by particularly high

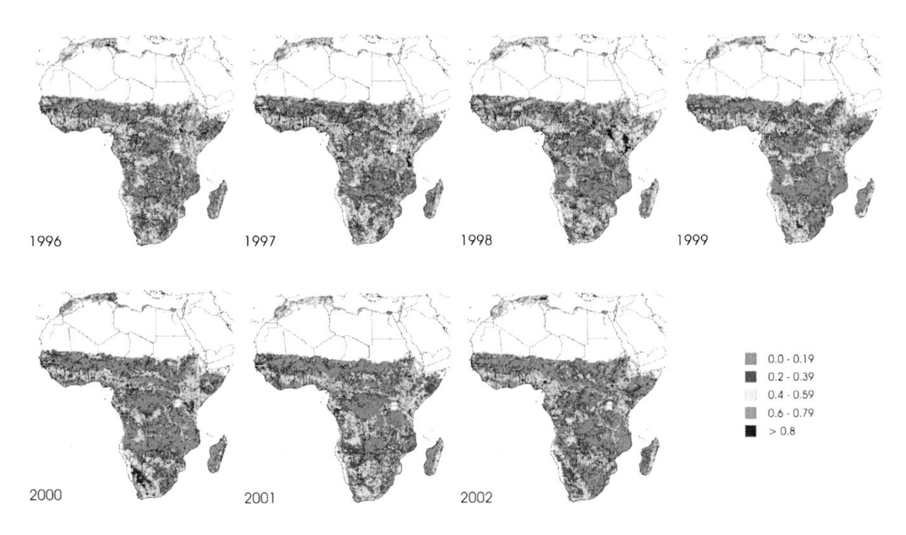

Fig. 4 Maps of Anomaly Indicator (AI) showing inter-annual variation for October during the seven years period 1996 - 2002.

values (AI > 0.8). Climatologists identified the same regions as the most sensitive to El Niño Southern Oscillation-effects over the continent.

To test the flexibility and robustness of the approach, experiments were carried out to analyse the effects of both distinct fusion strategies on the AI, and missing data by excluding one of the input factors [8]. In the first experiment, all the four contributing factors were considered, but they were aggregated by OWA operators implementing three distinct strategies (representing cautious, neutral and alarming attitudes). Not surprisingly, when aggregating the factors with an alarmist fusion the AI values are generally higher than by the neutral attitude, and they are always higher than those by the cautious fusion. More interesting was to analyse the correlation between distinct contributing factors and the AI values. The cautious fusion criterion produced an AI trend that was deeply influenced by 'peak of greenness', due to the high level of importance of this factor, derived automatically. It is interesting to observe that high anomalous values (> 0.8) were indicated by all approaches for the year 1998, supporting the results of the first test (see also the maps of figure 4), and showing that anomalous conditions were experienced in that period, whatever the fusion attitude. The comparison of the AI maps that were generated by applying distinct soft fusion strategies was useful to identify areas with stable AI values, thus reinforcing the accumulation of evidence approach.

In the second experiment the factor 'peak of greenness' was excluded and the remaining factors were aggregated with a neutral attitude. No modification of the score functions was necessary, and the importance degrees were automatically re-normalised by the system. Also in this case the system was able to produce an AI with a trend similar to the neutral output obtained by aggregating four factors. Moreover, it was still able to highlight the strong anomaly that occurred in 1998.

In conclusion, the approach proved to be flexible, robust with respect to missing data, open to the interaction with the analyst; it does not rely on disputed models, and it can be operationally implemented. It is based on the concept of the reinforcement of evidence to highlight areas where anomalies occur and can therefore help the analyst identify areas for further investigation with higher resolution data, ground field campaigns and/or more locally available models or expert knowledge.

4 Consensual Fusion of Imprecise Spatial Data by a Generalized OWA

Problem Statement

This section summarizes the definition of a flexible consensual fusion function based on a generalization of the OWA operator and its application for the computation of seismic hazard maps.

We have several decision maps, represented by grids of pixels, possibly with imprecise or fuzzy values, generated by n competitive models, software tools, or human experts (the sources), each one characterized by a distinct trust score (representing its reliability, presumed credit), and we want to fuse their possibly contradictory values so as to achieve a more robust consensual decision map.

We consider n grids with the same spatial reference and resolution. In the following we indicate with $v_1,...v_n$ the n values in the pixel with same coordinates in the n grids.

We assume that $v_1...,v_n$ have the same basic domain, that can be either numeric discrete, numeric continuous, or ordinal. Further, each value v_i can be an imprecise value, i.e., an interval on the basic domain, or a fuzzy value, i.e., a convex possibility distribution. This is a very common situation of environmental data, that are affected by measurement or systematic errors.

We want to model the following fusion criteria:

- the greater the trust score of the source, the more the respective data must determine the consensual result;
- the greater the spatial agreement of a source within a specified neighborhood of each pixel with the other sources, the more the source contributes to determine the consensual result;
- the consensual result must be affected at most by a maximum uncertainty level specified by the decision maker;
- the fusion strategy should not be rigid and fixed once for all, but flexibly tunable depending on the needs of the application so as to model decision attitudes with distinct trade-offs between risk-taken and risk-adverse.

We represent the fusion strategy by modeling a decision attitude as a quantified-guided function by a monotone non decreasing linguistic quantifier Q defined by a fuzzy set μ_Q:[0,1]→[0,1] specified by a triple (a,b,c) with $a,b\in[0,1]$ and $c>0$ with the meaning depicted in Figure 5 [43] [42].

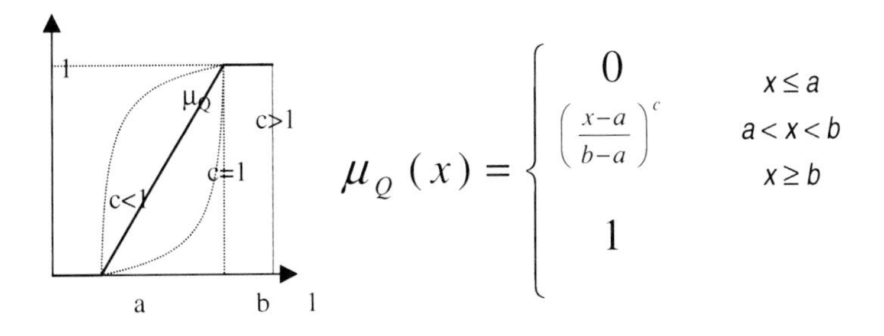

Fig. 5 Membership function of a relative monotone non decreasing linguistic quantifier specifying a fusion strategy

Q=all means that the pixel values in the consensual map must reflect the common decision of all the sources. In the case in which the values in the input maps are proportional to an alarm or anomaly condition, by specifying *all* one wants to model a risk-taken map: all the experts/models must agree on the need to issue the alarm on a given position of the map or to point at the anomaly in order to set an alarm in the consensual map for that position. *Q=at least 1* means that the pixel values in the consensual map must reflect the highest value. In the case in which the value is proportional to an alarm or anomaly, by selecting *at least 1* one models a risk-adverse map: one chooses the most alarming model. This can be useful in making precautionary decisions. *Q=most* means that the consensual map must reflect the shared decision of a fuzzy majority; this models a trade-off decision attitude between the two extreme cases.

Definition of the Generalized OWA Operator

The generalized $\underline{\text{OWA}}_Q$ operator of dimension n and weighting vector W_Q, with $\sum_{i=1,\ldots n} w_i = 1$, aggregates n imprecise values $[v_{1,m}, v_{1,M}]$, ..., $[v_{n,m}, v_{n,M}]$, $v_{1,m}, \ldots v_{1,M} \ldots, v_{n,m}, \ldots v_{n,M} \in D$ (D is a continuous domain) and $v_{n,m} \leq v_{n,M}$, and computes an imprecise value $[c_{1,m}, c_{1,M}]$ of D. This operator is defined as follows:

$$\underline{\text{OWA}}_Q : R(D)^n \rightarrow R(D)$$

where $R(D)$ is the set of all intervals on D and:

$$[c_{1,m}, c_{1,M}] = \underline{\text{OWA}}_Q([v_{1,m}, v_{1,M}], \ldots, [v_{n,m}, v_{n,M}])$$

$$\underline{\text{OWA}}_Q([v_{1,m}, v_{1,M}], \ldots, [v_{n,m}, v_{n,M}]) = \sum_{i=1,\ldots n} w_i * [g_{i,m}, g_{i,M}] \quad (1)$$

in which $[g_{i,m}, g_{i,M}]$ is the i-th largest interval of the $[v_{1,m}, v_{1,M}]$,..., $[v_{n,m}, v_{n,M}]$ such that:

Order: $[a_1, a_2] > [b_1, b_2]$ if $(a_1+a_2) > (b_1+b_2) \vee ((a_1+a_2) = (b_1+b_2) \wedge (b_2 - b_1) \geq (a_2 - a_1))$

$[a_1, a_2] < [b_1, b_2]$ otherwise

Addition: $[a_1, a_2] + [b_1, b_2] = [a_1+b_1, a_2+b2]$
Product: $[a_1, a_2] * [b_1, b_2] = [a_1* b_1, a_2* b_2]$

The weighting vector W_Q of the $\underline{\text{OWA}}_Q$ operator is derived as for the classic OWA operator starting from the definition of μ_Q [39].

$$w_i = \mu_Q\left(\frac{1}{e}\sum_{j=1}^{i} e_j\right) - \mu_Q\left(\frac{1}{e}\sum_{j=0}^{i-1} e_j\right) \quad with \quad e = \sum_{i=1}^{n} e_i = \sum_{j=1}^{n} i_j \tag{2}$$

where e_j is the importance degree i_j associated with the j-th largest value to aggregate. This way w_k, i.e., the increment in satisfaction in having k non-null values with respect to k-1 increases with e_k. The values with no importance play no role.

When all values to be aggregated are precise, the $\underline{\text{OWA}}_Q$ reduces to the usual OWA_Q definition [40]. If the values to fuse are defined on a discrete domain D we have to apply a further rounding function to the result of $\underline{\text{OWA}}_Q([v_{1,m},v_{1,M}],\ldots,[v_{n,m},v_{n,M}])$ so as to yield an interval $[c_{1,m},c_{1,M}]$ defined on the same discrete domain D. In the case in which the data to fuse are ordinal values, several proposals have been defined in the literature for the definition of the fusion operation [14][16]. Finally, in the case in which the values to fuse are fuzzy values, represented by convex possibility distributions μ_v, we apply the $\underline{\text{OWA}}_Q$ operator to their u–cut(μ_v)={x | μ_v(x)> u}, where u is the maximum uncertainty level (specified by the decision maker) that can be tolerated in the consensual result. In fact, if we apply an u–cut to a possibility distribution representing some real variable we can say that the values in the u–cut are affected by at most an uncertainty degree equal to u, i.e., we cannot be completely sure that the real value of the variable is in the set u–cut, unless u=0. Thus, if we apply the fusion to the u–cut affected by an uncertainty u we obtain a fused imprecise value affected at most by the same uncertainty degree.

Definition of the Consensual Fusion by the Generalized OWA Operator

The consensual fusion function associated with a quantifier Q and aggregating possibly imprecise values, i.e., the intervals $[v_{1,m},v_{1,M}],\ldots,[v_{n,m},v_{n,M}]$ on a real domain D, is defined by the generalized $\underline{\text{OWA}}_Q$ operator, defined in formula (1) as an extension of the standard OWA operator [40].

The values to aggregate $[v_{1,m},v_{1,M}],\ldots,[v_{n,m},v_{n,M}]$ belong to the pixels in the n maps (n number of sources) with same x and y coordinates. We indicate by p the common position of the n pixels values to aggregate within the n maps. They are weighted by importance degrees $i_1(p),\ldots, i_n(p) \in [0,1]$ that can vary depending on the location of the pixel p in the n maps. These importance degrees are computed by taking into account both the trust scores of the sources, that we represent by values $t_k \in [0,1]$ with k=1,…,n, and the agreements of the sources themselves within a varying neighborhood C_p of the pixels located in p. The agreement is defined by means of either a compatibility measure or a distance measure:

$$i_i(p) = \alpha * t_i + \beta * \left(Agreement\,(i, C_p)\right) \quad with \quad \alpha + \beta = 1 \quad and$$

$$Agreement\,(i, C_p) = \frac{1}{\left|C_p\right|} \sum_{\forall (x,y) \in C_p} \frac{\sum_{k=1, k \neq i}^{n} f(v_i(x,y), v_k(x,y))}{max_{i=1,\dots,n} \sum_{k=1, k \neq i}^{n} f(v_i(x,y), v_k(x,y))} \tag{3}$$

$$with \qquad f = compatibility \quad or \ f = 1 - normalized\_distance$$

in which (x,y) are the coordinates of the pixels within an neighborhood C_p of location p, $|C_p|$ is the number of pixels in C_p, and f can be chosen as either a compatibility measure between values v_i, and v_j [42] or the complement of a normalized distance measure [13]. The values in a map are important if they belong to a trusted map or, if they are in agreement with the correspondent values in the other maps. The parameters α and β control the relative influence of the trust of a map, and of the agreement degree. For example, by choosing β=1 and α=0, the influence of the values in the fusion strategy will be totally dependent on their agreement values. Notice that the agreement degree $Agreement(i,C_p)$ of a source i with the other n-1 sources is computed with respect to all the values of the pixels in the areas C_p centered in p. If C_p. consists of a single pixel p the agreement is defined locally and does not depend on larger areas of the maps. If C_p. covers the whole maps then the agreement between the sources is global. This introduces further flexibility in the model since it allows considering data variability locally or globally. The function f determines the choice for a strong or a weak agreement. By choosing a *compatibility* function we require a strong agreement among the values since two imprecise values having no overlapping are considered as totally disagreeing. By choosing the complement of a normalized distance we are more tolerant of the differences among the values.

Consensual Fusion to Model Seismic Hazard

As a first example of application of the consensual fusion strategy we discuss the generation of a consensual seismic hazard map based on the fusion of six maps produced independently by applying distinct input models. The six hazard maps are referred to the same area (Calabria region, Southern Italy) and each one is associated with a trust score (in this case a ground motion value g, which is a positive real number). In the classical approach the fused map is generated as the weighted average of the maps, where weights are generally assigned by an expert [28].

In figure 6 we have depicted the six maps of ground motions independently computed by the six input models. The grey level represents the difference between the ground motion values of the first map (high on the left) produced by the first model with respect to the others. It can be noticed that the models mostly disagree in their estimation of the ground motion values in the central area of the maps.

The proposed consensual fusion function defined in (1) is applied to generate the consensual ground motion map relative to a specified fuzzy majority of the trusted models. We applied our approach by modeling two distinct fusion strategies: a risk-taken fusion defined by the quantifier *most* (a=0.6, b=0.9, c=1 – See Figure 5) and a risk-adverse fusion defined by the quantifier *some* (a=0.0, b=0.3, c=1).

We take into account the imprecision of the models in generating their ground motion maps by representing the pixel values through fuzzy numbers (g_m, g, g_M) in which g is the ground motion value computed by the model in the current pixel, and $g_m, < g < g_M$ are defined to capture the approximations applied by the models. The imprecise values of ground motion to fuse by applying definition (8) are derived as 0-cuts of the fuzzy numbers of ground motion, i.e., $0\text{-}cut((g_m, g, g_M))= [g_m, g_M]$. This way we require maximum certainty on the fused result.

In computing the importance of a value from a source we considered a local definition of the agreement among the models according to formula (3): this way the agreement of a model with the others is computed independently for each pixel of the fused map.

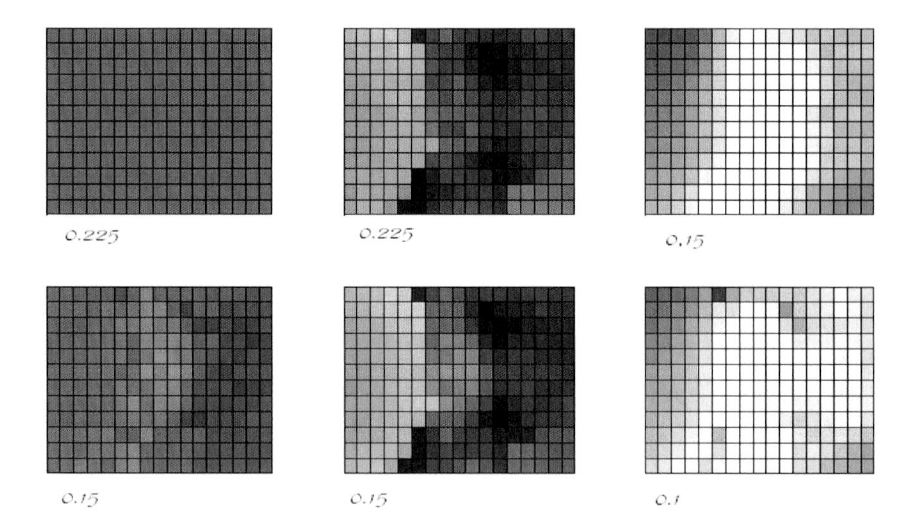

Fig. 6 Ground motion maps computed by the six models: the trust score is indicated below each map. The grey level represents the difference of the ground motion value with respect to the upper left map.

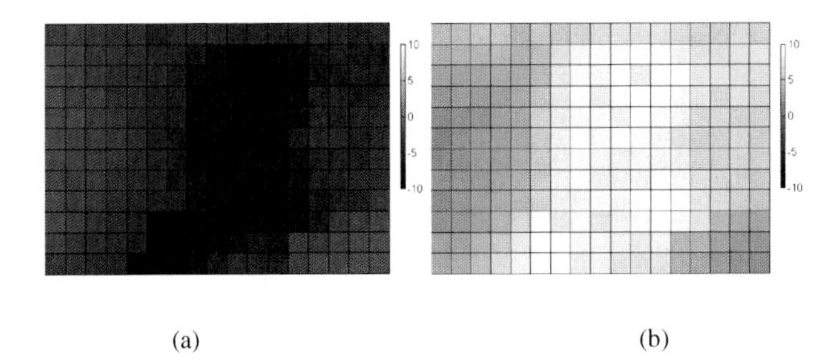

(a) (b)

Fig. 7 Seismic hazard maps obtained using the proposed consensual fusion model: the grey level represents the difference with respect to the map produced with the classic weighted average. (a) fusion based on *most* (a=0.6, b=0.9, c=1); (b) fusion based on *some* (a=0.0, b=0.3, c=1)

Figure 7.a and 7.b depict the maps obtained by the difference of the two consensual fusion strategies specified by *most* and *some* with respect to the classic weighted mean. It can be observed that the most precautionary strategy corresponding with *some* in figure 7.b produces as expected higher ground motion values than the weighted mean while the opposite occurs for the most risk-taken strategy specified by *most* depicted in figure 7.a.

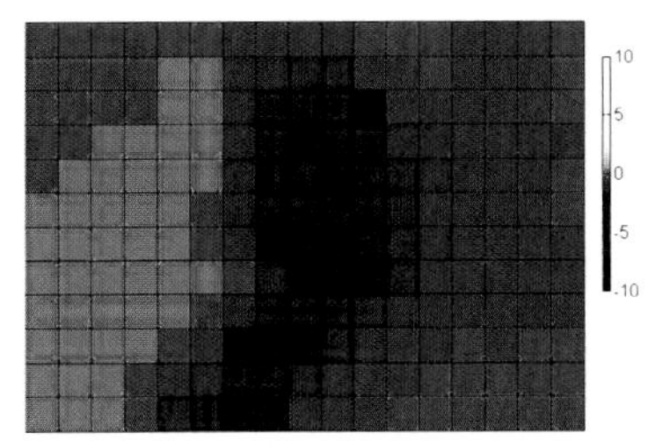

Fig. 8 Map of the differences between the ground motion maps obtained by the consensual fusion based on *average* (a=0.0, b=1.0, c=1.0) and the classic weighted mean

To show the influence of the consensual dynamics on the results we show in Figure 8 the map obtained by the difference of the classic weighted mean map and the consensual map corresponding with the quantifier *average* (a=0,b=1,c=1) that models an average (arithmetic mean) of all the models. It can be observed that the effect of the consensual dynamics is more evident in the central region of the map where there is the lowest agreement among the original maps in figure 6. Specifically, in this area only two out of the six models determine high ground motion values, while the other four models agree for lower values. The consensual ground motion values in this area are then in accordance with the majority of the models, i.e., the ground motion values are lower with respect to those produced by the classic approach.

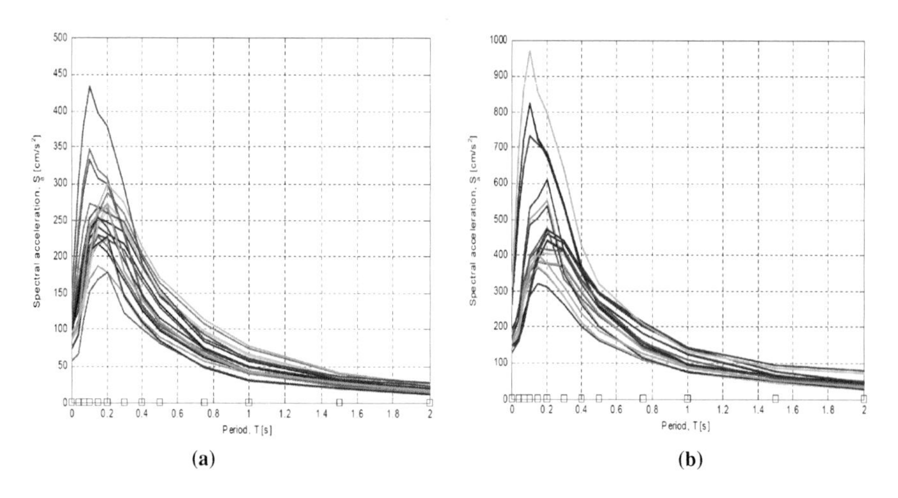

Fig. 9 The **(a)** panel shows the isoprobable response spectra with 10% probability of exceedance in 50 years—5% damping, computed for the test-site, while the **(b)** panel reports the isoprobable spectra with 2% probability of exceedance in 50 years—5% damping, for the same site.

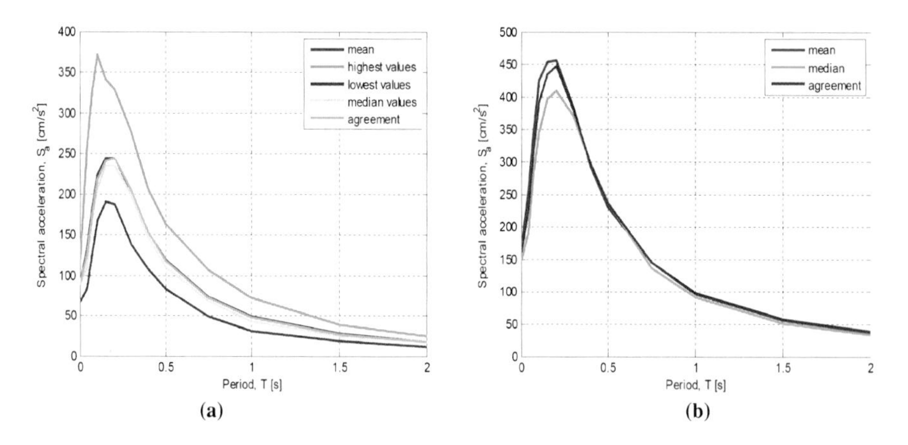

Fig. 10 The left **(a)** panel shows single iso-probable spectra (10% probability of exceedance in 50 years—5% damping) computed for the test-site using different fusion strategies: classic approach based on mean and median estimates, mean of the highest (at least a few, close to OR fusion) and lowest values (almost all, close to AND fusion), and consensual mean fusion. The right **(b)** panel show single isoprobable spectra (5% probability of exceedance in 50 years—5% damping) computed for the test-site using mean, median and consensual mean fusions.

A second experiment based on the application of the proposed consensual fusion was for the estimation of consensual iso-probable response spectra in seismic hazard analysis [23].

To this purpose, we performed a Probabilistic Seismic Hazard Aalalysis for a village placed along the Po river (Northern Italy). The classic analysis followed the structure of the logic-tree defined for the computation of the seismic hazard of the Italian national territory [15]. Figure 9 shows the 20 isoprobable spectra (5% damping) characterized by an exceedance probability in 50 years of 10% and 2%, respectively obtained with the classic approach.

Successively we applied the aggregation of these isoprobable spectra by first using the OWA aggregation operator without consensus, with several attitudinal vectors, i.e., linguistic quantifiers, that correspond to the classical statistical operators mean and median, and to the max and min aggregations.

Besides this experiment we applied the consensual fusion with the same attitudinal vectors, mean and median, by considering the consensus among the isoprobable spectra. In this respect we considered a local agreement (i.e. an interval of periods). The trust weights have been set as the product of the weights on each path of the logic tree from the root to the leaves in the classic approach. The fused spectra are shown in figure 10. The examples we show highlight some interesting aspects especially in the case of the 2475yr RP (see Figure 9(b)). In this situation the consensual mean fusion gives sensibly different results with respect to the ones obtained with a classical mean estimate.

5 Conclusions

In this chapter we analyzed the problems involved in modeling environmental ill-known phenomena. We discussed how multisource data fusion can be exploited to compute environmental indicators of occurrence of such phenomena. To this end the OWA operator is useful to define flexible fusion strategies. A generalization of the OWA to aggregate imprecise values, that often characterize observations of variables, has been introduced, and further its application to define a consensual fusion function has been formalized. Finally, two applications of the OWA operator and of its generalization have been described.

References

[1] Adeh, L.A.: On the validity of Dempster's rule of combination of evidence, Memo M79/24, Univ. of California, Berkeley (1979)

[2] Bloch, I., Maître, H.: Information combination operators for data fusion: A comparative review with classification. IEEE Transactions on Systems, Man, and Cybernetics, part A: systems and humans 26(1), 52–67 (1996)

[3] Bone, C., Dragicevic, S., Roberts, A.: Integrating high resolution remote sensing, GIS and fuzzy set theory for identifying susceptibility areas of forest insect infestations. International Journal of Remote Sensing 26(21), 4809–4828 (2005)

[4] Bordogna, G., Pagani, M., Pasi, G.: A Flexible Decision support approach to model ill-defined knowledge in GIS. Presented at the NATO Workshop on Environmental Impact Assement, Kiev (June 2006)

[5] Breiman, L., Friedman, J., Olshen, R.A., Stone, C.J.: Classification and Regression Trees. Wadsworth, Belmont (1984)

[6] Brivio, P.A., Boschetti, M., Carrara, P., Stroppiana, D., Bordogna, G.: Fuzzy integration of satellite data for detecting environmental anomalies across Africa. In: Hill, J., Roeder, A. (eds.) Advances in Remote Sensing and Geoinformation Processing for Land Degradation Assessment. Taylor & Francis, London (2006)

[7] Burrough, P.A., McDonnel, R.A.: Principles of Geographical Information Systems. Oxford University Press, Oxford (1998)

[8] Carrara, P., Bordogna, G., Stroppiana, D., Brivio, P.A., Nelson, A., Boschetti, M.: A flexible multi-source spatial data fusion system for environmental status assessment at continental scale. Journal of Geographic Information Science 22(7) (2008)

[9] Chanussot, J., Mauris, G., Lambert, P.: Fuzzy fusion techniques for linear features detection in multitemporal sar images. IEEE Trans. on Geoscience and Remote Sensing 37(3), 1292–1305 (1999)

[10] Chen, H., Meer, P.: Robust Fusion of Uncertain Information. IEEE Transactions on Systems, Man and Cybernetics, Part B 35(3), 578–586 (2005)

[11] DeCETI Project, Multi-sources information fusion for satellite image classification, electronic Report of the DeCETI Project, Leonardo da Vinci Programme, Strand II, Measure II.1.1.C, Contract No: GR/1996/II/0953/PI/II.1.1.c/FPC (2000), `http://www.survey.ntua.gr/main/labs/rsens/DeCETI/IRIT/MSI-FUSION/` (accessed 20/12/2006)

[12] Dempster, P.: A generalization of the Bayesian inference. Journal of Royal Statistical Society 30, 205–447 (1968)

[13] Bordogna, G., Pagani, M., Pasi, G.: Consensual Fusion of Uncertain Multisource Spatial data. In: Proc. of the FUZZIEEE 2007, London, July 24-27 (2007)

[14] Gogo, L., Torra, V.: On Aggregation Operators for Ordinal Qualitative Information. IEEE Trans. on Fuzzy Systems 8(2), 143–153 (2000)

[15] di Lavoro, G.: Redazione della Mappa di pericolosità sismica prevista dall'ordinanza PCM 3274 del 20 March 2003. Rapporto conclusivo per il Dipartimento di Protezione Civile, INGV, Milano-Roma, appendixes (April 2004), http://zonesismiche.mi.ingv.it/

[16] Herrera, F., Herrera-Viedma, E.: Linguistic decision analysis: Steps for solving decision problems under linguistic information. Fuzzy Sets Syst. 115(1), 67–82 (2000)

[17] Jeon, B., Landgrebe, D.A.: Decision fusion approach for multitemporal classification. IEEE Transactions on Geoscience and Remote Sensing 37(3), 1227–1233 (1999)

[18] Jiang, H., Eastman, J.R.: Application of fuzzy measures in multi-criteria evaluation in GIS. International Journal of Geographical Information Science 14(2), 173–184 (2000)

[19] Kacprzyk, J.: Group decision making with a fuzzy linguistic majority. Fuzzy Sets and Systems 18, 105–118 (1986)

[20] Kam, M.: geometric interpretation for decision fusion with memory. IEEE Trans. on Systems, Man and Cybernetics, Part A 29(1), 52–62 (1999)

[21] Keenan, P.B.: Spatial Decision Support Systems: An coming of age. Control and Cybernetics 35, 9–27 (2006)

[22] Lenz, R., Malkina-Pykh, I.G., Pykh, Y.: Introduction and overview. Ecological Modelling 130, 1–11 (2000)

[23] Pagani, M., Pagani, M., Bordogna, G., Marcellini, A.: About the use of the OWA operator in case of a PSHA based on a logic tree, xxx

[24] Malczewski, J.: GIS-based multicriteria decision analysis: a survey of the literature. International Journal of Geographical Information Science 20(7), 703–726 (2006)

[25] Malczewski, J., Rinner, C.: Exploring multicriteria decision strategies in GIS with linguistic quantifiers: A case study of residential quality evaluation. Journal of Geographical Sytems 7(2), 249–268 (2005)
[26] Metternicht, G.: Assessing temporal and spatial changes of salinity using fuzzy logic, remote sensing and GIS. Foundations of an expert system. Ecol. Model. 144, 163–179 (2001)
[27] Morris, A., Jankowski, P.: Fuzzy techniques for multiple criteria decision making in GIS. In: Joint 9th IFSA World Congress and 20th NAFIPS International Conference, Vancouver, CA, July 25-28, pp. 2446–2451 (2001) (CD ROM proceedings)
[28] Rabinowitz, N., Steinberg, D.M., Leonard, G.: Logic Trees, sensitivity analysis and data reduction in probabilistic seismic hazard assessment. Earthquake spectra 14(1), 189–201 (1998)
[29] Robinson, P.B.: A perspective on the fundamentals of fuzzy sets and their use in Geographic Information Systems. Transactions in GIS 7(1), 3–30 (2003)
[30] Ruger, N., Schluter, M., Matthies, M.: A fuzzy habitat suitability index for Populus euphratica in the Northern Amudarya delta (Uzbekistan). Ecol. Model. 184, 313–328 (2005)
[31] Shafer, G.: A mathematical theory of evidence. Princeton University Press, London (1976)
[32] Silvert, W.: Ecological impact classification with fuzzy sets. Ecological Modelling 96, 1–10 (1997)
[33] Solaiman, B.: Multisensor data fusion using fuzzy concepts: application to land-cover classification using ERS-1/JERS-1 SAR composites. IEEE Trans. on Geoscience and Remote Sensing 37(3), 1316–1326 (1999)
[34] Stroppiana, D., Brivio, P.A., Boschetti, M., Carrara, P., Bordogna, G.: A fuzzy anomaly indicator for environmental monitoring at continental Scale. Ecological Indicators 9, 92–106 (2009)
[35] Tran, L.T., Knight, C.G., O'Neill, R.V., Smith, E.R., Riitters, K.H., Wickham, J.: Environmental assessment, fuzzy decision analysis of integrated environmental vulnerability assessment of the Mid-Atlantic region. Environmental Monitoring 29(6), 845–859 (2002)
[36] Valet, L., Mauris, G., Bolon, P.: A statistical overview of recent literature in information fusion. IEEE AESS Systems Magazine 1, 7–14 (2001)
[37] Wald, L.: Some terms of reference in data fusion. IEEE Trans. on Geoscience and Remote Sensing 37(3), 1190–1193 (1999)
[38] Wan, W., Fraser, D.: Multisource data fusion with multiple self-organizing maps. IEEE Trans. on Geoscience and Remote Sensing 37(3), 1344–1349 (1999)
[39] Yager, R.R.: Interpreting Linguistically Quantified Propositions. International Journal of Intelligent Systems 9, 541–569 (1994)
[40] Yager, R.R.: On ordered weighted averaging aggregation operators in multi-criteria decision making. IEEE Trans. on Systems, Man and Cybernetics 18, 183–190 (1988)
[41] Yager, R.R.: Quantifier guided aggregation using OWA operators. International Journal of Intelligent Systems 11, 49–73 (1996)
[42] Yager, R.R.: A framework for multi-source data fusion. Information Sciences 163, 175–200 (2004)
[43] Zadeh, L.A.: A computational approach to fuzzy quantifiers in natural languages. Computers and Mathematics with Applications 9, 149–184 (1983)

Decision Making with Dempster-Shafer Theory Using Fuzzy Induced Aggregation Operators

José M. Merigó and Montserrat Casanovas

Abstract. We develop a new approach for decision making with Dempster-Shafer theory of evidence where the available information is uncertain and it can be assessed with fuzzy numbers. With this approach, we are able to represent the problem without losing relevant information, so the decision maker knows exactly which are the different alternatives and their consequences. For doing so, we suggest the use of different types of fuzzy induced aggregation operators in the problem. Then, we can aggregate the information considering all the different scenarios that could happen in the analysis. As a result, we get new types of fuzzy induced aggregation operators such as the belief structure – fuzzy induced ordered weighted averaging (BS-FIOWA) and the belief structure – fuzzy induced hybrid averaging (BS-FIHA) operator. We study some of their main properties. We further generalize this approach by using fuzzy induced generalized aggregation operators. We also develop an application of the new approach in a financial decision making problem about selection of financial strategies.

1 Introduction

The Dempster-Shafer (D-S) theory of evidence (Dempster, 1967; Shafer, 1976) provides a unifying framework for representing uncertainty because it includes the situations of risk and ignorance as special cases. Since its appearance, it has been studied in a wide range of situations (Le et al., 2007; Reformat and Yager, 2008; Srivastava and Mock, 2002; Yager et al., 1994; Yager and Liu, 2008).

Usually, when using the D-S theory it is assumed that the available information are exact numbers (Engemann et al., 1996; Merigó and Casanovas, 2008; 2009a;

José M. Merigó · Montserrat Casanovas
Department of Business Administration, University of Barcelona,
Av. Diagonal 690, 08034 Barcelona, Spain
e-mails: jmerigo@ub.edu, mcasanovas@ub.edu

R.R. Yager et al. (Eds.): Recent Developments in the OWA Operators, STUDFUZZ 265, pp. 209–228.
springerlink.com

Yager, 1992; 2004a). However, this may not be the real situation found in the decision making problem because often, the available information is vague or imprecise and it is not possible to analyze it with exact numbers. Then, a better approach may be the use of fuzzy numbers (FNs) because it considers the best and worst possible scenarios and a lot of other ones that could occur. Note that this problem has already been studied by Casanovas and Merigó (2007) by using fuzzy OWA operators. When using FNs, we will follow the ideas of (Chang and Zadeh, 1972; Zadeh, 1975; Dubois and Prade, 1980; Kaufmann and Gupta, 1985). Note that the OWA operator (Yager, 1988) is an aggregation operator that provides a parameterized family of aggregation operators between the maximum and the minimum. For further research on the OWA operator, see for example (Beliakov et al., 2007; Calvo et al., 2002; Chiclana et al., 2007; Merigó, 2008; Torra and Narukawa, 2007; Wang, 2008; Yager, 1993; 1996; 2003; 2007; 2008; Yager and Filev, 1999; Yager and Kacprzyk, 1997; Zarghami et al., 2008).

Going a step further, the aim of this paper is to suggest the use of different types of fuzzy induced aggregation operators for aggregating the information in decision making with D-S theory. The reason for using various types of aggregation operators is that we want to show that the fuzzy decision making problem with D-S theory can be modeled in different ways depending on the interests of the decision maker. We will use the fuzzy induced ordered weighed averaging (FIOWA) operator (Chen and Chen, 2003) and the fuzzy induced hybrid averaging (FIHA) operator because they provide a parameterized family of aggregation operators that include the fuzzy maximum, the fuzzy minimum, the fuzzy average (FA), the fuzzy weighted average (FWA) and the fuzzy OWA (FOWA), among others. Then, we will get new aggregation operators that we will call the belief structure - FIOWA (BS-FIOWA) and the belief structure - FIHA (BS-FIHA) operator. We will study some of their main properties and we will develop different families of FIOWA and FIHA operators that could be used in the analysis such as the step-FIOWA, the olympic-FIOWA, the centered-FIOWA, the S-FIOWA, etc.

We will further generalize this approach by using generalized aggregation operators such as the generalized OWA (GOWA) and the Quasi-OWA operator (Beliakov, 2005; Fodor et al., 1995; Karayiannis, 2000; Yager, 2004b). In this paper we will focus on the use of the order inducing variables. Therefore, we will follow the induced generalized OWA (IGOWA) and the Quasi-IOWA operator developed by Merigó and Gil-Lafuente (2009a). By using FNs, we will use the fuzzy generalized OWA (FGOWA) (Merigó and Casanovas, 2007), the fuzzy induced generalized OWA (FIGOWA) (Merigó and Gil-Lafuente, 2009b), the fuzzy induced generalized hybrid averaging (FIGHA) (Merigó and Casanovas, 2009b) and its corresponding quasi-arithmetic versions.

We also study the applicability of this new approach developing an application of this new model in a business decision making problem about selection of financial strategies. We see that depending on the particular type of aggregation operator used, the results may lead to different decisions.

In order to do so, the remainder of the paper is organized as follows. In Section 2, we briefly describe some basic concepts such as the FNs, the FIOWA and the FIHA operator. Section 3 briefly comments the main concepts of the D-S theory.

In Section 4 and 5, we present the new approach about using fuzzy induced aggregation operators in decision making with D-S theory. Finally, in Section 6 we develop an application of the new approach in a decision making problem. In this case, we consider a problem about selection of financial strategies.

2 Preliminaries

In this Section, we briefly review some basic concepts about the FNs, the FIOWA, the FIHA, the FIGOWA, the Quasi-FIOWA, the FIGHA and the Quasi-FIGHA operators.

2.1 Fuzzy Numbers

The FN was introduced by (Chang and Zadeh, 1972; Zadeh, 1975). Since then, it has been studied and applied by a lot of authors such as (Dubois and Prade, 1980; Kaufmann and Gupta, 1985).

A FN is a fuzzy subset (Zadeh, 1965) of a universe of discourse that is both convex and normal (Kaufmann and Gupta, 1985). Note that the FN may be considered as a generalization of the interval number (Moore, 1966) although it is not strictly the same because the interval numbers may have different meanings.

In the literature, we find a wide range of FNs (Dubois and Prade, 1980; Kaufmann and Gupta, 1985). For example, a trapezoidal FN (TpFN) A of a universe of discourse R can be characterized by a trapezoidal membership function $A = (\underline{a}, \overline{a})$ such that

$$\begin{aligned} \underline{a}(\alpha) &= a_1 + \alpha(a_2 - a_1), \\ \overline{a}(\alpha) &= a_4 - \alpha(a_4 - a_3). \end{aligned} \tag{1}$$

where $\alpha \in [0, 1]$ and parameterized by (a_1, a_2, a_3, a_4) where $a_1 \leq a_2 \leq a_3 \leq a_4$, are real values. Note that if $a_1 = a_2 = a_3 = a_4$, then, the FN is a crisp value and if $a_2 = a_3$, the FN is represented by a triangular FN (TFN). Note that the TFN can be parameterized by (a_1, a_2, a_4).

In the following, we are going to review the FN arithmetic operations as follows. Let A and B be two TFN, where $A = (a_1, a_2, a_3)$ and $B = (b_1, b_2, b_3)$. Then:

1) $A + B = (a_1 + b_1, a_2 + b_2, a_3 + b_3)$
2) $A - B = (a_1 - b_3, a_2 - b_2, a_3 - b_1)$
3) $A \times k = (k \times a_1, k \times a_2, k \times a_3)$; for $k > 0$.

Note that other operations could be studied but in this paper we will focus on these ones.

2.2 The Fuzzy Induced OWA Operator

The FIOWA (or FN-IOWA) operator was introduced by S.J. Chen and S.M. Chen (2003). It is an extension of the OWA operator (Calvo et al., 2002; Yager, 1988; 1993; 1996; 2007; Yager and Kacprzyk, 1997) that uses uncertain information represented by FNs for the situations where it is not possible to use the classical numbers. Moreover, it also uses a reordering process different from the values of the arguments. In this case, the reordering step is based on order inducing variables. It can be defined as follows.

Definition 1. *Let Ψ be the set of FN. A FIOWA operator of dimension n is a mapping FIOWA: $\Psi^n \rightarrow \Psi$ that has an associated weighting vector W of dimension n such that $w_j \in [0, 1]$ and $\sum_{j=1}^{n} w_j = 1$, then:*

$$FIOWA\left(\langle u_1,\tilde{a}_1\rangle, \ldots, \langle u_n,\tilde{a}_n\rangle\right) = \sum_{j=1}^{n} w_j b_j \tag{2}$$

where b_j is the $\tilde{a}_i$ value of the FIOWA pair $\langle u_i, \tilde{a}_i\rangle$ having the jth largest u_i, u_i is the order inducing variable and $\tilde{a}_i$ is the argument variable represented in the form of a FN.

Note that from a generalized perspective of the reordering step it is possible to distinguish between descending (DFIOWA) and ascending (AFIOWA) orders. Note also that this operator provides a parameterized family of aggregation operators that includes the fuzzy maximum, the fuzzy minimum and the fuzzy average (FA), among others. For more information, see (Merigó, 2008).

When using order inducing variables, it is not necessary to establish a criterion for ranking FNs because the reordering process is carried out according to the order inducing variables. However, in the final results, sometimes it is necessary to rank FNs. Among the different methods existing in the literature for ranking FNs, we recommend the use of the methods commented by Merigó (2008) such as the use of the value found in the highest membership level ($\alpha = 1$) and if it is an interval, the average of the interval.

2.3 The Fuzzy Induced Hybrid Averaging Operator

The fuzzy induced hybrid averaging (FIHA) operator is an extension of the HA operator (Xu and Da, 2003; Xu, 2006) that uses uncertain information represented in the form of FNs and order inducing variables in the reordering of the arguments. It uses in the same formulation the fuzzy weighted average (FWA) and the FIOWA operator. Then, with this operator we can represent the subjective probability and the attitudinal character of a decision maker in the same problem. It can be defined as follows.

Definition 2. *Let Ψ be the set of FN. A FIHA operator of dimension n is a mapping FIHA: $\Psi^n \rightarrow \Psi$ that has an associated weighting vector W of dimension n such that $w_j \in [0, 1]$ and $\sum_{j=1}^{n} w_j = 1$, then:*

$$FIHA\,(\langle u_1,\tilde{a}_1\rangle, \ldots, \langle u_n,\tilde{a}_n\rangle) = \sum_{j=1}^{n} w_j b_j \tag{3}$$

where b_j is the $\hat{a}_i$ ($\hat{a}_i = n\omega_i\tilde{a}_i$, $i = 1,2,\ldots,n$) value of the FIHA pair $\langle u_i,\tilde{a}_i\rangle$ having the jth largest u_i, u_i is the order inducing variable, $\omega = (\omega_1, \omega_2, \ldots, \omega_n)^T$ is the weighting vector of the $\tilde{a}_i$, with $\omega_i \in [0, 1]$ and the sum of the weights is 1, and the $\tilde{a}_i$ are FNs.

From a generalized perspective of the reordering step we can distinguish between the descending FIHA (DFIHA) and the ascending FIHA (AFIHA) operators.

The FIHA operator is monotonic and idempotent. It is not bounded by the maximum and the minimum because we may find some situations where the aggregation gives higher and lower results than the maximum and the minimum, respectively.

Different families of FIHA operators are found by using a different manifestation of the weighting vector such as the FA, the FWA, the FOWA, the FIOWA, the step-FIHA, the olympic-FIHA, the median-FIHA, the window-FIHA, the S-FIHA, the centered-FIHA, etc.

2.4 The Fuzzy Induced Generalized OWA Operator

The fuzzy induced generalized OWA (FIGOWA) operator (Merigó and Gil-Lafuente, 2009b) is an extension of the GOWA operator that uses uncertain information in the aggregation represented in the form of FNs. Thus, it is able to include a wide range of particular cases in its formulation. The reason for using this operator is that sometimes, the uncertain factors that affect our decisions are not clearly known and in order to assess the problem we need to use FNs. This operator also uses a reordering process based on order inducing variables in order to assess complex reordering processes. It can be defined as follows.

Definition 3. *Let Ψ be the set of FNs. A FIGOWA operator of dimension n is a mapping FIGOWA: $\Psi^n \rightarrow \Psi$ that has an associated weighting vector W of dimension n such that $w_j \in [0, 1]$ and $\sum_{j=1}^{n} w_j = 1$, then:*

$$FIGOWA\,(\langle u_1,\tilde{a}_1\rangle, \langle u_2,\tilde{a}_2\rangle \ldots \langle u_n,\tilde{a}_n\rangle) = \left(\sum_{j=1}^{n} w_j b_j^{\lambda} \right)^{1/\lambda} \tag{4}$$

where b_j is the $\tilde{a}_i$ value of the FIGOWA pair $\langle u_i,\tilde{a}_i\rangle$ having the jth largest u_i, u_i is the order inducing variable, $\tilde{a}_i$ is the argument variable represented in the form of FN and λ is a parameter such that $\lambda \in (-\infty, \infty)$.

The FIGOWA can be further generalized by using quasi-arithmetic means. The result is the Quasi-FIOWA operator. It can be defined as follows.

Definition 4. *Let Ψ be the set of FNs. A Quasi-FIOWA operator of dimension n is a mapping f: $\Psi^n \rightarrow \Psi$ that has an associated weighting vector W of dimension n such that $w_j \in [0, 1]$ and $\sum_{j=1}^{n} w_j = 1$, then:*

$$f(\langle u_1,\tilde{a}_1\rangle, \langle u_2,\tilde{a}_2\rangle \ldots, \langle u_n,\tilde{a}_n\rangle) = g^{-1}\left(\sum_{j=1}^{n} w_j g(b_j)\right) \tag{5}$$

where b_j is the $\tilde{a}_i$ value of the Quasi-FIOWA pair $\langle u_i,\tilde{a}_i\rangle$ having the jth largest u_i, u_i is the order inducing variable, $\tilde{a}_i$ is the argument variable represented in the form of FN and g(b) is a strictly continuous monotone function.

As we can see, when $g(b) = b^\lambda$, we get the FIGOWA operator. Note that it is also possible to distinguish between descending (Quasi-DFIOWA) and ascending (Quasi-AFIOWA) orders.

Note that different types of FNs could be used in the aggregation of the FIGOWA and the Quasi-FIOWA such as TFNs, TpFNs, L-R FNs, interval-valued FNs, intuitionistic FNs and more complex structures.

Moreover, both the FIGOWA and the Quasi-FIOWA includes a wide range of particular cases such as the FIOWA operator, the fuzzy induced ordered weighted geometric averaging (FIOWGA) operator, the fuzzy induced ordered weighted quadratic averaging (FIOWQA) operator and the fuzzy induced ordered weighted harmonic averaging (FIOWHA) operator.

2.5 Fuzzy Induced Generalized Hybrid Averaging Operator

The FIGHA operator (Merigó and Casanovas, 2009b) is a unified aggregation model between the FIGOWA operator and the fuzzy weighted generalized mean for situations where we want to deal with more complex reordering processes in the aggregation of the FNs. It uses order-inducing variables in the reordering of the FNs and it includes the fuzzy generalized hybrid averaging (FGHA) operator as a particular case. It can be defined as follows.

Definition 5. *Let Ψ be the set of FNs. A FIGHA operator of dimension n is a mapping FIGHA: $\Psi^n \rightarrow \Psi$ that has an associated weighting vector W of dimension n with $\sum_{j=1}^{n} w_j = 1$ and $w_j \in [0, 1]$, such that:*

$$FIGHA\ (\langle u_1,\tilde{a}_1\rangle, \langle u_2,\tilde{a}_2\rangle \ldots, \langle u_n,\tilde{a}_n\rangle) = \left(\sum_{j=1}^{n} w_j b_j^\lambda\right)^{1/\lambda} \tag{6}$$

where b_j is the $\hat{a}_i$ value ($\hat{a}_i = n\omega_i\tilde{a}_i$, $i = 1,2,\ldots,n$) of the IOWA pair $\langle u_i, \tilde{a}_i\rangle$ having the jth largest u_i, u_i is the order inducing variable, $\omega = (\omega_1, \omega_2, \ldots, \omega_n)^T$ is the *weighting vector of the* $\tilde{a}_i$, *with* $\omega_i \in [0, 1]$ *and the sum of the weights is* 1, *the* $\tilde{a}_i$ *are FNs, and* λ *is a parameter such that* $\lambda \in (-\infty, \infty)$.

The FIGHA can be further generalized by using quasi-arithmetic means, obtaining the Quasi-FIHA operator. This operator is defined as follow.

Definition 6. *Let* Ψ *be the set of FN. A Quasi-FIHA operator of dimension n is a mapping QFIHA:* $\Psi^n \to \Psi$ *that has an associated weighting vector W of dimension n such that* $\sum_{j=1}^{n} w_j = 1$ *and* $w_j \in [0, 1]$, *then*:

$$\text{Quasi-FIHA}\,(\langle u_1,\tilde{a}_1\rangle, \langle u_2,\tilde{a}_2\rangle \ldots, \langle u_n,\tilde{a}_n\rangle) = g^{-1}\left(\sum_{j=1}^{n} w_j g(b_j)\right) \tag{7}$$

where b_j *is the* $\hat{a}_i$ *value* ($\hat{a}_i = n\omega_i\tilde{a}_i$, $i = 1,2,\ldots,n$) *of the IOWA pair* $\langle u_i, \tilde{a}_i\rangle$ *having the jth largest* u_i, u_i *is the order inducing variable,* $\omega = (\omega_1, \omega_2, \ldots, \omega_n)^T$ *is the weighting vector of the* $\tilde{a}_i$, *with* $\omega_i \in [0, 1]$ *and the sum of the weights is* 1, *the* $\tilde{a}_i$ *are FNs, and g(b) is a strictly continuous monotonic function.*

Note that in this case we can also use a wide range of FNs in the aggregation process and they include a wide range of particular cases such as the FIHA, the fuzzy induced hybrid geometric averaging (FIHGA), the fuzzy induced hybrid quadratic averaging (FIHQA) and the fuzzy induced hybrid harmonic averaging (FIHHA) operator.

3 The Dempster-Shafer Theory of Evidence

The D-S theory provides a unifying framework for representing uncertainty as it can include the situations of risk and ignorance as special cases. Note that the case of certainty is also included as it can be seen as a particular case of risk and ignorance.

Definition 7. *A D-S belief structure defined on a space X consists of a collection of n nonnull subsets of X,* B_j *for* $j = 1,\ldots,n$, *called focal elements and a mapping m, called the basic probability assignment, defined as, m:* $2^X \to [0, 1]$ *such that*:

$$\begin{aligned} &(1) \quad m(B_j) \in [0, 1]. \\ &(2) \quad \textstyle\sum_{j=1}^{n} m(B_j) = 1. \\ &(3) \quad m(A) = 0, \forall A \neq B_j. \end{aligned} \tag{8}$$

As said before, the cases of risk and ignorance are included as special cases of belief structure in the D-S framework. For the case of risk, a belief structure is called Bayesian belief structure if it consists of n focal elements such that $B_j = \{x_j\}$, where each focal element is a singleton. Then, we can see that we are in a situation of decision making under risk environment as $m(B_j) = P_j = \text{Prob}\ \{x_j\}$.

The case of ignorance is found when the belief structure consists in only one focal element *B*, where *m(B)* essentially is the decision making under ignorance environment as this focal element comprises all the states of nature. Thus, *m(B)* = 1. Other special cases of belief structures such as the consonant belief structure or the simple support function are studied in (Shafer, 1976). Note that two important evidential functions associated with these belief structures are the measures of plausibility and belief (Shafer, 1976).

Two important evidential functions associated with these belief structures are the measures of plausibility and belief. In the following, we provide a definition of these two measures as developed by Shafer[3].

Definition 8. *The plausibility measure Pl is defined as, Pl:* $2^X \rightarrow [0, 1]$ *such that*:

$$\mathrm{Pl}(A) = \sum_{A \cap B_j \neq \varnothing} m(B_j) \tag{9}$$

Definition 9. *The belief measure Bel is defined as Bel:* $2^X \rightarrow [0, 1]$ *such that*:

$$\mathrm{Bel}(A) = \sum_{B_j \subseteq A} m(B_j) \tag{10}$$

Bel(A) represents the exact support to A and Pl(A) represents the possible support to A. With these two measures we can form the interval of support to A as [Bel(A),Pl(A)]. This interval can be seen as the lower and upper bounds of the probability to which A is supported such that $\mathrm{Bel}(A) \leq \mathrm{Prob}(A) \leq \mathrm{Pl}(A)$. From this we see that $\mathrm{Pl}(A) \geq \mathrm{Bel}(A)$ for all A. Another interesting feature about these two measures is that they are connected by $\mathrm{Bel}(A) = 1 - \mathrm{Pl}(\bar{A})$ or $\mathrm{Pl}(A) = 1 - \mathrm{Bel}(\bar{A})$, where $\bar{A}$ is the complement of A.

4 Fuzzy Induced Aggregation Operators in D-S Theory

In this Section, we describe the process to follow when using fuzzy induced aggregation operators in decision making with D-S theory. We divide it in three subsections. In the first one, we comment the decision process. In the second one, we analyze the aggregation used in the problem. And in the third one, we study different types of fuzzy induced aggregation operators that could be used in the aggregation.

4.1 Decision Making Approach

A new method for decision making with D-S theory is possible by using FN aggregation operators in the problem. This problem has been studied in (Merigó and Casanovas, 2007). Going a step further, we see that it is possible to use fuzzy induced aggregation operators such as the FIOWA and the FIHA operator. Note it is also possible to consider other cases such as the use of different types of fuzzy induced generalized means and fuzzy induced quasi-arithmetic means. The motivation for using FNs appears because sometimes, the available information is not clear and it is necessary to assess it with another approach such as the use of FNs. Although the information is uncertain and it is difficult to take decisions with it, at least we can represent the best and worst possible scenarios and the possibility that the internal values of the fuzzy interval will occur. The decision process can be summarized as follows.

Assume we have a decision problem in which we have a collection of alternatives $\{A_1, ..., A_q\}$ with states of nature $\{S_1, ..., S_n\}$. $\tilde{a}_{ih}$ is the uncertain payoff, given in the form of FNs, to the decision maker if he selects alternative A_i and the state of nature is S_h. The knowledge of the state of nature is captured in terms of a belief structure m with focal elements $B_1, ..., B_r$ and associated with each of these focal elements is a weight $m(B_k)$. The objective of the problem is to select the alternative which gives the best result to the decision maker. In order to do so, we should follow the following steps:

Step 1: Calculate the uncertain payoff matrix.

Step 2: Calculate the belief function m about the states of nature.

Step 3: Calculate the attitudinal character (or degree of orness) of the decision maker $\alpha(W)$ (Yager, 1988).

Step 4: Calculate the collection of weights, w, to be used in the FIOWA aggregation for each different cardinality of focal elements. Note that it is possible to use different methods depending on the interests of the decision maker (Merigó, 2008; Merigó and Casanovas, 2009a; Merigó and Gil-Lafuente, 2009a; Xu, 2005; Yager, 1988; 1993; 1996; 2003; 2007).

Step 5: Determine the uncertain payoff collection, M_{ik}, if we select alternative A_i and the focal element B_k occurs, for all the values of i and k. Hence $M_{ik} = \{a_{ih} \mid S_h \in B_k\}$.

Step 6: Calculate the fuzzy induced aggregated payoff, V_{ik} = FIOWA (M_{ik}), using Eq. (2), for all the values of i and k.

Step 7: For each alternative, calculate the generalized expected value, C_i, where:

$$C_i = \sum_{r=1}^{r} V_{ik} m(B_k) \tag{11}$$

Step 8: Select the alternative with the largest C_i as the optimal.

From a generalized perspective of the reordering step, it is possible to distinguish between ascending and descending orders in the FIOWA aggregation.

The procedure to follow if we use the AIOWA operator in the aggregation step is the same than the procedure used for the FIOWA or DFIOWA operator with the following differences.

In *Step 3*, when calculating the inducing variables we should consider that in these cases, the lowest inducing variable is the first result in the reordering of the arguments.

In *Step 4*, when calculating the collection of weights, we should consider that the reordering will now be different so that we might associate each weight correctly with its corresponding position.

In *Step 6*, when calculating the aggregated payoff, we should use V_{ik} = AIOWA(M_{ik}), for all the values of i and k.

4.2 Using FIOWA Operators in Belief Structures

Analyzing the aggregation in *Steps* 6 and 7 of the previous subsection, it is possible to formulate in one equation the whole aggregation process. We will call this process the belief structure – FIOWA (BS-FIOWA) aggregation. It can be defined as follows.

Definition 10. *A BS-FIOWA operator is defined by*

$$C_i = \sum_{k=1}^{r} \sum_{j_k=1}^{q_k} m(B_k) w_{j_k} b_{j_k} \tag{12}$$

where w_{j_k} is the weighting vector of the kth focal element such that $\sum_{j=1}^{n} w_{j_k} = 1$ and $w_{j_k} \in [0, 1]$, b_{j_k} is the j_kth largest of the $\tilde{a}_{i_k}$ and the $\tilde{a}_{i_k}$ are FNs, and $m(B_k)$ is the basic probability assignment.

Note that q_k refers to the cardinality of each focal element and r is the total number of focal elements.

The BS-FIOWA operator is monotonic, commutative, bounded and idempotent. We can prove these properties with the following theorems.

Theorem 1 (Commutativity). *Assume f is the BS-FIOWA operator, then*

$$f(\langle u_{1_1}, \tilde{a}_{1_1} \rangle, \ldots, \langle u_{q_r}, \tilde{a}_{q_r} \rangle) = f(\langle u_{1_1}^*, \tilde{a}_{1_1}^* \rangle, \ldots, \langle u_{q_r}^*, \tilde{a}_{q_r}^* \rangle) \tag{13}$$

where $(\langle u_{1_1}^*, \tilde{a}_{1_1}^* \rangle, \ldots, \langle u_{q_r}^*, \tilde{a}_{q_r}^* \rangle)$ is any permutation of $(\langle u_{1_1}, \tilde{a}_{1_1} \rangle, \ldots, \langle u_{q_r}, \tilde{a}_{q_r} \rangle)$ for each focal element k.

Theorem 2 (Monotonicity). *Assume f is the BS-FIOWA operator, if $\tilde{a}_{i_k} \geq \tilde{a}_{i_k}^*$ for all i, then,*

$$f(\langle u_{1_1}, \tilde{a}_{1_1} \rangle, \ldots, \langle u_{q_r}, \tilde{a}_{q_r} \rangle) \geq f(\langle u_{1_1}^*, \tilde{a}_{1_1}^* \rangle, \ldots, \langle u_{q_r}^*, \tilde{a}_{q_r}^* \rangle) \tag{14}$$

Theorem 3 (Boundedness). *Assume f is the BS-FIOWA operator, then*

$$\min\{\tilde{a}_i\} \leq f(\langle u_{1_1}, \tilde{a}_{1_1} \rangle, \ldots, \langle u_{q_r}, \tilde{a}_{q_r} \rangle) \leq \max\{\tilde{a}_i\} \tag{15}$$

Theorem 4 (Idempotency). *Assume f is the BS-FIOWA operator, if* $\tilde{a}_{i_k} = \tilde{a}$ *for all* $i \in N$, *then*

$$f(\langle u_{1_1}, \tilde{a}_{1_1} \rangle, ..., \langle u_{q_r}, \tilde{a}_{q_r} \rangle) = \tilde{a} \quad (16)$$

From a generalized perspective of the reordering step, it is possible to distinguish between descending and ascending orders by using $w_j = w^*_{n-j+1}$, where w_j is the *j*th weight of the DFIOWA and w^*_{n-j+1} the *j*th weight of the AFIOWA operator. Then, we obtain the BS-DFIOWA and the BS-AFIOWA operators.

4.3 Families of BS-FIOWA Operators

By choosing a different manifestation in the weighting vector of the FIOWA operator, we are able to develop different families of FIOWA and BS-FIOWA operators. As it can be seen in Definition 10, each focal element uses a different weighting vector in the aggregation step with the FIOWA operator. Therefore, the analysis needs to be done individually.

Remark 1. For example, it is possible to obtain the fuzzy maximum, the fuzzy minimum, the FA, the FWA and the FOWA operator.

- The fuzzy maximum is obtained if $w_p = 1$ and $w_j = 0$, for all $j \neq p$, and $u_p = \text{Max}\{\tilde{a}_i\}$.
- The fuzzy minimum is obtained if $w_p = 1$ and $w_j = 0$, for all $j \neq p$, and $u_p = \text{Min}\{\tilde{a}_i\}$.
- The FA is found when $w_j = 1/n$, for all $\tilde{a}_i$.
- The FWA is obtained if $u_i > u_{i+1}$, for all i.
- The FOWA operator is obtained if the ordered position of u_i is the same than the ordered position of b_j such that b_j is the *j*th largest of $\tilde{a}_i$.

Other families of FIOWA operators could be used in the BS-FIOWA operator such as the step-FIOWA, the S-FIOWA, the olympic-FIOWA, the window-FIOWA and the centered-FIOWA operator, among others. Note that we find in the literature a wide range of methods for determining OWA weights that could be applied for the FIOWA. But in this subsection, we simply give a general overview commenting some basic cases that are applicable in the FIOWA operator. For more information on these families, see (Merigó, 2008).

Remark 2. The step-FIOWA operator is found when $w_k = 1$ and $w_j = 0$, for all $j \neq k$. Note that the median-FIOWA can be seen as a particular case of this situation when the number of arguments is odd.

Remark 3. The olympic-FIOWA operator is found if $w_1 = w_n = 0$, and for all others $w_j = 1/(n - 2)$. Note that the window-FIOWA operator can be seen as a generalization of this case.

Remark 4. A further interesting family is the S-FIOWA operator. In this case, we can distinguish between three types: the "orlike", the "andlike", and the "generalized" S-FIOWA operator. The generalized S-FIOWA operator is obtained when $w_1 = (1/n)(1 - (\alpha + \beta)) + \alpha$, $w_n = (1/n)(1 - (\alpha + \beta)) + \beta$, and $w_j = (1/n)(1 - (\alpha + \beta))$ for all $j = 2$ to $n - 1$ where $\alpha, \beta \in [0, 1]$ and $\alpha + \beta \leq 1$. Note that if $\alpha = 0$, we get the andlike S-FIOWA and if $\beta = 0$, the orlike S-FIOWA. Also note that if $\alpha + \beta = 1$, we get the fuzzy induced Hurwicz criteria.

Remark 5. The centered-FIOWA operator is found if the aggregation is symmetric, strongly decaying and inclusive. It is symmetric if $w_j = w_{j+n-1}$. It is strongly decaying when $i < j \leq (n + 1)/2$, then $w_i < w_j$ and when $i > j \geq (n + 1)/2$ then $w_i < w_j$. It is inclusive if $w_j > 0$. Note that it is possible to consider different particular situations of this operator by softening the second condition with $w_i \leq w_j$ instead of $w_i < w_j$ and by removing the third condition.

Remark 6. Finally, if we assume that all the focal elements use the same weighting vector, then, we can refer to these families as the BS-fuzzy maximum, the BS-fuzzy minimum, the BS-FA, the BS-FWA, the BS-S-FIOWA, the BS-olympic-FIOWA, the BS-window-FIOWA, the BS-centered-FIOWA, etc.

5 Fuzzy Induced Hybrid Averaging Operators in D-S Theory

In some situations, the decision maker could prefer to use another type of fuzzy aggregation operator such as the FIHA operator. The main advantage of this operator is that it uses the characteristics of the FWA and the FIOWA in the same aggregation. Then, if we introduce this operator in decision making with D-S theory, we are able to develop a unifying framework that includes in the same formulation probabilities, FWAs and FIOWAs.

In order to use this type of aggregation in D-S framework we should consider that now in *Step* 3, when calculating the collection of weights to be used in the aggregation, we are using two weighting vectors because we are mixing in the same problem the FWA and the FIOWA.

In *Step* 5, when calculating the fuzzy aggregated payoff, we should use the FIHA operator instead of the FIOWA operator by using Eq. (3).

In this case, it is also possible to formulate in one equation the whole aggregation process. We will call it the BS-FIHA operator.

Definition 11. *A BS-FIHA operator is defined by*

$$C_i = \sum_{k=1}^{r} \sum_{j_k=1}^{q_k} m(B_k) w_{j_k} b_{j_k} \tag{17}$$

where w_{j_k} is the weighting vector of the kth focal element such that $\sum_{j=1}^{n} w_{j_k} = 1$ and $w_{j_k} \in [0, 1]$, b_{j_k} is the $\hat{a}_{i_k}$ ($\hat{a}_{i_k} = n\omega_i \tilde{a}_{i_k}$, $i = 1,2,\ldots,n$) value of the BS-FIHA pair

$\langle u_{i_k}, \tilde{a}_{i_k} \rangle$ having the j_kth largest u_{i_k}, u_{i_k} is the order inducing variable, $\omega = (\omega_1, \omega_2, \ldots, \omega_n)^T$ is the weighting vector of the $\tilde{a}_{i_k}$, with $\omega_i \in [0, 1]$ and the sum of the weights is 1, and the $\tilde{a}_{i_k}$ are FNs, and $m(B_k)$ is the basic probability assignment.

As we can see, the focal weights are aggregating the results obtained by using the FIHA operator. Note that if $\omega_i = 1/n$ for all i, then, Eq. (17) is transformed in Eq. (12).

In this case, we could also study different properties and particular cases of the BS-FIHA operator, such as the distinction between descending (BS-DFIHA) and ascending (BS-AFIHA) orders.

When aggregating the collection of fuzzy payoffs of each focal element, it is also possible to consider a wide range of families of FIHA operators. For example, we could mention the fuzzy hybrid maximum, the fuzzy hybrid minimum, the FA, the FWA, the FOWA and the FIOWA operator. These operators are obtained in a similar way as explained in subsection 4.3 excepting for the FWA and the FIOWA. Note that the FWA is found when $w_j = 1/n$, for all j, and the FIOWA operator when $\omega_i = 1/n$, for all i, respectively.

Other families of FIHA operators that could be used are the step-FIHA operator, the window-FIHA, the olympic-FIHA, the S-FIHA, the median-FIHA, etc. Note that these families follow a similar methodology as it has been explained for the FIOWA operator.

Finally, if we use the same family of FIHA operator for all the focal elements, then, we can refer to the aggregation as the BS-fuzzy hybrid maximum, the BS-fuzzy hybrid minimum, the Hurwicz BS-fuzzy hybrid criteria, the BS-step-FIHA, the BS-olympic-FIHA, the BS-S-FIHA, the BS-centered-FIHA, etc.

6 Fuzzy Induced Generalized Aggregation Operators in D-S Theory of Evidence

A more general formulation of the previous methods can be developed by using generalized aggregation operators. In this case, this would mean that we use the FIGOWA, the Quasi-FIOWA, the FIHA and the Quasi-FIHA operator. If we introduce these fuzzy induced generalized aggregation operators in the decision making process with Dempster-Shafer belief structure, we have to make the following changes to the previous procedures explained in Section 4 and 5.

In *Step* 3, when calculating the collection of weights to be used in the aggregation, we have to adapt the problem of calculating the OWA weights to the particular type of aggregation operator that we are using.

In *Step* 5, when calculating the fuzzy aggregated payoff, we should use the FIGOWA, th Quasi-FIOWA, the FIGHA and the Quasi-FIHA operator by using Eq. (4), Eq. (5), Eq. (6) and Eq (7), respectively.

In this case, it is also possible to formulate in one equation the whole aggregation process. We will call it the BS-FIGOWA, the BS-Quasi-FIOWA, the BS-FIGHA and the BS-Quasi-FIHA operator.

Definition 12. *A BS-FIGOWA operator is defined by*

$$C_i = \sum_{k=1}^{r} m(B_k) \left(\sum_{j_k=1}^{q_k} w_{j_k} b_{j_k}^{\lambda} \right)^{1/\lambda} \quad (18)$$

where w_{j_k} *is the weighting vector of the kth focal element such that* $\sum_{j=1}^{n} w_{j_k} = 1$ *and* $w_{j_k} \in [0, 1]$, b_{j_k} *is the* j_k*th largest of the* $\tilde{a}_{i_k}$ *and the* $\tilde{a}_{i_k}$ *are FNs, and* $m(B_k)$ *is the basic probability assignment and* λ *is a parameter such that* $\lambda \in (-\infty, \infty)$.

Definition 13. *A BS-FIGHA operator is defined by*

$$C_i = \sum_{k=1}^{r} m(B_k) \left(\sum_{j_k=1}^{q_k} w_{j_k} b_{j_k}^{\lambda} \right)^{1/\lambda} \quad (19)$$

where w_{j_k} *is the weighting vector of the kth focal element such that* $\sum_{j=1}^{n} w_{j_k} = 1$ *and* $w_{j_k} \in [0, 1]$, b_{j_k} *is the* $\hat{a}_{i_k}$ $(\hat{a}_{i_k} = n\omega_i \tilde{a}_{i_k}$, $i = 1,2,\ldots,n)$ *value of the BS-FIGHA pair* $\langle u_{i_k}, \tilde{a}_{i_k} \rangle$ *having the* j_k*th largest* u_{i_k}, u_{i_k} *is the order inducing variable,* $\omega = (\omega_1, \omega_2, \ldots, \omega_n)^T$ *is the weighting vector of the* $\tilde{a}_{i_k}$, *with* $\omega_i \in [0, 1]$ *and the sum of the weights is* 1, *and the* $\tilde{a}_{i_k}$ *are FNs, and* $m(B_k)$ *is the basic probability assignment and* λ *is a parameter such that* $\lambda \in (-\infty, \infty)$.

Note that the BS-Quasi-FIOWA and the BS-Quasi-FIHA operator have the same definition with the only difference that now we replace the parameter λ by a strictly continuous monotonic function $g(b)$.

In this case, we could study a wide range of particular cases that are formed as particular cases of the FIGOWA, the Quasi-FIOWA, the FIGHA and the Quasi-FIHA operators. Specially, it is worth noting the BS-FIOWGA, the BS-FIOWQA, the BS-FIOWHA, the BS-FIHGA, the BS-FIHQA and the BS-FIHHA operators.

7 Application in Financial Decision Making

In the following, we are going to develop an application of the new approach in a decision making problem. We will analyze a problem about selection of financial strategies where an enterprise is looking for its optimal financial strategy the next year. Note that other decision making applications could be developed such as the selection of financial products, the selection of human resources, etc.

We will develop the analysis considering a wide range of particular cases of fuzzy induced aggregation operators such as the FA, the FWA, the FOWA, the FIOWA and the FIHA operator.

Assume an enterprise is planning its financial strategy for the next year and they consider 5 possible financial strategies to follow.

- A_1 = Invest in the Asian market.
- A_2 = Invest in the South American market.
- A_3 = Invest in the African market.
- A_4 = Invest in all the three continents.
- A_5 = Do not develop any investment.

In order to evaluate these financial strategies, the group of experts considers that the key factor is the economic situation of the company for the next year. After careful analysis, the experts have considered five possible situations that could happen in the future: S_1 = Very bad, S_2 = Bad, S_3 = Regular, S_4 = Good, S_5 = Very good.

Depending on the uncertain situations that could happen in the future, the experts establish the payoff matrix. As the future states of nature are very imprecise, the experts cannot determine exact numbers in the payoff matrix. Instead, they use FNs to calculate the future benefits of the enterprise depending on the state of nature that happens in the future and the financial strategy selected. The results are shown in Table 1.

Table 1 Fuzzy payoff matrix

	S_1	S_2	S_3	S_4	S_5
A_1	(50,60,70)	(30,40,50)	(30,40,50)	(60,70,80)	(40,50,60)
A_2	(10,20,30)	(20,30,40)	(50,60,70)	(50,60,70)	(80,90,100)
A_3	(30,40,50)	(50,60,70)	(40,50,60)	(40,50,60)	(40,50,60)
A_4	(60,70,80)	(40,50,60)	(30,40,50)	(30,40,50)	(30,40,50)

After careful analysis of the information, the experts have obtained some probabilistic information about which state of nature will happen in the future. This information is represented by the following belief structure about the states of nature.

Focal element

$$B_1 = \{S_1, S_2, S_3\} = 0.3$$
$$B_2 = \{S_3, S_4, S_5\} = 0.3$$
$$B_3 = \{S_2, S_3, S_4, S_5\} = 0.4$$

The attitudinal character of the enterprise is very complex because it involves the opinion of different members of the board of directors. Therefore, the experts use order inducing variables for analyzing the attitudinal character of the enterprise. The results are shown in Table 2.

Table 2 Inducing variables

	S_1	S_2	S_3	S_4	S_5
A_1	7	6	4	9	2
A_2	1	5	7	9	3
A_3	4	3	8	6	5
A_4	2	5	6	7	8

The experts establish the following weighting vectors for both the FWA and the FIOWA.

Weighting vector

$$W = (0.3, 0.3, 0.4)$$

$$W = (0.2, 0.2, 0.3, 0.3)$$

$$W = (0.1, 0.2, 0.2, 0.2, 0.3)$$

With this information, we can obtain the fuzzy aggregated payoffs. The results are shown in Table 3.

Table 3 Fuzzy aggregated results

	FA	FWA	FOWA	FIOWA	FIHA
V_{11}	(36.6,46.6,56.6)	(36,46,56)	(36,46,56)	(36,46,56)	(28.5,37,45.5)
V_{12}	(43.3,53.3,63.3)	(43,53,63)	(42,52,62)	(43,53,63)	(51,63,75)
V_{13}	(40,50,60)	(42,52,62)	(38,48,58)	(39,49,59)	(45,56.5,68)
V_{21}	(26.6,36.6,46.6)	(29,39,49)	(25,35,45)	(25,35,45)	(23,31,39)
V_{22}	(60,70,80)	(62,72,82)	(59,69,79)	(62,72,82)	(78,90,102)
V_{23}	(50,60,70)	(53,63,73)	(47,57,67)	(50,60,70)	(62,73.5,85)
V_{31}	(40,50,60)	(40,50,60)	(39,49,59)	(41,51,61)	(36.5,45,53.5)
V_{32}	(40,50,60)	(40,50,60)	(40,50,60)	(40,50,60)	(48,60,72)
V_{33}	(42.5,52.5,62.5)	(42,52,62)	(42,52,62)	(43,53,63)	(49,60.5,72)
V_{41}	(43.3,53.3,63.3)	(42,52,62)	(42,52,62)	(45,55,65)	(33,41,49)
V_{42}	(30,40,50)	(30,40,50)	(30,40,50)	(30,40,50)	(34.5,46,57.5)
V_{43}	(32.5,42.5,52.5)	(32,42,52)	(32,42,52)	(33,43,53)	(36,47,58)

Once we have the aggregated results, we have to calculate the fuzzy generalized expected value. The results are shown in Table 4.

Table 4 Fuzzy generalized expected value

	FA	FWA	FOWA	FIOWA	FIHA
A_1	(40,50,60)	(40.5,50.5,60.5)	(38.6,48.6,58.6)	(39.3,49.3,59.3)	(41.8,52.6,63.3)
A_2	(46,56,66)	(48.5,58.5,68.5)	(44,54,64)	(46.1,56.1,66.1)	(55.1,65.7,76.3)
A_3	(41,51,61)	(40.8,50.8,60.8)	(40.5,50.5,60.5)	(41.5,51.5,61.5)	(44.9,55.7,66.4)
A_4	(35,45,55)	(34.4,44.4,54.4)	(34.4,44.4,54.4)	(35.7,45.7,55.7)	(34.6,44.9,55.1)

As we can see, depending on the fuzzy aggregation operator used, the results and decisions may be different. Note that in this case, our optimal choice is the same for all the aggregation operators but in other situations we may find different decisions between each aggregation operator.

A further interesting issue is to establish an ordering of the financial strategies. Note that this is very useful when the decision maker wants to consider more than one alternative. The results are shown in Table 5.

Table 5 Ordering of the financial strategies

	Ordering
FA	$A_2 \succ A_3 \succ A_1 \succ A_4$
FWA	$A_2 \succ A_3 \succ A_1 \succ A_4$
FOWA	$A_2 \succ A_3 \succ A_1 \succ A_4$
FIOWA	$A_2 \succ A_3 \succ A_1 \succ A_4$
FIHA	$A_2 \succ A_3 \succ A_1 \succ A_4$

As we can see, depending on the aggregation operator used, the ordering of the financial strategies may be different. Note that in this example the results are clear being A_2 the optimal choice.

8 Conclusions

We have studied the D-S theory of evidence in decision making with uncertain information represented in the form of FNs. With this approach, we have been able to assess the information in a more complete way because in this model we consider the different scenarios that could happen in the problem. For doing so, we have used different types of fuzzy induced aggregation operators in the decision process such as the FIOWA and the FIHA operators. Then, we have obtained new aggregation operators: the BS-FIOWA and the BS-FIHA operator. We have studied some of their main properties and different particular cases.

We have further generalized this approach by using generalized aggregation operators that use generalized and quasi-arithmetic means. We have used the FIGOWA, the Quasi-FIOWA, the FIGHA and the Quasi-FIHA operators obtaining the BS-FIGOWA, the BS-Quasi-FIOWA, the BS-FIGHA and the BS-Quasi-FIHA operators, respectively.

We have also developed an application of the new approach in a business decision making problem about selection of financial strategies. We have seen the usefulness of this approach about using probabilities, FWAs and FIOWAs in the same problem. We have also seen that depending on the fuzzy induced aggregation operator used the results may lead to different decisions.

In our future research, we expect to develop further extensions to this approach by adding new characteristics in the problem such as the use of more complete aggregation operators such as the ones explained by Merigó (2008) and applying it to other decision making problems.

References

Beliakov, G.: Learning Weights in the Generalized OWA Operators. Fuzzy Optimization and Decision Making 4, 119–130 (2005)

Beliakov, G., Calvo, T., Pradera, A.: Aggregation Functions: A Guide for Practitioners. Springer, Berlin (2007)

Calvo, T., Mayor, G., Mesiar, R.: Aggregation Operators: New Trends and Applications. Physica-Verlag, New York (2002)

Casanovas, M., Merigó, J.M.: Using fuzzy OWA operators in decision making with Dempster-Shafer belief structure. In: Proceedings of the AEDEM International Conference, Krakow, Poland, pp. 475–486 (2007)

Chang, S.S.L., Zadeh, L.A.: On fuzzy mapping and control. IEEE Transactions on Systems, Man and Cybernetics 2, 30–34 (1972)

Chen, S.J., Chen, S.M.: A new method for handling multi-criteria fuzzy decision making problems using FN-IOWA operators. Cybernetics and Systems 34, 109–137 (2003)

Chiclana, F., Herrera-Viedma, E., Herrera, F., Alonso, S.: Some induced ordered weighted averaging operators and their use for solving group decision-making problems based on fuzzy preference relations. European Journal of Operational Research 182, 383–399 (2007)

Dempster, A.P.: Upper and lower probabilities induced by a multi-valued mapping. Annals of Mathematical Statistics 38, 325–339 (1967)

Dubois, D., Prade, H.: Fuzzy Sets and Systems: Theory and Applications. Academic Press, New York (1980)

Engemann, K.J., Miller, H.E., Yager, R.R.: Decision making with belief structures: an application in risk management. International Journal of Uncertainty, Fuzziness and Knowledge-Based Systems 4, 1–26 (1996)

Fodor, J., Marichal, J.L., Roubens, M.: Characterization of the ordered weighted averaging operators. IEEE Transactions on Fuzzy Systems 3, 236–240 (1995)

Karayiannis, N.: Soft Learning Vector Quantization and Clustering Algorithms Based on Ordered Weighted Aggregation Operators. IEEE Transactions on Neural Networks 11, 1093–1105 (2000)

Kaufmann, A., Gupta, M.M.: Introduction to fuzzy arithmetic. Publications Van Nostrand, Rheinhold (1985)

Le, C.A., Huynh, V.N., Shimazu, A., Nakamori, Y.: Combining classifiers for word sense disambiguation based on Dempster-Shafer theory and OWA operators. Data & Knowledge Engineering 63, 381–396 (2007)

Merigó, J.M.: New extensions to the OWA operator and its application in decision making. PhD Thesis, Department of Business Administration, University of Barcelona (2008) (in Spanish)

Merigó, J.M., Casanovas, M.: The fuzzy generalized ordered weighted averaging operator. In: Proceedings of the 14th SIGEF Congress, Poiana-Brasov, Romania, pp. 504–517 (2007)

Merigó, J.M., Casanovas, M.: Decision making with Dempster-Shafer theory of evidence using geometric operators. International Journal of Computational Intelligence 4, 261–268 (2008)

Merigó, J.M., Casanovas, M.: Induced aggregation operators in decision making with Dempster-Shafer belief structure. International Journal of Intelligent Systems 24, 934–954 (2009a)

Merigó, J.M., Casanovas, M.: Fuzzy generalized hybrid aggregation operators and its application in fuzzy decision making, International Journal of Fuzzy Systems (2009b) (submitted for publication)
Merigó, J.M., Gil-Lafuente, A.M.: The induced generalized OWA operator. Information Sciences 179, 729–741 (2009a)
Merigó, J.M., Gil-Lafuente, A.M.: The fuzzy induced generalized OWA operator and its application in business decision making. In: Proceeding of the IFSA-EUSFLAT International Conference, Lisbon, Portugal, pp. 1661–1666 (2009b)
Moore, R.E.: Interval Analysis. Prentice Hall, Englewood Cliffs (1966)
Reformat, M., Yager, R.R.: Building ensemble classifiers using belief functions and OWA operators. Soft Computing 12, 543–558 (2008)
Shafer, G.A.: Mathematical Theory of Evidence. Princeton University Press, Princeton (1976)
Srivastava, R.P., Mock, T.: Belief Functions in Business Decisions. Physica-Verlag, Heidelberg (2002)
Torra, V., Narukawa, Y.: Modelling Decisions: Information Fusion and Aggregation Operators. Springer, Berlin (2007)
Wang, X.: Fuzzy number intuitionistic fuzzy arithmetic aggregation operators. International Journal of Fuzzy Systems 10, 104–111 (2008)
Xu, Z.S.: An overview of methods for determining OWA weights. International Journal of Intelligent Systems 20, 843–865 (2005)
Xu, Z.S.: A Note on Linguistic Hybrid Arithmetic Averaging Operator in Multiple Attribute Group Decision Making with Linguistic Information. Group Decision and Negotiation 15, 593–604 (2006)
Xu, Z.S., Da, Q.L.: An overview of operators for aggregating the information. International Journal of Intelligent Systems 18, 953–969 (2003)
Yager, R.R.: On Ordered Weighted Averaging Aggregation Operators in Multi-Criteria Decision Making. IEEE Transactions on Systems, Man and Cybernetics B 18, 183–190 (1988)
Yager, R.R.: Decision Making Under Dempster-Shafer Uncertainties. International Journal of General Systems 20, 233–245 (1992)
Yager, R.R.: Families of OWA operators. Fuzzy Sets and Systems 59, 125–148 (1993)
Yager, R.R.: Quantifier guided aggregation using OWA operators. International Journal of Intelligent Systems 11, 49–73 (1996)
Yager, R.R.: Induced aggregation operators. Fuzzy Sets and Systems 137, 59–69 (2003)
Yager, R.R.: Uncertainty modeling and decision support. Reliability Engineering and System Safety 85, 341–354 (2004a)
Yager, R.R.: Generalized OWA Aggregation Operators. Fuzzy Optimization and Decision Making 3, 93–107 (2004b)
Yager, R.R.: Centered OWA operators. Soft Computing 11, 631–639 (2007)
Yager, R.R.: Using trapezoids for representing granular objects: Applications to learning and OWA aggregation. Information Sciences 178, 363–380 (2008)
Yager, R.R., Fedrizzi, M., Kacprzyk, J.: Advances in the Dempster-Shafer theory of evidence. John Wiley & Sons, New York (1994)
Yager, R.R., Filev, D.P.: Induced ordered weighted averaging operators. IEEE Transaction on Systems, Man and Cybernetics 29, 141–150 (1999)
Yager, R.R., Kacprzyk, J.: The Ordered Weighted Averaging Operators: Theory and Applications. Kluwer Academic Publishers, Norwell (1997)

Yager, R.R., Liu, L.: Classic Works of the Dempster-Shafer Theory of Belief Functions. Springer, Berlin (2008)

Zadeh, L.A.: Fuzzy Sets. Information and Control 8, 338–353 (1965)

Zadeh, L.A.: The Concept of a Linguistic Variable and its application to Approximate Reasoning, Part 1, Information Sciences, 8, pp. 199-249; Part 2, Information Sciences, 8, pp. 301-357; Part 3, Information Sciences, 9, pp. 43-80 (1975)

Zarghami, M., Szidarovszky, F., Ardakanian, R.: A fuzzy-stochastic OWA model for robust multi-criteria decision making. Fuzzy Optimization and Decision Making 7, 11–15 (2008)

Two Methods for Image Compression/Reconstruction Using OWA Operators

H. Bustince, D. Paternain, B. De Baets, T. Calvo, J. Fodor,
R. Mesiar, J. Montero, and A. Pradera

Abstract. In this chapter we address image compression by means of two alternative algorithms. In the first algorithm, we associate to each image an interval-valued fuzzy relation, and we build an image which is n times smaller than the original

H. Bustince · D. Paternain
Universidad Pública de Navarra, Campus Arrosadía s/n, 31006, Pamplona, Spain
e-mail: bustince@unavarra.es

B. De Baets
Dept. of Applied Mathematics, Biometrics and Process Control,
Universiteit Gent Coupure links 653, 9000 Gent, Belgium
e-mail: bernard.debaets@ugent.be

T. Calvo
Universidad de Alcalá, Spain
e-mail: tomasa.calvo@uah.es

J. Fodor
Institute of Intelligent Engineering Systems, Budapest Tech, Bécsi út 96/b,
H-1034 Budapest, Hungary
e-mail: fodor@bmf.hu

R. Mesiar
Department of Mathematics and Descriptive Geometry, Slovak University of Technology,
Radlinskho11,81368 Bratislava, Slovakia
e-mail: mesiar@math.sk

J. Montero
Facultad de Matemáticas, Universidad Complutense, 28040 Madrid, Spain
e-mail: monty@mat.ucm.es

A. Pradera
Departamento de Ciencias de la Computación, Universidad Rey Juan Carlos 28933 Móstoles,
Madrid, Spain
e-mail: ana.pradera@urjc.es

R.R. Yager et al. (Eds.): Recent Developments in the OWA Operators, STUDFUZZ 265, pp. 229–253.
springerlink.com

one, by using two-dimensional OWA operators. The experimental results show that, in this case, best results are obtained with ME-OWA operators. In the second part of the work, we describe a reduction algorithm that replaces the image by several eigen fuzzy sets associated with it. We obtain these eigen fuzzy sets by means of an equation that relates the OWA operators we use and the relation (image) we consider. Finally, we present a reconstruction method based on an algorithm which minimizes a cost function, with this cost function built by means of two-dimensional OWA operators.

1 Introduction

The goal of algorithms of compression in image processing is to reduce the size of the considered images with a loss of information as small as possible (see [13]).

In the literature, there exist (see [16, 17, 18]) different techniques to compress and reconstruct images. In [3, 4, 9, 12], it is proved that Fuzzy Sets Theory provides a good tool for image processing. Besides, OWA operators were introduced by Yager in 1988 as an aggregation technique (see [23, 24]). Moreover, in 1983 Atanassov presented a class of operators that associate with each interval a point inside it (see [2]). We will see that under appropriate conditions, we can build two-dimensional OWA operators from Atanassov's operators (see [6]). Such two-dimensional OWA operators should play, as claimed out in [7], a key role in order to build up families of OWA operators that allow an operational and consistent reckoning [8, 1] of arbitrary number of information units, a problem indeed related to the structural issue in aggregation recently stressed in [14].

In this chapter we present two methods for image compression. They have in common the following: *They use two-dimensional OWA operators (constructed from Atanassov's operators) and interval-valued fuzzy sets (IVFSs)* (see [5, 11, 19]). Nevertheless, there are big differences between both methods, as, for instance:

A) In the first algorithm we go from an image of dimension $N \times M$ to another n times smaller.
B) In the second algorithm, which is a generalization of the algorithm proposed by Nobuhara et al. in [17], we go from an image to a set of eigen fuzzy sets associated with it.

To know if an image compression method is good, we must reconstruct the image and compare it to the original one. In this chapter we propose two methods of reconstruction. The first one consists of surrounding each pixel with a set of pixels of the same intensity; the second method is a generalization of the one in [16, 17], and it is characterized by the use of two-dimensional OWA operators and of the gradient algorithm to minimize an appropriate cost function. We will see that, due to the nature of our compressions, the method proposed in item A) provides better reconstructions than those obtained by means of B).

We have organized this work in the following way: In Section 2 we start with some preliminary definitions which are necessary for this chapter. Next, in

Section 3, we show the relation between OWA and Atanassov's operators. In Section 4 we present our first method of compression, which is based on IVFS. In Section 5 we consider a second algorithm of compression, based on eigen fuzzy sets. We compare results by reconstructing the images, and we conclude with a final comments section.

2 Preliminary Definitions

Let $X = \{0,1,\ldots,N-1\}$ and $Y = \{0,1,\ldots,M-1\}$ be two finite, non empty referential sets . We will denote by $FS(X)$ the set of all fuzzy sets defined over the referential X.

Definition 1. A fuzzy relation R over the referential $X \times Y$ is a fuzzy set given by

$$R = \{(x,y), \mu_R(x,y) = R(x,y))|(x,y) \in X \times Y\},$$

where $FR(X \times Y)$ denotes the set of all fuzzy relations defined on $X \times Y$.

For us, an image Q of $N \times M$ pixels is going to be a fuzzy relation R of $N \times M$ elements. Each element of the relation has as membership value the normalized intensity of the corresponding pixel. We normalize by dividing the value of the intensity of each pixel by $L-1$ (with L the number of gray levels starting from 0).

We will denote by $L([0,1])$ the set of all closed subintervals of the unit interval $[0,1]$, that is,

$$L([0,1]) = \{\mathbf{x} = [\underline{x},\overline{x}]|(\underline{x},\overline{x}) \in [0,1]^2 \text{ and } \underline{x} \leq \overline{x}\}. \tag{1}$$

$L([0,1])$ is a partially ordered set with respect to the order relationship $\leq_L$ defined in the following way: given $\mathbf{x},\mathbf{y} \in L([0,1])$

$$\mathbf{x} \leq_L \mathbf{y} \text{ if and only if } \underline{x} \leq \underline{y} \text{ and } \overline{x} \leq \overline{y}. \tag{2}$$

With this order relationship, $(L([0,1]), \leq_L)$ is a complete lattice (see [5, 10]), where the smallest element is $0_L = [0,0]$ and the largest is $1_L = [1,1]$.

Definition 2. An interval-valued fuzzy set ($IVFS$) A on the universe $U \neq \emptyset$ is a mapping $\mathbf{A} : U \to L([0,1])$.

For us $\mathbf{A}(u) = [\underline{A}(u), \overline{A}(u)] \in L([0,1])$ will be the membership degree of $u \in U$, with $\underline{A}(u)$, $\overline{A}(u) \in [0,1]$ denoting the lower bound and the upper bound respectively of the membership associated to u.

In 1983, Atanassov proposed the following operator:

Definition 3. Let $\alpha \in [0,1]$. The operator $K_\alpha : L([0,1]) \to [0,1]$ is defined as a convex combination of the bounds of its argument by

$$K_\alpha(\mathbf{x}) = \underline{x} + \alpha(\overline{x} - \underline{x})$$

for all $\mathbf{x} \in L([0,1])$.

Clearly, the following properties hold:

1. $K_0(\mathbf{x}) = \underline{x}$ for all $\mathbf{x} \in L([0,1])$,
2. $K_1(\mathbf{x}) = \overline{x}$ for all $\mathbf{x} \in L([0,1])$,
3. $K_\alpha(\mathbf{x}) = K_\alpha([K_0(\mathbf{x}), K_1(\mathbf{x})]) = K_0(\mathbf{x}) + \alpha(K_1(\mathbf{x}) - K_0(\mathbf{x}))$ for all $\mathbf{x} \in L([0,1])$.

Observe that, if we denote by K the system of operators $(K_\alpha)_{\alpha \in [0,1]}$, then K can be regarded as an operator on $[0,1] \times L([0,1])$. In [5, 6] a generalization of this operator is proposed.

3 K_α and OWA Operators

3.1 Relation between Atanassov's Operators and OWA Operators

Let us remind here that Yager ([21]) defined OWA operators in the following way.

Definition 4. A mapping $F : [0,1]^n \to [0,1]$ is called an OWA operator of dimension n if there exists a weighting vector W, $W = (w_1, w_2, \ldots, w_n) \in [0,1]^n$ with $\sum_i w_i = 1$ and such that

$$F(a_1, a_2, \ldots, a_n) = \sum_{j=1}^{n} w_j b_j$$

with b_j the j-th largest of the a_i.

If we reduce ourselves to consider two-dimensional OWA operators using as weighting vector $W = (\alpha, 1-\alpha)$, we can think of applying these operators to the bounds of an interval. In this case, the numerical result of the action of the OWA operator over the bounds of an interval and the numerical result of the K_α operator acting over that interval are the same. Nevertheless, these two operators are very different conceptually. Whereas K_α operators act over intervals defined in $L([0,1])$, the domain of two-dimensional OWA operators is $[0,1] \times [0,1]$. For this reason, to relate both concepts, OWA operators of dimension two require a reordering operation, ensuring that the two points they are acting over are actually the bounds of an interval in $L([0,1])$. This is given by the following theorem.

We define (see [6]) a new operator $\mathbb{K}_\alpha$ by composing the K_α operator with the map

$$\begin{aligned} i : [0,1]^2 &\to L([0,1]) \\ (x,y) &\to [\min(x,y), \max(x,y)] \end{aligned} \tag{3}$$

Theorem 1. *1. Let $\alpha \in [0,1]$ and $\mathbb{K}_\alpha = K_\alpha \circ i$ where K_α is the operator given in Definition 3. Then, if F is the OWA operator (of dimension 2) defined by the weighting vector $W = (\alpha, 1-\alpha)$, we have that*

$$\mathbb{K}_\alpha(x,y) = F(x,y) \text{ for all } x,y \in [0,1].$$

2. Let F be an OWA operator (of dimension 2) with weighting vector $W = (w_1, w_2)$. Then for any $(x,y) \in [0,1]^2$ we have that

$$F(x,y) = \mathbb{K}_\alpha(x,y), \text{ with } \alpha = w_1.$$

From now on in this chapter, we will take: $\mathbb{K}_\alpha = K_\alpha \circ i$.

3.2 Calculation of the Coefficient α by Means of Families of OWA Operators

Now, we study how to calculate the α coefficient using different families of OWA operators, ME-OWA and BADD-OWA operators.

First of all, we recall the definition of orness of an OWA operator ([23]).

Definition 5. Let F be an OWA operator and W its weighting vector. The orness measure is defined as

$$orness(F) = \frac{1}{(n-1)} \sum_{i=1}^{n} (n-i) w_i.$$

Proposition 1. $orness(\mathbb{K}_\alpha) = \alpha$

Definition 6. F is called a ME-OWA operator if F is an OWA operator such that given a desired value of orness β, it maximizes the entropy. In particular we solve the following problem:

$$\begin{array}{ll} \text{Max} & -\sum_{i=1}^{n} w_i ln w_i \\ \text{subject to} & \beta = \frac{1}{(n-1)} \sum_{i=1}^{n} (n-i) w_i \\ \text{where} & \sum_{i=1}^{n} w_i = 1,\ w_i \in [0,1]. \end{array}$$

Proposition 2. *The operator $\mathbb{K}_\alpha$ is a ME-OWA for all $\alpha \in [0,1]$, with $orness(\mathbb{K}_\alpha) = \alpha$.*

Definition 7. F is called a BADD-OWA operator if F is an OWA operator where

$$w_i = \frac{b_i^\beta}{\sum_{j=1}^{n} b_j^\beta}$$

with $\beta \geq 0$ and being b_i as in Definition 4.

Proposition 3. *Let* $\beta \geq 0$. *If we take*

$$\alpha = \frac{(\mathbb{K}_1(x,y))^\beta}{(\mathbb{K}_0(x,y))^\beta + (\mathbb{K}_1(x,y))^\beta}$$

with $x,y \in [0,1]$, *then* $\mathbb{K}_\alpha$ *is a BADD-OWA operator.*

4 First Algorithm for Compression and Reconstruction

The algorithm that we propose consists of two steps:

1. Associate with each image of size $N \times M$ an interval-valued fuzzy set of $\frac{N}{n} \times \frac{M}{n}$ elements, with $n = 2,3,\cdots, n \leq N$ and $n \leq M$ in case N and M are multiples of n. If this is not the case, we will eliminate as many rows and/or columns as it is necessary to have that n exactly divides N and M.
2. Build an image of dimension $\frac{N}{n} \times \frac{M}{n}$ such that the intensity of each pixel is obtained by applying the two-dimensional OWA operator we are considering to the interval that represents the membership of that pixel to the $IVFS$ we have built in the previous step, obviously multiplied by $L-1$ (with L the number of gray levels in the image).

In the following subsections we detail in a very precise way this compression algorithm, as well as a reconstruction algorithm and the corresponding experimental results. All along the paper we are going to use as test images : Lena and Camera (see Figure 1).

Fig. 1 Original Lena and Camera images

4.1 Specification of the First Algorithm of Compression

We consider an image (relation) Q of dimension $N \times M$. The intensity or gray level of the pixel (or the membership of each element of the relation) located in (i,j) is denoted by q_{ij}, with $0 \leq q_{ij} \leq 1$ and $i = 1,\ldots,N$ and $j = 1,\ldots,M$.

1. Divide the image Q in blocks of size $n \times n$. If M or N are not multiple of n, we delete the minimum number of rows/columns in the boundary of the image until the new size of the image satisfies the property.
2. Associate each block with an interval in the following way: the lower bound of the interval is given by the minimum of the intensities in the block and the upper bound by the maximum.
3. Use Propositions 1,2 and 3 to calculate the parameter α for the operator K_α.
4. Associate with each interval the number obtained after applying the operator K_α.

Example: Let Q be an image of dimension 6×6 and let $n = 3$

$$\begin{pmatrix} q_{1,1} & q_{1,2} & q_{1,3} & q_{1,4} & q_{1,5} & q_{1,6} \\ q_{2,1} & q_{2,2} & q_{2,3} & q_{2,4} & q_{2,5} & q_{2,6} \\ q_{3,1} & q_{3,2} & q_{3,3} & q_{3,4} & q_{3,5} & q_{3,6} \\ q_{4,1} & q_{4,2} & q_{4,3} & q_{4,4} & q_{4,5} & q_{4,6} \\ q_{5,1} & q_{5,2} & q_{5,3} & q_{5,4} & q_{5,5} & q_{5,6} \\ q_{6,1} & q_{6,2} & q_{6,3} & q_{6,4} & q_{6,5} & q_{6,6} \end{pmatrix}$$

Then, the interval-valued fuzzy set associated to Q will be formed by 4 elements:

$$\begin{pmatrix} \left[\bigwedge_{\substack{i=1,2,3 \\ j=1,2,3}} q_{i,j} \quad \bigvee_{\substack{i=1,2,3 \\ j=1,2,3}} q_{i,j} \right] & \left[\bigwedge_{\substack{i=1,2,3 \\ j=4,5,6}} q_{i,j} \quad \bigvee_{\substack{i=1,2,3 \\ j=4,5,6}} q_{i,j} \right] \\ \left[\bigwedge_{\substack{i=4,5,6 \\ j=1,2,3}} q_{i,j} \quad \bigvee_{\substack{i=4,5,6 \\ j=1,2,3}} q_{i,j} \right] & \left[\bigwedge_{\substack{i=4,5,6 \\ j=4,5,6}} q_{i,j} \quad \bigvee_{\substack{i=4,5,6 \\ j=4,5,6}} q_{i,j} \right] \end{pmatrix}$$

Remark: Symbols $\wedge$ and $\vee$ stand for minimum and maximum respectively. We take $\alpha \in [0,1]$ and $K_\alpha([\underline{x},\overline{x}]) = \alpha\overline{x} + (1-\alpha)\underline{x}$. Then we have:

$$(L-1) \begin{pmatrix} \alpha(\bigvee_{\substack{i=1,2,3 \\ j=1,2,3}} q_{i,j}) + (1-\alpha)(\bigwedge_{\substack{i=1,2,3 \\ j=1,2,3}} q_{i,j}) & \alpha(\bigvee_{\substack{i=1,2,3 \\ j=4,5,6}} q_{i,j}) + (1-\alpha)(\bigwedge_{\substack{i=1,2,3 \\ j=4,5,6}} q_{i,j}) \\ \alpha(\bigvee_{\substack{i=4,5,6 \\ j=1,2,3}} q_{i,j}) + (1-\alpha)(\bigwedge_{\substack{i=4,5,6 \\ j=1,2,3}} q_{i,j}) & \alpha(\bigvee_{\substack{i=4,5,6 \\ j=4,5,6}} q_{i,j}) + (1-\alpha)(\bigwedge_{\substack{i=4,5,6 \\ j=4,5,6}} q_{i,j}) \end{pmatrix}$$

Once the interval-valued fuzzy set associated to the image has been obtained, we build a fuzzy set. Applying the operator K_α to each interval we get the reduced image of size 2×2.

Next we show two methods for calculating the value of the α coefficient following the theoretical developments in Subsection 3.2.

4.1.1 Compression with Constant α

The aim of this subsection is to apply the algorithm that we have developed in the previous section using a constant parameter α for the whole image. To calculate that fixed value we use ME-OWA operators.

As we have seen in Proposition 2, the construction of an ME-OWA operator is straight. We are going to consider only the following cases: $\alpha = 0$, $\alpha = 0.5$ and $\alpha = 1$. The motivations for these choices are the following: With $\alpha = 0$, we associate the lower bound of the interval to each block, with $\alpha = 0.5$, we take the mean point of the interval and finally, with $\alpha = 1$, we associate the upper bound of the interval.

Fig. 2 Compression of Lena and Camera images with ME-OWA operators and compression block size $n = 2$ and $n = 3$

Obviously, the higher the value of α, the higher the membership degree and therefore the intensity of each pixel. The image is darker with $\alpha = 0$ than with $\alpha = 0.5$, which is also darker than with $\alpha = 1$.

4.1.2 Compression with Variable α

In this subsection, we develop the previous algorithm by taking for each block a different value of α. We do the calculation of the values of α using BADD-OWA operators. Hence, we calculate the value of α in terms of the bounds of the interval we are considering.

As we see in Definition 7, to construct a BADD-OWA operator it is necessary to select a value of $\beta \geq 0$. We know that if we take $\beta = 0$, then we get $\alpha = 0.5$, already studied in Subsection 4.1.1. In other case, if we take $\beta = 1$, the value of α is calculated as follows:

$$\alpha = \frac{\overline{x}}{\overline{x} + \underline{x}}$$

being $[\underline{x}, \overline{x}] \in L([0,1])$ the interval representing each block. We also know that $\alpha \geq 0.5$, and if we increase the value of β, $\alpha \to 1$. That is, the result tends to the upper bound of the interval. For this reason, with a high value of β, we get reduced images similar as the images analyzed in third column of Figure 2.

4.2 *First Algorithm of Reconstruction*

The algorithm that we propose is the following:

1. Create a a block of size $n \times n$ for each pixel of the compressed image, centered on the considered pixel.
2. Copy the value of the central pixel for each of the elements of the $n \times n$ block we have constructed around it.

With this construction, we get an image of the same size of the original, so we can compare them. To evaluate the quality of the reconstructed image we are going to use the Mean Squared Error (MSE); that is,

$$MSE(Q, Q') = \frac{1}{MN} \sum_{i=1}^{M} \sum_{j=1}^{N} (q_{ij} - q'_{ij})^2 \tag{4}$$

with q_{ij} the intensity of pixel (i, j) in the original image and q'_{ij} the intensity of the same pixel in the reconstructed image.

4.3 *First Experimental Results*

In this section we show the results that we obtain when we apply the compression/reconstruction algorithms that we have studied in Subsections 4.1 and 4.2.

In the first two rows of Figure 2 we show the result of the algorithm of compression when we apply it to images in Figure 1, taking $n = 2$ and α constant, in the following way: $\alpha = 0$ for the first column, $\alpha = 0.5$ for the second column and $\alpha = 1$ for the third column. In the last two rows of Figure 2 we show the result of the algorithm when we take $n = 3$.

In Figures 4 and 5 we show the results of reconstructing the images in Figure 2 using in all of the cases the algorithm developed in Subsection 4.2.

In Table 1 we study the MSE of the reconstructions with respect to the original image for some values of α (constant for the whole image once it has been chosen) and different sizes of the compression blocks. Notice that, in order to analyze in a more detailed way the error in our method, we have extended the experiment and considered more than three values of α and n, as it is clear from the Tables.

From the analysis of Table 1, we deduce experimentally the following:

1. We obtain the smallest error when $\alpha = 0.5$.
2. Error increases as α goes to zero or one.
3. In all of the cases, we obtain the maximum error for $\alpha = 1$.

Clearly, taking into account the nature of our algorithm, error increases if the window size n becomes bigger.

In the first two rows of Figure 3 we show the result of applying the algorithm of compression to images of Figure 1 taking $n = 2$ and α variable in terms of the parameter β associated with the BADD-OWA operator. We have taken $\beta = 1, 10, 20, 100$ for the first, second, third and last column, respectively. In the last two rows of Figure 3 we show the result with $n = 3$.

In Figures 6,7,8 and 9 we present the results of reconstructing the images of Figure 3 using in all of the cases the algorithm we have described in Subsection 4.2.

In Table 2 we show in a detailed way the MSE for Lena and Camera images, varying the parameter β of the BADD-OWA operator and the size of the compression block.

Table 1 Error in the reconstruction of Lena (left) and Camera (right) images using the first algorithm of compression and ME-OWA operators

	$n=2$	$n=3$	$n=4$	$n=5$
$\alpha=0$	284.15	655.38	1012.35	1414.89
$\alpha=0.25$	156.42	338.2	504.14	699.13
$\alpha=0.375$	125.88	262.75	389.36	531.46
$\alpha=0.5$	116.81	242.71	364.85	490.61
$\alpha=0.625$	129.21	278.11	430.6	576.56
$\alpha=0.75$	163.1	368.92	586.61	789.32
$\alpha=1$	295.29	716.83	1169.4	1595.65

	$n=2$	$n=3$	$n=4$	$n=5$
$\alpha=0$	452.08	947.14	1379.99	1723.91
$\alpha=0.25$	254.87	507.65	720.08	893.78
$\alpha=0.375$	207.99	408.26	576.36	720.23
$\alpha=0.5$	194.41	389.11	553.2	707.7
$\alpha=0.625$	214.14	450.21	650.62	856.19
$\alpha=0.75$	267.18	591.55	868.61	1165.65
$\alpha=1$	473.17	1114.94	1666.29	2267.65

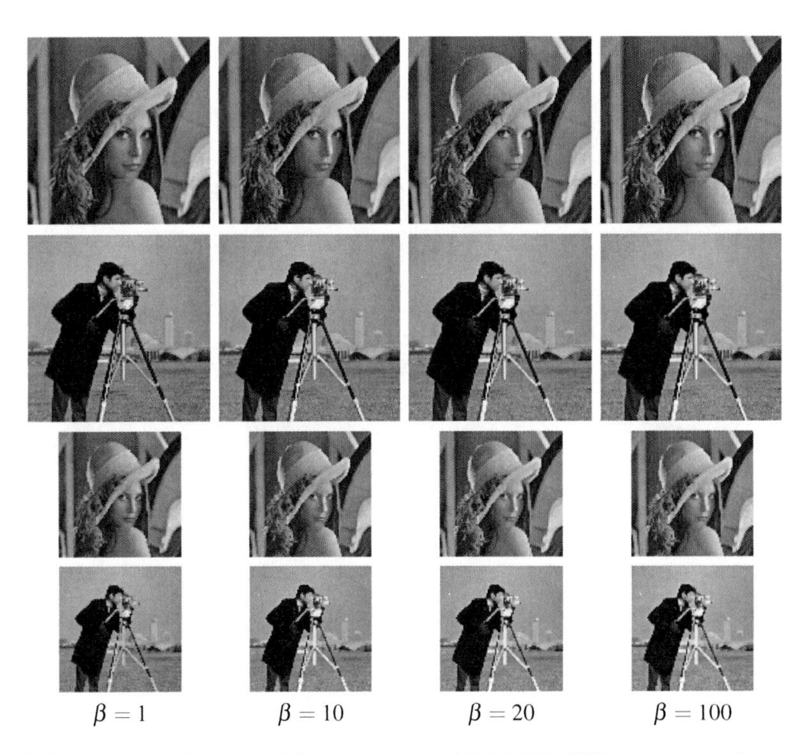

Fig. 3 Compression of Lena and Camera images with BADD-OWA operators and compression block size $n = 2$ and $n = 3$

Table 2 Error in the reconstruction of Lena (left) and Camera (right) images using the first algorithm of compression and BADD-OWA operators

	$n = 2$	$n = 3$	$n = 4$	$n = 5$
$\beta = 1$	164.27	424.05	736.22	1044.08
$\beta = 10$	281.73	699.23	1150.36	1575.22
$\beta = 20$	290.23	711.01	1163.52	1589.69
$\beta = 100$	295.07	716.56	1169.33	1595.22

	$n = 2$	$n = 3$	$n = 4$	$n = 5$
$\beta = 1$	302.96	771.07	1211.37	1693.29
$\beta = 10$	457.74	1097.41	1648.69	2249.31
$\beta = 20$	468.23	1109.37	1660.32	2261.06
$\beta = 100$	472.75	1114.56	1665.99	2267.43

From the analysis of Table 2 we deduce experimentally the following:

1. We obtain the smallest error if we take $\beta = 1$.
2. Error increases if β increases.

From the comparison of Tables 1 and 2 we deduce that we achieve the best results taking α constant and equal to 0.5.

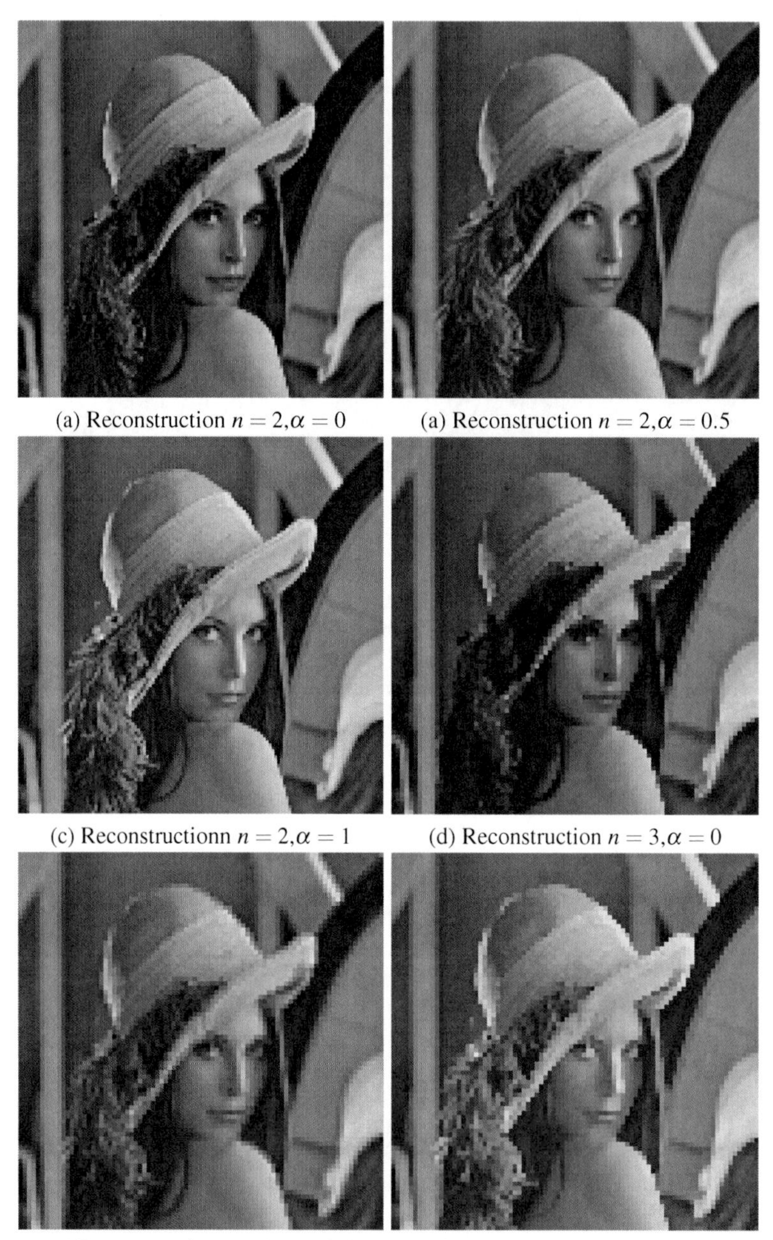

(a) Reconstruction $n = 2, \alpha = 0$ (a) Reconstruction $n = 2, \alpha = 0.5$

(c) Reconstructionn $n = 2, \alpha = 1$ (d) Reconstruction $n = 3, \alpha = 0$

(e) Reconstruction $n = 3, \alpha = 0.5$ (f) Reconstruction $n = 3, \alpha = 1$

Fig. 4 Reconstructions of Lena image using the first algorithm of compression and ME-OWA operators

(a) Reconstruction $n = 2, \alpha = 0$

(a) Reconstruction $n = 2, \alpha = 0.5$

(c) Reconstruction $n = 2, \alpha = 1$

(d) Reconstruction $n = 3, \alpha = 0$

(e) Reconstruction $n = 3, \alpha = 0.5$

(f) Reconstruction $n = 3, \alpha = 1$

Fig. 5 Reconstructions of Camera image using the first algorithm of compression and ME-OWA operators

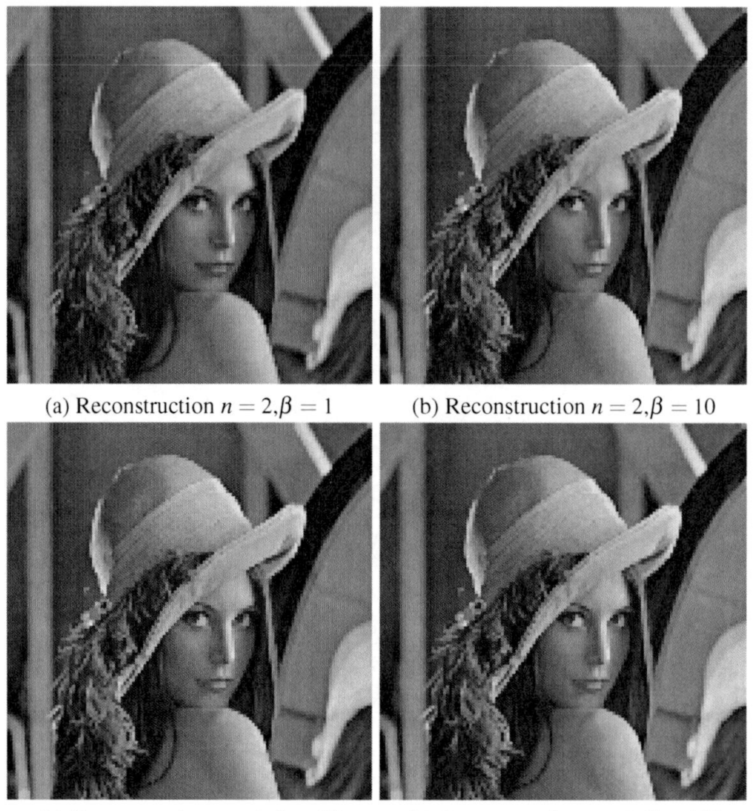

(a) Reconstruction $n = 2, \beta = 1$ (b) Reconstruction $n = 2, \beta = 10$

(a) Reconstruction $n = 2, \beta = 20$ (b) Reconstruction $n = 2, \beta = 100$

Fig. 6 Reconstructions of Lena image using the first algorithm of compression and BADD-OWA operators. Compression block size $n = 2$.

5 Eigen Fuzzy Sets and Compression/Reconstruction of Images

In [17], Nobuhara et al. use the concept of eigen fuzzy set associated with a fuzzy relation R to formulate two algorithms: one for image compression and another for image reconstruction. In the compression algorithm the original image is replaced by a set of eigen fuzzy sets of it (see [20, 21]). In the following, we recall the concepts of greatest and smallest eigen fuzzy set associated with a fuzzy relation R. Actually, Nobuhara et al. present two algorithms of compression: one using the smallest and the greatest eigen fuzzy sets, and the other using a family of eigen fuzzy sets obtained as convex linear combination of the smallest and the greatest eigen fuzzy sets. We will see that the latter algorithm can be generalized by means of Theorem 1; that is, using operators $\mathbb{K}_\alpha = K_\alpha \circ i$.

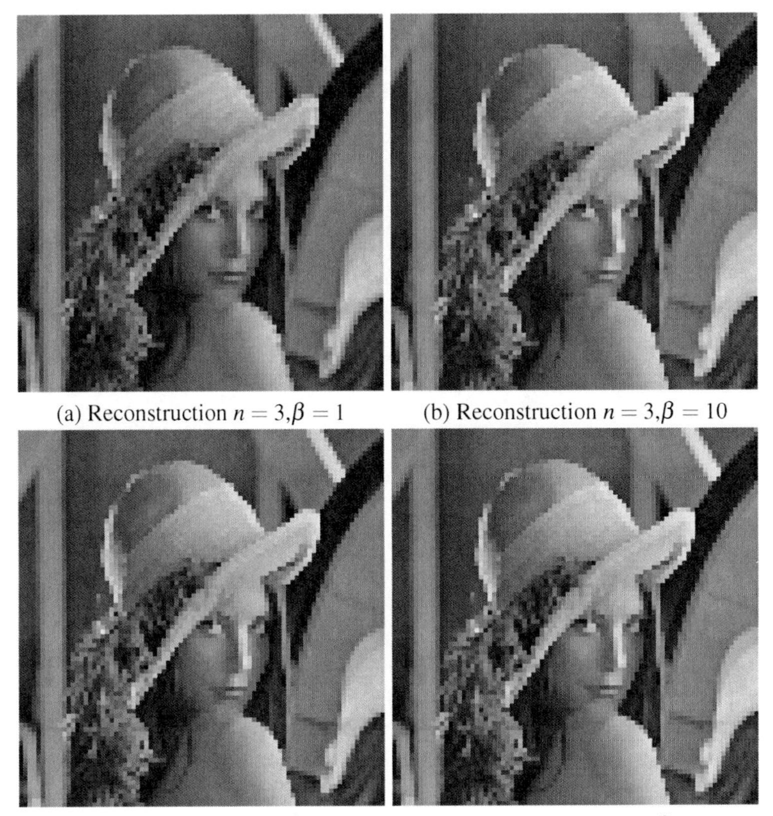

(a) Reconstruction $n = 3, \beta = 1$ (b) Reconstruction $n = 3, \beta = 10$

(a) Reconstruction $n = 3, \beta = 20$ (b) Reconstruction $n = 3, \beta = 100$

Fig. 7 Reconstructions of Lena image using the first algorithm of compression and BADD-OWA operators. Compression block size $n = 3$.

Definition 8. Let $R \in FR(X \times X)$ be a fuzzy relation over $X \times X$; and let $A \in FS(X)$ be a fuzzy set. A is an eigen fuzzy set-(max − min) associated with R if A satisfies:

$$A \circ R = A \tag{5}$$

where

$$A \circ R(x') = \bigvee_{x \in X} A(x) \wedge R(x, x') \text{ for all } x' \in \mathbf{X} \tag{6}$$

In [17], it can be found an iterative algorithm to find a solution of Equation (5), and it is proved that the eigen fuzzy set-(max − min) obtained is the biggest of the eigen fuzzy sets associated with R verifying Equation (5) using (6). This eigen fuzzy set-(max − min) is denoted A_G.

(a) Reconstruction $n = 2, \beta = 1$ (b) Reconstruction $n = 2, \beta = 10$

(a) Reconstruction $n = 2, \beta = 20$ (b) Reconstruction $n = 2, \beta = 100$

Fig. 8 Reconstructions of Camera image using the first algorithm of compression and BADD-OWA operators. Compression block size $n = 2$.

Definition 9. Let $R \in FR(X \times X)$ be a fuzzy relation over $X \times X$; and let $A \in FS(X)$ be a fuzzy set. A is an eigen fuzzy set-(min − max) associated with R if A satisfies:

$$A \bullet R = A \tag{7}$$

where

$$A \bullet R(x') = \bigwedge_{x \in X} A(x) \vee R(x, x') \ \forall x' \in \mathbf{X} \tag{8}$$

In [17], it can also be found an iterative algorithm to find a solution of Equation (7) and it is proved that the eigen fuzzy set-(min − max) obtained is the smallest of the eigen fuzzy sets associated with R verifying Equation (7) using (8). This eigen fuzzy set is denoted A_S.

(a) Reconstruction $n = 3, \beta = 1$ (b) Reconstruction $n = 3, \beta = 10$

(a) Reconstruction $n = 3, \beta = 20$ (b) Reconstructionn $n = 3, \beta = 100$

Fig. 9 Reconstructions of Camera image using the first algorithm of compression and BADD-OWA operators. Compression block size $n = 3$.

Proposition 4. *Let $R \in FR(X \times X)$ and let $A_S, A_G \in FS(X)$ be the smallest and greatest of the eigen fuzzy sets associated with R, respectively. Then, the set*

$$\boldsymbol{A}(x_i) = [A_S(x_i), A_G(x_i)]$$

is an $IVFS$ on the referential X.

Proof: Direct. □.

Proposition 4 allows us to associate to each square relation R over $X \times X$, and, hence, to each square image, an interval-valued fuzzy set build using the eigen fuzzy sets A_S and A_G. If the original image is not squared, we must supress the appropriate number of rows and/or columns to get a squared one.

From the *IVFS* built as stated and the operator K_α we can construct a family of fuzzy sets associated with the relation R in the following way:

$$K_\alpha(x) = \alpha A_G(x) + (1-\alpha)A_S(x) \text{ with } \alpha \in [0,1] \text{ and for all } x \in X \tag{9}$$

Basically, the algorithm of Nobuhara et al. starts from a fixed fuzzy set (the seed) and proceeds iteratively to create a convergent sequence of fuzzy sets. We should remark that if we use the fuzzy sets generated by means of 9 as seed to generate eigen fuzzy sets associated to R in the algorithm proposed by Nobuhara et al., it may happen that:

$$\begin{array}{ll} K_\alpha(A) = A_0 & \quad K_\alpha(A) = A_0 \\ A_0 \circ R = A_1 & \quad A_0 \bullet R = A_1 \\ A_1 \circ R = A_2 & \quad A_1 \bullet R = A_2 \\ \dots & \quad \dots \end{array}$$

are not convergent sequences, and, therefore, we can not actually generate eigen fuzzy sets. This fact has led us to obtain eigen fuzzy sets by means of operators $\mathbb{K}_\lambda$. We will see later that these operators give raise to convergent sequences.

In fact, in [17] Nobuhara et al. propose three different algorithms for compression. In the first one, they exclusively use A_S, in the second one, A_G, and in the third one, both of them. They show experimentally that the best results are obtained when the two eigen fuzzy sets are used. For this reason, the authors claim that results improve when the number of eigen fuzzy sets increases. In particular, this led them to propose a new algorithm using P eigen fuzzy sets. In this chapter, we present a generalization of this algorithm.

Take $R \in FR(X \times X)$ and $A \in FS(X)$. It is easy to verify that $A \circ R$ and $A \bullet R$ are not comparable; that is, in general, we can not assert which of the two compositions is greater. This observation has led us to study equations of the type

$$\lambda(\vee(A \circ R, A \bullet R)) + (1-\lambda)(\wedge(A \circ R, A \bullet R)) = A \text{ with } \lambda \in [0,1], \tag{10}$$

instead of the equations considered by Nobuhara et al. in [17], that is, $\lambda(A \circ R) + (1-\lambda)(A \bullet R) = A$. We must stress that we can write Equation (10) using Theorem 1 as follows:

$$\mathbb{K}_\lambda(A \circ R, A \bullet R) = A$$

We will prove that for any given $\lambda \in [0,1]$ and any $R \in FR(X \times X)$ there exists at least a fixed point, that is, there exists at least one fuzzy set $A \in FS(X)$ satisfying Equation (10).

Lemma 1. *Let $A, B \in FS(X)$ such that $A \leq B$. Then, for all $\lambda \in [0,1]$ and for all $R \in FR(X \times X)$ the following inequality holds:*

$$\mathbb{K}_\lambda(A \circ R, A \bullet R) \leq \mathbb{K}_\lambda(B \circ R, B \bullet R)$$

Proof: Clearly, $A \circ R(x') = \bigvee_{x \in X} A(x) \wedge R(x,x') \leq \bigvee_{x \in X} B(x) \wedge R(x,x') = B \circ R(x')$ and $A \bullet R(x') = \bigwedge_{x \in X} A(x) \vee R(x,x') \leq \bigwedge_{x \in X} B(x) \vee R(x,x') = B \bullet R(x')$. Bearing in mind that $\wedge$, $\vee$ and $K_\lambda(x,y) = \lambda x + (1-\lambda)y$ are increasing functions, the result follows. □

Theorem 2. *Let $\lambda \in [0,1]$. Then, for any relation $R \in FR(X \times X)$ there exists at least one $A \in FS(X)$ such that:*

$$\mathbb{K}_\lambda(A \circ R, A \bullet R) = A$$

Proof: We need to find a fixed point of:

$$F_R : FS(X) \longrightarrow FS(X)$$
$$F_R(A) = \mathbb{K}_\lambda(A \circ R, A \bullet R)$$

Clearly, F_R is continuous, since it is a composition of continuous mappings. To get the result, it is enough to find an iterated, convergent sequence, that is, a convergent sequence $\{A_n\} \subset FS(X)$ such that

$$A_{n+1} = F_R(A_n) = \mathbb{K}_\lambda(A_n \circ R, A_n \bullet R) ,$$

since, by continuity, if $A_\infty = \lim_{n\to\infty} A_n$, then $A_\infty = \lim_{n\to\infty} A_{n+1} = \lim_{n\to\infty} F_R(A_n) = F_R(A_\infty)$.

Let $A_0 = \{(x,0)|x \in X\}$. Take

$$A_1 = \mathbb{K}_\lambda(A_0 \circ R, A_0 \bullet R) \geq A_0$$
$$A_2 = \mathbb{K}_\lambda(A_1 \circ R, A_1 \bullet R) \geq \mathbb{K}_\lambda(A_0 \circ R, A_0 \bullet R) = A_1 .$$

By induction, $A_{n+1} \geq A_n$, so the sequence is increasing. Moreover, $A_n \leq \{(x,1) |x \in X\}$ for all $n \geq 0$, so the sequence is also bounded. These two facts imply that the sequence is convergent and the proof is complete. □

5.1 Second Algorithm of Compression

The compression algorithm that we propose is based on replacing the relation (image) by P vectors (fuzzy sets) satisfying 10.

For a given relation R:

1.- Fix the number P of vectors;
2.- FOR i:=1 to P DO
    Fix $\lambda_i \in [0,1]$;
    Take the set $A_0^i(x) = 0$ for all $x \in X$;
    Build $A_1^i(x) = \mathbb{K}_{\lambda_i}((A_0^i \circ R)(x), (A_0^i \bullet R)(x))$ for all $x \in X$;
    Iterate until $A_{n+1}^i(x) = \mathbb{K}_{\lambda_i}((A_n^i \circ R)(x), (A_n^i \bullet R)(x)) = A_n^i(x)$ for all $x \in X$;
  ENDFOR;

Note that this compression is stronger than the one provided by our first algorithm. This is due to the fact that we are keeping a set of vectors instead of a smaller image; that is, we are keeping less information. Regarding the algorithm itself, observe that the choice of both P and λ_i are completely arbitrary, depending only on the needs of the user.

In Figure 10 we show the vector that results of applying our second algorithm of compression to Lena image with $P = 4$ eigen fuzzy sets and $\lambda = 0, \frac{1}{3}, \frac{2}{3}, 1$. In the horizontal axis we represent the components of the vector, whereas in the vertical axis we represent the intensity of each of the components.

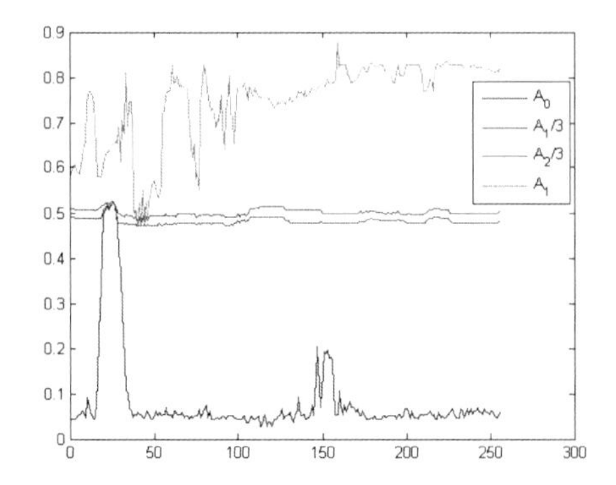

Fig. 10 Compression of Lena with 4 eigen fuzzy sets

5.2 *Second Algorithm of Reconstruction*

The algorithm that we propose is an optimization algorithm minimizing an appropriate cost function.

1. Let $\lambda_i \in [0,1]$, with $i = 1,\cdots,P$, and the corresponding $A_i \in FS(X)$ such that:

$$\mathbb{K}_{\lambda_i}(A_i \circ R, A_i \bullet R) = A_i \ .$$

   Observe that we are considering a family of fuzzy sets obtained by means of the second algorithm of compression.
2. Take a random fuzzy relation $\widetilde{R} \in FR(X \times X)$.
3. Build the cost function:

$$Q(\widetilde{R}) = \sum_{i=1}^{P} \sum_{x \in X} \{A_i(x) - \mathbb{K}_{\lambda_i}((A_i \circ \widetilde{R})(x), (A_i \bullet \widetilde{R})(x)))\}^2 \qquad (11)$$

4. Build in an iterative way the final relation, using the gradient algorithm as in [18, 16].

As the gradient algorithm is iterative, we denote by "iter" the iteration step. The gradient optimization method is given by:

$$\widetilde{R}^{(iter+1)}(l',x') = \widetilde{R}^{(iter)}(l',x') - \alpha \frac{\partial Q^{(iter)}}{\partial \widetilde{R}^{(iter)}(l',x')} \tag{12}$$

where:
$\widetilde{R}^{(iter)}$ is the fuzzy relation that results after iter iterations
$Q^{(iter)} = Q(\widetilde{R}^{(iter)})$
$\alpha =$ is a learning parameter ($\in (0,+\infty)$) fixed beforehand (see [17])
The derivative is calculated as follows.

$$\frac{\partial Q^{(iter)}}{\partial \widetilde{R}^{(iter)}(l',x')} = -2\sum_{n=1}^{P}(A_n(x') - \mathbb{K}_{\lambda_n}(A_n \circ \widetilde{R}^{(iter)}, A_n \bullet \widetilde{R}^{(iter)})) \cdot (Z_1 + Z_2)$$

where Z_1 is given by

$$\begin{aligned} Z_1 =& \lambda_n \varphi(A_n \circ \widetilde{R}^{(iter)}(x'), A_n \bullet \widetilde{R}^{(iter)}(x')) + (1-\lambda_n)\psi(A_n \circ \widetilde{R}^{(iter)}(x'), A_n \bullet \widetilde{R}^{(iter)}(x')) \\ & \cdot \varphi(A_n(l') \wedge \widetilde{R}^{(iter)}(l',x'), \bigvee_{y \in X \setminus \{l'\}} A_n(y) \wedge \widetilde{R}^{(iter)}(y,x'))\psi(A_n(l'), \widetilde{R}^{(iter)}(l',x')) \end{aligned}$$

and Z_2 by

$$\begin{aligned} Z_2 =& \lambda_n \psi(A_n \circ \widetilde{R}^{(iter)}(x'), A_n \bullet \widetilde{R}^{(iter)}(x')) + (1-\lambda_n)\varphi(A_n \circ \widetilde{R}^{(iter)}(x'), A_n \bullet \widetilde{R}^{(iter)}(x')) \\ & \cdot \psi(A_n(l') \vee \widetilde{R}^{(iter)}(l',x'), \bigwedge_{y \in X \setminus \{l'\}} A_n(y) \vee \widetilde{R}^{(iter)}(y,x'))\varphi(A_n(l'), \widetilde{R}^{(iter)}(l',x')) \end{aligned}$$

with

$$\varphi(x,y) = \begin{cases} 1 & \text{if } x > y \\ 0 & \text{otherwise} \end{cases} \quad \text{and } \psi(x,y) = \begin{cases} 1 & \text{if } x < y \\ 0 & \text{otherwise} \end{cases}$$

These minimization processes are performed until the inequality $Q^{(iter+1)} - Q^{(iter)} < \varepsilon$ is satisfied, where ε is a fixed threshold.

5.3 *Experimental Results for the Second Compression and Reconstruction Algorithms*

In this subsection we show the results of the application of the second algorithm of compression and the second algorithm of reconstruction using eigen fuzzy sets.

We show in Table 3 the mean reconstruction error over a set of 5000 images of size 10×10. The first column corresponds to 2500 blocks of size 10×10 which

Table 3 Mean error in the reconstruction of test images by means of eigen fuzzy sets

	Real	Random
GEFS	13.7818	16.6867
SEFS+GEFS	13.7617	16.6867
SEFS+$A_{1/3}$+$A_{2/3}$+GEFS	13.729	16.6402

have been obtained from real images. In the second column, we have generated the 2500 test images randomly. We have compressed the image by calculating the greatest eigen fuzzy set (first row), the greatest and the smallest eigen fuzzy set (second row) and four eigen fuzzy sets obtained with the algorithm developed in Subsection 5.1 (third row), with $P = 4$ and $A_{1/3}, A_{2/3}$ denoting the fuzzy sets which results from the algorithm taking $\lambda = 1/3$ and $\lambda = 2/3$, respectively.

In Table 3 we observe that the error in reconstructions decreases when the number of eigen fuzzy sets we are using in the compression increases. This fact can be visually corroborated in Figure 11. We have made the experiment with the image of Lena and taking $P = 2, 8, 16$ eigen fuzzy sets. We must remark that the main drawback of this algorithm is its poor time performance.

In Table 4 we show the mean error in image reconstruction when we use the algorithm proposed in [17] (first row) and when we use our second algorithm of reconstruction (second row). We have made the test with random images of sizes 3×3, 4×4 and 5×5. We have chosen 3 eigen fuzzy sets with $\lambda = \{0, \frac{1}{2}, 1\}$ for the 3×3 images; 4 eigen fuzzy sets with $\lambda = \{0, \frac{1}{3}, \frac{2}{3}, 1\}$ for the 4×4 images, and 5 eigen fuzzy sets with $\lambda = \{0, \frac{1}{4}, \frac{2}{4}, \frac{3}{4}, 1\}$ for the 5×5 images. Results are very similar for both algorithms, although in some cases, as those of 4×4 and 5×5 images, the results of our algorithm are slightly better.

Table 4 Comparison mean error in the original algorithm and in our second algorithm of reconstruction

	3×3	4×4	5×5
Algorithm proposed in [12]	1.4839	2.6271	4.2188
Second algorithm of reconstruction	1.4991	2.6131	4.1981

6 Final Comments

The possibility of building two-dimensional OWA operators using Atanassov's operators (Section 3) has allowed us to develop an algorithm of image reduction (first algorithm of compression), and later, an algorithm of image reconstruction (first algorithm of reconstruction). Moreover, we have experimentally showed that, if we

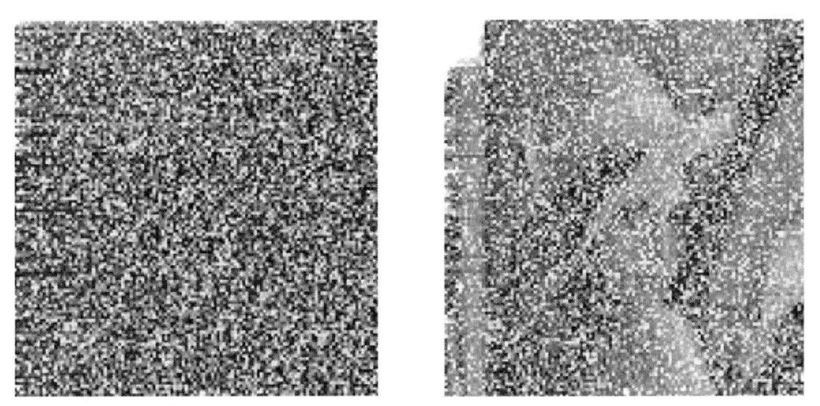

Reconstruction with $P = 2$ Eigen Fuzzy Sets Reconstruction with $P = 8$ Eigen Fuzzy Sets

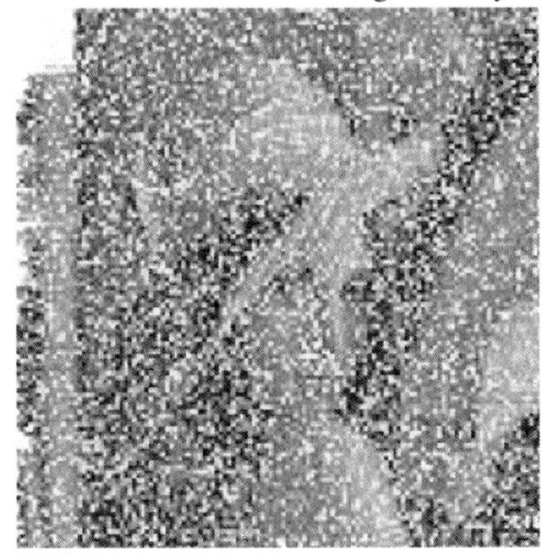

Reconstruction with $P = 16$ Eigen Fuzzy Sets

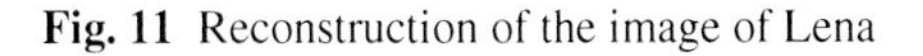

Fig. 11 Reconstruction of the image of Lena

reconstruct the images reduced with the first algorithm of compression, then we obtain the best results when we take ME-OWA operators with $\alpha = 0.5$ and constant for the whole image. Another experimentally established fact is that the time complexity of both the algorithm of compression and the algorithm of reconstruction is very low. That is, they are very efficient from the point of view of the time that they require to be run.

The concept of eigen fuzzy set has allowed us to replace an image by a set of vectors. Furthermore, the expression in Theorem 1, $\mathbb{K}_\alpha = K_\alpha \circ i$ has enabled us to calculate new eigen fuzzy sets that have been later used in our second algorithm of reconstruction. The algorithms that use eigen fuzzy sets provide worse results than those algorithms that do not use them, since the latter do not use random relations for reconstruction and keep more information than the former. The main disadvantage of the methods of reconstruction based on eigen fuzzy sets and the gradient algorithm lays on the high time expenses.

Nevertheless, it is extremely relevant to stress how our two approaches assume a global view of the image, so its structural information is someway taken into account, as claimed in [14]. Not taking into account information of surrounding pixels suggests that we are assuming independency between pixels, and in this context, a comparison of the quality of results limited to the direct analysis of mean error values can lead to seriously wrong conclusions, as dramatically shown in [15].

As future research lines, a comparison of our algorithms with other algorithms that can be found in the literature is necessary. Moreover, we would also like to specify in which cases our algorithms improves the results on the algorithm of Nobuhara et al, and why this improvement takes place. Finally, an interesting point is to reduce the temporal waste of the algorithms.

Acknowledgements. H. Bustince and D. Paternain have been supported by project TIN2007-65981. T Calvo was supported by the Spanish projects MTM2006-08322 and TIN2009-07901. Also by the European project 143423-2008-LLP-ES-KA3-KA3MP. J. Fodor was supported within the frames of Bilateral Romanian-Hungarian S&T Co-operation Project conducted by the Research and Technology Fund in Hungary, by OTKA K063405, and by COST Actions IC0602 and IC0702. R. Mesiar has been supported by grants APVV-0012-07, VEGA 1/4209/07. J. Montero has been supported by project TIN2006-06190.

References

1. Amo, A., Montero, J., Molina, E.: Representation of consistent recursive rules. Eur. Jour. Oper. Res. 130, 29–53 (2001)
2. Atanassov, K.: Intuitionistic fuzzy sets. Fuzzy Sets and Systems 20, 87–96 (1986)
3. Bezdek, J.C., Keller, J., Krisnapuram, R., Pal, N.R.: Fuzzy Models and algorithms for pattern recognition and image processing. In: Dubois, D., Prade, H. (eds.) The Handbooks of Fuzzy Sets Series. Kluwer Academic Publishers, Dordrecht (1999)
4. Bustince, H., Barrenechea, E., Pagola, M.: Image thresholding using restricted equivalence functions and maximizing the measures of similarity. Fuzzy Sets and Systems 158, 496–516 (2007)
5. Bustince, H., Barrenechea, E., Pagola, M.: Generation of interval-valued fuzzy and Atanassov's intuitionistic fuzzy connectives from fuzzy conectives and from K_α operators. Laws for conjunctions and disjunctions. Amplitude. International Journal of Intelligent Systems 23, 680–714 (2008)
6. Bustince, H., Orduna, R., Fernandez, J., et al.: Relation between OWA operator of dimension two and Atanassov's operators. Construction. In: Proceedings of the 8th International Conference on Fuzzy Logic and Intelligent Technologies in Nuclear Science, September 21-24, vol. 1, pp. 519–524 (2008)
7. Cutello, V., Montero, J.: Hierarchical aggregation of OWA operators, basic measures and related computational problems. Uncert., Fuzz. and Know. -Based Syst. 3, 17–26 (1995)
8. Cutello, V., Montero, J.: Recursive connective rules. Int. J. Intelligent Systems 14, 3–20 (1999)
9. De Baets, B.: Generalized Idempotence in Fuzzy Mathematical Morphology. In: Kerre, E.E., Nachtegael, M. (eds.) Fuzzy Techniques in Image Procesing, pp. 58–73. Springer, Berlin (2000)

10. Deschrijver, G., Cornelis, C., Kerre, E.E.: On the representation of intuitionistic fuzzy T-norms and T-conorms. IEEE Transactions on Fuzzy Systems 12(1), 45–61 (2004)
11. Gorzalczany, M.B.: A method of inference in approximate reasoning based on interval-valued fuzzy sets. Fuzzy Sets and Systems 21, 1–17 (1987)
12. Kerre, E.E., et al. (eds.): Fuzzy Filters for Image Processing. Springer, Berlin (2003)
13. Loia, V., Sessa, S.: Fuzzy relation equations for coding/decoding processes of images and videos. Information Sciences 171, 145–172 (2005)
14. Montero, J., Gómez, D., Bustince, H.: On the relevance of some families of fuzzy sets. Fuzzy Sets And Systems 158, 2429–2442 (2007)
15. Nachtegael, M., Mélange, T., Kerre, E.E.: The Possibilities of Fuzzy Logic in Image Processing. In: Ghosh, A., De, R.K., Pal, S.K. (eds.) PReMI 2007. LNCS, vol. 4815, pp. 198–208. Springer, Heidelberg (2007)
16. Nobuhara, N., Hirota, K., Pedrycz, W., Sessa, S.: Two iterative methods of decomposition of a fuzzy relation for image compression/decomposition processing. Soft Computing 8, 698–704 (2004)
17. Nobuhara, H., Bede, B., Hirota, K.: On various eigen fuzzy sets and their application to image reconstruction. Information Sciences 176, 2988–3010 (2006)
18. Pedrycz, W., Hirota, K., Sessa, S.: A decomposition of fuzzy relations. IEEE Transaction on Systems, Man, and Cybernetics Part B 31(4), 657–663 (2001)
19. Sambuc, R.: Function Φ-Flous, Application a l'aide au Diagnostic en Pathologie Thyroidienne, These de Doctorat en Medicine, University of Marseille (1975)
20. Sanchez, E.: Resolution of eigen fuzzy sets equations. Fuzzy Sets and Systems 1, 69–74 (1978)
21. Sanchez, E.: Eigen fuzzy sets and fuzzy relations. Journal of Mathematical Analysis and Applications 81, 399–421 (1981)
22. Tarski, A.: A lattice-theoretical fixpoint theorem and its applications. Pacific Journal of Mathematics 5, 285–309 (1995)
23. Yager, R.R.: On ordered weighted averaging aggregation operators in multicriteria decision making. IEEE Trans. Syst. Man Cybern. 18, 183–190 (1988)
24. Yager, R.R.: Families of OWA operators. Fuzzy Sets and Systems 59, 125–148 (1993)

OWA-Based Fuzzy m-ary Adjacency Relations in Social Network Analysis

Matteo Brunelli, Mario Fedrizzi, and Michele Fedrizzi

Abstract. In this paper we propose an approach to Social Network Analysis (SNA) based on fuzzy m-ary adjacency relations. In particular, we show that the dimension of the analysis can naturally be increased starting from the traditional two–dimensional case and interesting results can be derived. Therefore, fuzzy m-ary adjacency relations can be computed starting from fuzzy binary relations and introducing OWA-based aggregations. The behavioral assumptions derived from the measure and the exam of individual propensity to connect with other suggest that OWA operators can be considered particularly appealing tools in characterizing such relations.

1 Introduction

Social Network Analysis (SNA) is a relatively new and still developing subject that focuses on the study of social relations [30, 35] as a branch of the broader discipline named network analysis [26] whose main object is studying the relationships between objects belonging to one or more universal sets. SNA focuses its attention on social objects and has principally concerned with the structure and effects of relations between people, groups or organizations, rather than on individual psychological attributes. Nevertheless, as pointed out in [31], psychological attributes and behavioral issues are likely to influence the dynamics of networks of individuals.

Matteo Brunelli
IAMSR and TUCS, Åbo Akademi University, Joukahainengatan 3-5 A,
FIN-20520 Åbo, Finland
e-mail: matteo.brunelli@abo.fi

Mario Fedrizzi · Michele Fedrizzi
Department of Computer and Management Science,
University of Trento Via Inama 5, 38100 Trento, Italy
e-mail: mario.fedrizzi@unitn.it, michele.fedrizzi@unitn.it

R.R. Yager et al. (Eds.): Recent Developments in the OWA Operators, STUDFUZZ 265, pp. 255–267.
springerlink.com

For instance, the role of individual differences in shaping organizational networks has been examined from several points of view [3], compelling as well the study of how similarity in personal relationships and social context affect each other [23].

For better understanding of roles played by actors in social networks the so called centrality indices have been introduced, accordingly a member is viewed as central whenever she or he has a high number of connections with a high number of different co-members [2, 4, 7, 15]. Since paths play a central role in the functioning of most of the networks, it is not surprising that a relevant number of centrality measures quantify importance with respect to the sharing of paths in the network. Betweenness centrality, as a measure of how many geodesic paths cross a given vertex, is one of the most popular and was introduced [14] to quantify the control of a given actor over the flow of information in the network. Therefore, this measure can be used to provide an ordering of the vertices in terms of their individual importance, but it does not provide any description of the way in which subsets of vertices influence the network as a whole. As pointed out in [21], vertex betweenness centrality can be naturally extended to sets of vertices either defining the betweenness of a set in terms of geodesic paths that pass through at least one of the vertices in the set, or in terms of geodesic paths that pass through all vertices in the set.

Everett and Borgatti [12] introduced the first type of extension and called it group betweenness centrality, the second type was introduced in [21] showing that the two notions are intimately related. The relationship between the two approaches has been mathematically characterized showing how the betweenness of a group of an arbitrary number of vertices can be bounded above and below by quantities involving only the betweenness of the individual vertices and the co-betweenness of pairs of these vertices. In this way a direct insight into the composition of subgroups of vertices is provided and it can be used in evaluating the robustness of potential coalitions and the deploying of consensual dynamics.

One of the most commonly used tool for representing social relationship among a set of actors in a network is the adjacency matrix representing a binary relation. The first limitation of binary relations is that they can be used only for representing pairwise adjacency, the second one is that the dichotomy is not suitable for shaping the strength of the adjacency relationship involving several social and individual attributes. One way to overcome the first limitation was introduced in [6] and it was based on the concept of multirelational systems [28]. The generalization of the definition of relation through the introduction of fuzziness opens the way to the extension of Social Network Analysis to contexts in which the network could be represented using fuzzy graphs [24], taking care of the vagueness influencing the relationships among the actors involved in the social dynamics and of the qualitative nature of the actors' attributes as well.

Fuzzy approaches to SNA provided so far are actually very few. In [25] a technique to model multi-modal social networks as fuzzy social networks is proposed. The technique is based on k-modal fuzzy graphs determined using the union operation on fuzzy graphs and a new operator called consolidation operator. The notion of regular equivalence [5] was generalized in [13] introducing the notion of regular similarity, represented by a fuzzy binary relation that describes the degree of similarity between actors in the social network. In [10] the problem of partitioning the nodes of a social network in overlapping groups allowing for multiple memberships and varied levels of membership was solved introducing the so called fuzzy groups. In [34], starting from the introduction of the natural connection between graph theory and granular computing, human-focused concepts associated with social networks are formalized using set-based relational network theory and fuzzy sets. A softening of the concept of node importance (centrality of a node) is provided, considering the number of close connections. Kokabu et al. [20] proposed a model for evaluating reciprocity of networks represented by means of fuzzy binary relations. In literature, it was also proposed [8] to use fuzzy relations for defining a characterization for fuzzy m-ary relations and therefore expand the dimension of the analysis for $m > 2$.

The approach proposed in this paper takes advantage of the ability of fuzzy relations [19, 36, 37] to model uncertainty permeating the relationships between the actors in the network, and of the OWA operators [32, 33] to move continuously from non-compensatory to full-compensatory situation and characterizing therefore the attitude of the actors to connect each other. The paper is outlined as follows. In section 2 we offer a presentation of SNA and adjacency matrix, which is the main tool to perform the analysis. In the same section we show that adjacency relations can be valued (cardinal) relations and that fuzzy adjacency relations are simply a special case of valued relations. Having presented that, in section 3 fuzzy m-ary adjacency relations are defined and a method based on aggregating functions for estimating them is presented. We claim, in section 4, that OWA operators satisfy some reasonable properties and that they can be employed as suitable aggregating functions to increase the dimension of the analysis. In section 5 we discuss an example and, finally, in section 6, we present our conclusions.

2 Crisp, Valued and Fuzzy Adjacency Relations in SNA

As already mentioned, SNA is the branch of network analysis devoted to studying and representing relationships between 'social' objects. To formalize, SNA mainly explores relationships between objects belonging to an universal set $X = \{x_1, \ldots, x_n\}$ and in order to achieve its aim, some mathematical properties of relations are utilized. More specifically, a binary relation on a single set, which is the most popular kind of relation used in the SNA, is a relation $A \subseteq X \times X$, whose characteristic function $\mu_A : X \times X \to \{0, 1\}$ is defined as

$$\mu_A(x_i, x_j) = \begin{cases} 1, & \text{if } x_i \text{ is related to } x_j \\ 0, & \text{if } x_i \text{ is not related to } x_j \end{cases}$$

By definition [19], adjacency relations satisfy properties of reflexivity, $\mu_A(x_i, x_i) = 1$, and symmetry, $\mu_A(x_i, x_j) = \mu_A(x_j, x_i)$. Note that no transitivity condition is required to hold. Moreover, if $a_{ij} := \mu_A(x_i, x_j)$ and X is reasonably not too large, then an adjacency matrix $\mathbf{A} = (a_{ij})_{n \times n}$ is a convenient way of representing a relation.

Some scholars in the field claim that **A** has its strong point in being a good synthesis of all the pairwise relations between elements of X. In contrast, according to some others, **A** is too poor of information, i.e. it does not contain information about the degree to which the relations between two elements hold. Therefore it may happen that it treats in the same way very different cases, without discriminating among situations where intensities of relationship may be very different. Indeed, many examples may be brought in order to support the latter point of view.

Some methods have already been proposed in order to overcome the problem related with the lack of information about the intensity of relationship between elements of a pair. For instance, a discrete scale can be adopted and a value be assigned to each entry a_{ij} to denote the intensity of relation between x_i and x_j. This approach, based on *valued adjacency relations*, is the most widely used in order to overcome the problem of unvalued relations.

Here, we want to propose an alternate approach based on *fuzzy sets* theory [36] in order to obtain a *fuzzy* adjacency relation. A binary fuzzy relation on a single set, $R_2 \subseteq X \times X$, is defined through the following membership function

$$\mu_{R2} : X \times X \rightarrow [0,1] \tag{1}$$

and also in this case, putting $r_{ij} := \mu_{R2}(x_i, x_j)$, a fuzzy relation can be conveniently represented by a matrix $\mathbf{R} = (r_{ij})_{n \times n}$ where the value of each entry is the degree to which the relation between x_i and x_j holds. In other words, the value of $\mu_{R2}(x_i, x_j)$ is the answer to the question: 'how strong is the relationship between x_i and x_j?'. Therefore, in the context of SNA

$$\mu_{R2}(x_i, x_j) = \begin{cases} 1, & \text{if } x_i \text{ has the strongest possible degree of relationship with } x_j \\ \gamma \in]0,1[& \text{if } x_i \text{ is, to some extent, related to } x_j \\ 0, & \text{if } x_i \text{ is not related with } x_j \end{cases}$$

Fuzzy adjacency relations, as well as crisp adjacency relations, are here assumed to be reflexive and symmetric. It is useful to spend some words about symmetry. A fuzzy binary relation is symmetric if and only if

$$\mu_{R2}(x_i, x_j) = \mu_{R2}(x_j, x_i) \;\; i, j = 1, \ldots, n. \tag{2}$$

Although the assumption of symmetry is a simplification, it is of great help for the model because, due to it, such relations can be represented by means of undirected graphs and problems related with the so-called combinatorial explosion are partially avoided. Furthermore, in many real-world cases, symmetry is spontaneously satisfied by the nature of the relationship.

At this point we remind that:

- A fuzzy relation contains more information than a crisp one and the former can overcome some drawbacks of the latter. See for example [19], where fuzzy adjacency relations are called fuzzy compatibility or proximity relations
- We can shift from the fuzzy approach to the crisp one thanks to the α-cuts. An α-cut is a crisp relation defined by

$$\mu_A(x_i,x_j) = \begin{cases} 1, \text{ if } \mu_{R2}(x_i,x_j) \geq \alpha \\ 0, \text{ if } \mu_{R2}(x_i,x_j) < \alpha . \end{cases}$$

For instance, given

$$\mathbf{R} = \begin{pmatrix} 1 & 0.7 & 0.3 & 0.7 \\ 0.7 & 1 & 0.1 & 0.8 \\ 0.3 & 0.1 & 1 & 0.2 \\ 0.7 & 0.8 & 0.2 & 1 \end{pmatrix}, \tag{3}$$

its α-cut with $\alpha = 0.5$ is

$$\mathbf{A} = \begin{pmatrix} 1 & 1 & 0 & 1 \\ 1 & 1 & 0 & 1 \\ 0 & 0 & 1 & 0 \\ 1 & 1 & 0 & 1 \end{pmatrix} \tag{4}$$

- Applying fuzzy relations to SNA, we can extend most of the techniques employed for analyzing crisp adjacency matrices. A significant example, which will be used later on in this discussion, is the normalized index of local centrality, that is

$$C(x_i) = \frac{1}{n-1} \sum_{\substack{j=1 \\ j \neq i}}^{n} r_{ij} \,. \tag{5}$$

  If $c_i := C(x_i)$ and $\mathbf{c} = (c_1, \ldots, c_n)$, then we can refer to $\mathbf{R}$ in (3) and find that $\mathbf{c} = (\frac{17}{30}, \frac{3}{5}, \frac{7}{20}, \frac{185}{300})$. This result is more informative than the same index computed on $\mathbf{A}$ in (4), i.e. $\mathbf{c} = (\frac{2}{3}, \frac{2}{3}, 0, \frac{2}{3})$
- It is possible to exploit already known indices for fuzzy sets as, for instance, a measure of fuzziness [11] which would estimate how much information we would have lost if we had used crisp relations instead of fuzzy ones
- Exploiting a fuzzy adjacency relation solves all the borderline cases, i.e. all the cases where it is difficult to establish whether x_i and x_j are related or not. To tell the truth, this property is, to some extent, shared with valued adjacency relations. However, a fuzzy relation is much easier to be interpreted from a logical point of view

Let's note that the whole issue can be addressed thanks to graph theory. In this case there are n nodes $x_1, x_2, \ldots, x_n$ and $\frac{n(n-1)}{2}$ edges connecting them. Hence, nodes are nothing else but elements of the universe set X, weights of edges are $\mu_{R2}(x_i,x_j)$ and the graph is $G = \langle X, R \rangle$. Therefore, the problem can also be addressed in a graphical way with $\mu_{R2}(x_i,x_j)$ representing the "thickness" of the edge between x_i and x_j.

One might wonder how it would be possible to define a fuzzy adjacency matrix starting from real-world information. In all those cases where it is difficult to define it directly we can derive it from valued adjacency matrices, e.g. some evidence under the form of numerical data about the relationships is available. Let us assume that a valued adjacency matrix, $\mathbf{V} = (v_{ij})_{n \times n}$, exists with $v_{ij} \in \mathbb{R}_{\geq}$. If it is possible to define a maximal level for the valued graph, say v^*, such that it represents the maximum possible value of relationship, then, with v^* playing the role of the upper bound for entries v_{ij}, we can rescale each v_{ij} into a r_{ij} thanks to a suitable mapping $r_{ij} = h(v_{ij})$, $h : [0, v^*] \rightarrow [0, 1]$.

3 Fuzzy *m*-ary Adjacency Relations and the Degree of Social Relationship

In this section we propose an extension of the analysis involving m-dimensional relations with $2 \leq m \leq n$. If we do so, then each element of the m-ary relation is the degree of social relationship among the m elements contained in the m-tuple which is taken into account. Analogously to the binary case, it is straightforward to define a fuzzy m-ary relation.

Definition 1. A fuzzy m-ary relation R_m on a single set X is a fuzzy subset of X^m defined by means of the membership function

$$\mu_{R_m} : X^m \rightarrow [0, 1] \tag{6}$$

Then, for $p_1, \ldots, p_m \in \{1, \ldots, n\}$, the membership function characterizing fuzzy m-ary relations is the following

$$\mu_{R_m}(x_{p_1}, \ldots, x_{p_m}) = \begin{cases} 1, & \text{if } x_{p_1}, \ldots, x_{p_m} \text{ are definitely related} \\ \gamma \in]0, 1[& \text{if } x_{p_1}, \ldots, x_{p_m} \text{ are, to some extent, related} \\ 0, & \text{if } x_{p_1}, \ldots, x_{p_m} \text{ are definitely not related} \end{cases}$$

The logic underlying the membership function remains substantially unchanged and therefore properties of reflexivity and symmetry are extended to the m-dimensional case in the following way. A fuzzy m-ary relation is reflexive if and only if

$$\mu_{R_m}(x_i, x_i, \ldots, x_i) = 1, \quad i = 1, \ldots, n.$$

A fuzzy m-ary relation is symmetric if and only if for any $p_1, p_2, \ldots, p_m \in \{1, \ldots, n\}$ it is

$$\mu_{R_m}(x_{p_1}, \ldots, x_{p_m}) = \mu_{R_m}(x_{q_1}, \ldots, x_{q_m})$$

where $(x_{q_1}, x_{q_2}, \ldots, x_{q_m})$ is any permutation of $(x_{p_1}, x_{p_2}, \ldots, x_{p_m})$.

A fuzzy m-ary relation satisfying the reflexivity and simmetry properties is called a fuzzy m-ary adjacency relation. At this point, having defined fuzzy m-ary adjacency relations, it is the case to highlight the difference between an element of a fuzzy m-ary adjacency relation and a clique [22, 29, 35]. Namely, a clique of a graph is a maximum complete subgraph whereas, if we deal with m-ary relations and the contrary is not made explicit, the value $\mu_{R_m}(x_{p_1}, \ldots, x_{p_m})$ simply states, by means of the bounded unipolar scale $[0, 1]$, the degree to which the relation holds, without taking into account any maximality condition.

However, problems arise when we try to elicit R_m in a direct way, as it is certainly not a trivial operation, especially when m is large enough. As we have seen in the previous section, in social network analysis adjacency relations in the form $R_2 \subseteq X \times X$ are often used and therefore degrees of relationship over pairs are known. That is why we propose an effective way to elicit R_m using the information embedded in the fuzzy binary adjacency relation on the same universal set X. We propose to recursively calculate the degree of relationship over m–tuples by means of the degree of relationship over pairs using aggregation functions $\rho_3, \ldots, \rho_m$ satisfying a fixed set of assumptions, as described in the next section. More precisely, from a given fuzzy binary adjacency relation R_2 we calculate the corresponding fuzzy 3–ary adjacency relation R_3, then from R_3 we calculate R_4, and so on. In general, to estimate $\mu_{R_m}(x_{p_1}, \ldots, x_{p_m})$ we construct it recursively in the following way

$$\mu_{R_k}(x_{p_1}, \ldots, x_{p_k}) = \rho_k(\mu_{R_{k-1}}(x_{p_1}, \ldots, x_{p_{k-1}}), \ldots, \mu_{R_{k-1}}(x_{p_2}, \ldots, x_{p_k})), \tag{7}$$

for $k = 3, \ldots, m$. The k arguments of the function ρ_k themselves are functions of $k-1$ variables. For example, the first argument of the function ρ_k in (7) is

$$\mu_{R_{k-1}}(x_{p_1}, \ldots, x_{p_{k-1}}) = \rho_{k-1}(\mu_{R_{k-2}}(x_{p_1}, \ldots, x_{p_{k-2}}), \ldots, \mu_{R_{k-2}}(x_{p_2}, \ldots, x_{p_{k-1}})) \,.$$

Note that to calculate $\mu_{R_k}(x_{p_1}, \ldots, x_{p_k})$ in (7) we need to aggregate precisely k values of $\mu_{R_{k-1}}(\cdot)$. Since we assume that the symmetric property holds, the order of the arguments in $\mu_{R_m}(x_{p_1}, \ldots, x_{p_m})$ is not relevant and we can assume, without loss of generality, $x_{p_1} \leq \cdots \leq x_{p_m}$. Therefore, a fuzzy m–ary adjacency relation on a set X requires $\binom{n+m-1}{m}$ relationship values to be completely defined, i.e. the number of combinations with repetition of size m from a set of n elements.

The most important case for applications is that of considering $\mu_{R_m}(x_{p_1}, \ldots, x_{p_m})$ with all different arguments, i.e. allowing no repetition. This corresponds to take into account only groups of distinct social objects and in the following we will focus on this case. Under this assumption, a fuzzy m–ary adjacency relation on a set X requires only $\binom{n}{m}$ relationship values to be completely defined.

Let us introduce the notation which will be used hereafter.

Definition 2. Given a finite non empty set $X = \{x_1, \ldots, x_n\}$, we denote by $\mathscr{F}_m(X)$ the family of subsets of X containing m elements,

$$\mathscr{F}_m(X) = \{A \subseteq X; |A| = m\} \tag{8}$$

Example 1. Given $X = \{x_1, x_2, x_3, x_4\}$, it is

$$\begin{aligned}
\mathscr{F}_1(X) &= \{\{x_1\}, \{x_2\}, \{x_3\}, \{x_4\}\} \\
\mathscr{F}_2(X) &= \{\{x_1, x_2\}, \{x_1, x_3\}, \{x_1, x_4\}, \{x_2, x_3\}, \{x_2, x_4\}, \{x_3, x_4\}\} \\
\mathscr{F}_3(X) &= \{\{x_1, x_2, x_3\}, \{x_1, x_2, x_4\}, \{x_1, x_3, x_4\}, \{x_2, x_3, x_4\}\} \\
\mathscr{F}_4(X) &= \{x_1, x_2, x_3, x_4\}.
\end{aligned}$$

In general, with $|X| = n$ and $m \leq n$, the cardinality of $\mathscr{F}_m(X)$ is

$$|\mathscr{F}_m(X)| = \binom{n}{m}.$$

It turns out that in order to calculate the degree of relationship among the m distinct objects of a set $\{x_{p_1},\dots,x_{p_m}\} \subseteq X$, we need to calculate first the degrees of relationship over all its subsets with cardinality greater than one and less than m. The total number of these subsets is

$$\binom{m}{2} + \cdots + \binom{m}{m-1} = 2^m - m - 2\,. \tag{9}$$

Let us now draw our attention again to the problem of aggregating relationship values in order to construct higher dimensional relations. The choice of aggregation functions $\rho_3, \rho_4, \dots, \rho_m$ plays clearly a crucial role in determining the fuzzy m–ary adjacency relation μ_{R_m}. In the following section, therefore, we will focus on the suitable properties we require for these functions.

4 Some Properties of the OWA-Based Aggregating Function

This section is devoted to present and justify the assumptions that we make regarding $\rho_3, \dots, \rho_m$. First of all, as we already said, we require $\rho_3, \dots, \rho_m$ to be 'aggregation functions'. We recall the corresponding definition [1].

Definition 3 (Aggregation function). An aggregation function is a function of $m > 1$ arguments that maps the (m-dimensional) unit cube onto the unit interval, $f : [0,1]^m \to [0,1]$, with the properties

- $f(0,\dots,0) = 0$ and $f(1,\dots,1) = 1$
- $\mathbf{a} \le \mathbf{b}$ implies $f(\mathbf{a}) \le f(\mathbf{b})$ for all $\mathbf{a}, \mathbf{b} \in [0,1]^m$ (monotonicity)

Moreover, we are going to propose some other properties which, in our opinion, should be satisfied by every $\rho_m,\ m = 3,\dots,n$.

1. idempotency $\rho_m(a,\dots,a) = a$. Therefore, if the objects of some set are pairwise related with degree a, then we assume that the intensity of relation computed on the tuple containing those objects has value a as well.
2. commutativity, $\rho_m(a_1,\dots,a_m) = \rho_m(a_{q_1},\dots,a_{q_m})$ where $(q_1,\dots,q_m)$ is any permutation of the indices. This property is required to hold because fuzzy adjacency relations are symmetrically defined for all $m = 2,\dots,n$.
3. *strict* monotonicity: $\rho_m(a_1,\dots,a_m) > \rho_m(b_1,\dots,b_m)$ if $a_i \ge b_i\ \forall i$ and there exists at least one j such that $a_j > b_j$. Strict monotonicity is asked to hold in order to overcome some evaluation problems which would arise if we used non-strictly monotonically increasing functions as, for instance, the geometric mean $g(\cdot)$. To give an example, substituting g to ρ_m we would have $g(1,\dots,1,0) = g(0,\dots,0)$, which is not a desirable result from the social analysis point of view
4. continuity. This is essentially a technical assumption.

These four assumptions lead us to choose within a restricted class of averaging operators. Namely, ρ_m should be an aggregating function respecting properties 1–4. It is easy to check that the geometric mean, as mentioned above, is excluded because it is not strictly monotone. The weighted arithmetic mean is also excluded because it is not commutative.

Conversely, provided that $0 < w_i < 1$, any OWA operator [32, 33] satisfies the listed properties [9]. Choosing between OWA operators would be anything but arbitrary as an index of orness is associated to each OWA and several approaches has been developed to find an OWA operator with a given level of orness and optimizing some other properties as, for instance, entropy and variance. Moreover, OWA operators cover a range of some well known aggregating functions, as they can be meant as trade offs between the min and the max operators.

Hence, in our case, as we use OWA operators, they should be defined such that $w_i \in]0,1[$ so that they are strictly monotonically increasing functions in all the terms. Let us therefore give the following modified definition of OWA.

Definition 4 (Strictly monotone OWA operator). A strictly monotone OWA operator of dimension m is a mapping $F : \mathbb{R}^m \to \mathbb{R}$, that has an associated weighting vector $\mathbf{w} = (w_1, \ldots, w_m)$ such that $0 < w_i < 1$ and

$$\sum_{i=1}^{m} w_i = 1$$

Furthermore

$$F(a_1, \ldots, a_m) = w_1 b_1 + \cdots + w_m b_m = \sum_{j=1}^{m} w_j b_j$$

where b_j is the j-th largest element of the bag $A = \langle a_1, \ldots, a_m \rangle$.

5 Example

A number of examples explaining the utility of m-ary relations can be brought. Let us, for example, consider the following fuzzy binary adjacency relation

$$\begin{pmatrix} 1 & 0.8 & 0.8 & 0.6 & 0.2 & 0.4 & 0.3 \\ 0.8 & 1 & 0.9 & 0.2 & 0.1 & 0.3 & 0.3 \\ 0.8 & 0.9 & 1 & 0.3 & 0.2 & 0.3 & 0.3 \\ 0.6 & 0.2 & 0.3 & 1 & 0.7 & 0.7 & 0.4 \\ 0.2 & 0.1 & 0.2 & 0.7 & 1 & 0.9 & 0.7 \\ 0.4 & 0.3 & 0.3 & 0.7 & 0.9 & 1 & 0.5 \\ 0.3 & 0.3 & 0.3 & 0.4 & 0.7 & 0.5 & 1 \end{pmatrix} \tag{10}$$

which we can suppose be representing of fuzzy adjacency relations between decision makers. Although we bring an example, it is easy to imagine several other possible applications indeed. Following our proposal, it is possible to estimate a ternary fuzzy adjacency relation simply by applying function ρ_3 according to (7). Let us further assume that, hereafter, function ρ_m is univocally determined as an

OWA operator of dimension m with maximal entropy and $orness(w) = 0.4$, which, in the special case with $m = 3$, is $w \simeq (0.238371, 0.323257, 0.438371)$. Thus, the result, would be

$$\begin{aligned}
\mu_{R3}(x_1,x_2,x_3) &= \rho_3(\mu_{R2}(x_1,x_2),\mu_{R2}(x_1,x_3),\mu_{R2}(x_2,x_3)) \simeq 0.823837 \\
\mu_{R3}(x_1,x_2,x_4) &= \rho_3(\mu_{R2}(x_1,x_2),\mu_{R2}(x_1,x_4),\mu_{R2}(x_2,x_4)) \simeq 0.472326 \\
\vdots &= \vdots \\
\mu_{R3}(x_4,x_6,x_7) &= \rho_3(\mu_{R2}(x_4,x_6),\mu_{R2}(x_4,x_7),\mu_{R2}(x_6,x_7)) \simeq 0.503837 \\
\mu_{R3}(x_5,x_6,x_7) &= \rho_3(\mu_{R2}(x_5,x_6),\mu_{R2}(x_5,x_7),\mu_{R2}(x_6,x_7)) \simeq 0.66
\end{aligned}$$

It could be particularly interesting to pick the element of $\mathscr{F}_m(X)$ such that the value of its membership function is maximal,

$$\max\{\mu_{R_m}(x_{p_1},x_{p_2},\ldots,x_{p_m})|\ p_1,\ldots,p_m \in \{1,\ldots,n\},\ p_1 < p_2 < \cdots < p_m\}\,, \quad (11)$$

In our case the maximum value of membership function is 0.823837 and it is achieved by the triplet (x_1,x_2,x_3).

A rather special case is that involving the $m^\star$-ary relation where $m^\star$ is defined as the integer part of $n/2+1$, more formally $m^\star = \lfloor n/2+1 \rfloor$, because the associated subset of X would be a minimum winning coalition. In our case $m^\star = 4$ and $w = (0.167087, 0.213266, 0.272208, 0.34744)$ with

$$\begin{aligned}
\mu_{R4}(x_1,x_2,x_3,x_4) &= \rho_4(\mu_{R3}(x_1,x_2,x_3),\ldots,\mu_{R3}(x_2,x_3,x_4)) \simeq 0.514996 \\
\mu_{R4}(x_1,x_2,x_3,x_5) &= \rho_4(\mu_{R3}(x_1,x_2,x_4),\ldots,\mu_{R3}(x_2,x_3,x_5)) \simeq 0.402686 \\
\vdots &= \vdots \\
\mu_{R4}(x_3,x_5,x_6,x_7) &= \rho_4(\mu_{R3}(x_3,x_5,x_6),\ldots,\mu_{R3}(x_5,x_6,x_7)) \simeq 0.41189 \\
\mu_{R4}(x_4,x_5,x_6,x_7) &= \rho_4(\mu_{R3}(x_4,x_5,x_6),\ldots,\mu_{R3}(x_5,x_6,x_7)) \simeq 0.595482
\end{aligned}$$

with the maximum being 0.595482, achieved by (x_4,x_5,x_6,x_7).

This latter proposal can be refined if we assume that every element $x_i \in X$ has a specific weight ω_i denoting its relative importance. Let us consider the weight vector

$$\omega = (\omega_1,\ldots,\omega_n) \quad \text{s.t.} \sum_{i=1}^{n} \omega_i = 1\,, \omega_i \geq 0\ \forall i. \quad (12)$$

Then, we can perform an analysis similar to that described above by assuming that parameter m is free, not necessarily equal to $m^\star$, and by requiring that the sum of the weights associated to the considered m elements is equal or greater than a given majority threshold $0 < t \leq 1$. In light of these observations, the optimization problem is

$$\max\{\mu_{R_m}(x_{p_1},\ldots,x_{p_m}) \mid p_1,\ldots,p_m \in \{1,\ldots,n\}, p_1 < \cdots < p_m, \\ \sum_{i=1}^{m} \omega_{p_i} > t,\ m = 2,\ldots,n-1\}\,. \tag{13}$$

Some comments on (13) could be useful to better understand the involved optimization. In (13) we are still interested in the strongest coalition, but the constraint of having a fixed number of elements is replaced by a constraint on a majority threshold t to be satisfied by the sum of the weights of the coalition's elements. That is, coalitions with different number m of elements are taken into account, provided that they fulfil threshold t. Note that large values of μ_{R_m} can be easily achieved if the number m of elements is small, while the constraint $\sum_{i=1}^{m} \omega_{p_i} > t$ is satisfied by the coalitions with a sufficiently large number of strong elements. Therefore, the optimal solution of (13) arises by taking into account the two conflicting criteria: power of the coalition and degree of relationship among the coalition's elements. We stress again that the number m of the coalition's elements is optimally determined only after having solved (13).

Although the example proposed here is not based on a real world case, the problem solved in (13) could be applied to economics and political sciences. In fact, it is possible to see vector ω as a collection of weights for political parties. At this point, if we are able to establish some distance measures between any two parties (i.e. a relationship degree), then we can apply (13) and find the strongest winning coalition.

Note that vector ω defining the relative importance of each $x_i \in X$ must not be confused with vector $\mathbf{w}$ of an OWA operator, which is used in this paper to assign weights to degrees of relationships among elements in X.

Another problem that can be addressed is that of maximizing the number m of elements in a subset satisfying a fixed majority threshold. Namely, let us fix a threshold $\delta \in [0,1]$ such that $\mu_{R_m}(x_{p_1},x_{p_2},\ldots,x_{p_m}) > \delta$ and leave the dimension m of our analysis free. In this way, progressively increasing m and calculating $\mu_{R_m}(x_{p_1},x_{p_2},\ldots,x_{p_m})$ at every stage, we can detect the largest $B \subseteq X$ such that $\mu_{R_m}(B) > \delta$. Let $\hat{m}$ denote this maximal cardinality,

$$\hat{m} = \max\{m \mid \mu_{R_m}(B) > \delta,\ m = |B|,\ B \subseteq X\}\,. \tag{14}$$

It may occur that set B is not unique, since there exist v different subsets B_j, $j = 1,\ldots,v$ satisfying inequality $\mu_{R_m}(B_j) > \delta$ with the same maximal cardinality $\hat{m}$. In this case, it is possible to define a winner as the subset B_i with the strongest degree of relationship, $\mu_{R_m}(B_i) \geq \mu_{R_m}(B_j)$, $j = 1,\ldots,v$. If again the solution is not unique, the multiple solutions are considered equivalent for our analysis.

The very last observation concerns the dimension of the analysis. If $m = n$, then the degree to which this particular relation holds is a measure of how strong the relation among all the $x_i \in X$ is. It can be interpreted as the degree of social relationship computed on the entire network.

6 Conclusions

We provided a new approach to the analysis of social networks based on m–ary fuzzy adjacency relations and OWA operators. Our aim was to show that from the combined use of these two mathematical tools, the vagueness pervading the relationships between the actors involved in the social network and their attitude to connect each other can be represented more effectively. Through the introduction of the strictly monotone OWA operator, we provided a representation of fuzzy m-ary adjacency relations using the information embedded in the fuzzy binary relations defined on the same universal set. Hopefully, starting from the results of this paper, it will be possible to provide further representations of the interactions characterizing the dynamics of social networks involving linguistically based evaluations as well.

References

1. Beliakov, G., Pradera, A., Calvo, T.: Aggregation Functions: A Guide for Practitioners. Studies in Fuzziness and Soft Computing Springer (2007)
2. Bonacich, P.: Power and centrality: A family of measures. Am. J. Sociol. 92, 1170–1182 (1987)
3. Borgatti, S.P., Foster, P.C.: The network paradigm in organizational research: A review and typology. Journal of Management 29, 991–1013 (2003)
4. Borgatti, S.P.: Centrality and network flow. Soc. Networks 27, 55–71 (2005)
5. Borgatti, S.P., Everett, M.: The class of all regular equivalences: Algebraic structure and computation. Soc. Networks 11, 65–88 (1989)
6. Borgatti, S.P., Everett, M.G.: Regular blockmodels of multiway, multimode matrices. Soc. Networks 14, 91–120 (1992)
7. Borgatti, S.P., Everett, M.G.: A graph-theoretic perspective on centrality. Soc. Networks 28, 466–484 (2006)
8. Brunelli, M., Fedrizzi, M.: A fuzzy approach to social network analysis. In: Proc. of the ASONAM 2009 Conference, Athens, Greece (2009)
9. Carlsson, C., Fullér, R.: Fuzzy Reasoning in Decision Making and Optimization. Studies in Fuzziness and Soft Computing, vol. 82, Springer, Heidelberg (2002)
10. Davis, G.B., Carley, K.M.: Clearing the FOG: Fuzzy overlapping groups for social networks. Soc. Networks 30, 201–212 (2008)
11. De Luca, A., Termini, S.: A Definition of a Nonprobabilistic Entropy in the Setting of Fuzzy Sets Theory. Information and Control 20, 301–312 (1972)
12. Everett, M.G., Borgatti, S.P.: The centrality of groups and classes. J. Math. Sociol. 23, 181–201 (1999)
13. Fan, T.-F., Liau, C.-J., Lin, T.-Y.: Positional Analysis in Fuzzy Social Networks. In: Proc. of IEEE International Conference on Granular Computing, pp. 423–428 (2007)
14. Freeman, L.C.: A set of measures of centrality based on betweenness. Sociometry 40, 35–41 (1977)
15. Freeman, L.C.: Centrality in social networks: Conceptual clarification. Soc. Networks 1, 215–239 (1979)
16. Fullér, R., Majlender, P.: An analytic approach for obtaining maximal entropy. OWA operator weights Fuzzy Sets Syst. 124, 53–57 (2001)

17. Fullér, R., Majlender, P.: On obtaining minimal variability OWA operator weights Fuzzy Sets Syst. 136, 203–215 (2003)
18. Hsie, M.-H., Magee, C.L.: An algorithm and metric for network decomposition from similarity matrices: Application to positional analysis. Soc. Networks 30, 146–158 (2008)
19. Klir, G.J., Yuan, B.: Fuzzy Sets and Fuzzy Logic: Theory and Applications. Pretience Hall, Englewood Cliffs (1995)
20. Kokabu, M., Katai, O., Shiose, T., Kawakami, H.: Design concept of community currency based on fuzzy network analysis Complexity International. vol. 11, pp. 102–112 (2003)
21. Kolaczyk, E.D., Chua, D.B., Barthélemy, M.: Group betweenness and co-betweenness: Inter-related notions of coalition centrality. Soc.Networks 31, 190–203 (2009)
22. Luce, R.D., Perry, A.D.: A method of matrix analysis of group structure. Psychometrika 14, 95–116 (1949)
23. Mollenhorst, G., Voelker, B., Flap, H.: Social contexts and personal relationships: The effect of meeting opportunities on similarity for relationships of different strength. Soc. Networks 30, 60–69 (2008)
24. Mordeson, J.N., Nair, P.S.: Fuzzy Graphs and Fuzzy Hypergraphs. Physica-Verlag, New York (2000)
25. Nair, P.S., Sarasamma, S.T.: Data mining through fuzzy social network analysis. In: Proc. Of the 26th International Conference of North American Fuzzy Information Processing Society, San Diego, California, pp. 251–255 (2007)
26. Newman, M., Barabasi, A.L., Watts, D.J.: The Structure and Dynamics of Networks. Princeton University Press, Princeton (2006)
27. O'Hagan, M.: Aggregating template or rule antecedents in real-time expert systems with fuzzy set logic. In: Proc. 22nd Annual IEEE Asilomar Conf. on Signals, Systems, Computers, pp. 681–689 (1988)
28. Pattison, P.E.: The analysis of semigroups of multirelational systems. J. of Math. Psych. 25, 87–118 (1982)
29. Peay, E.R.: Hierarchical Clique Structures. Sociometry 37, 54–65 (1974)
30. Scott, J.: Social Network Analysis. A Handbook. Sage, London (2000)
31. Todderdell, P., Holman, D., Hukin, A.: Social networks: Measuring and examining individual differences in propensity to connect with others. Soc. Networks 30, 283–296 (2008)
32. Yager, R.R.: Ordered weighted averaging operators in multicriteria decision making. IEEE T. Syst. Man Cy. 18, 183–190 (1988)
33. Yager, R.R., Kacprzyk, J.: The ordered weighted averaging operators: Theory and application. Kluwer Academic Publisher, Boston (1997)
34. Yager, R.R.: Intelligent Social Network Analysis Using Granular Computing. Int. J. of Intell. Syst. 23, 1197–1220 (2008)
35. Wasserman, S., Faust, K.: Social Networks Analysis: Methods and Applications. Cambridge University Press, Cambridge (1994)
36. Zadeh, L.A.: Fuzzy Sets. Information and Control 8, 338–353 (1965)
37. Zadeh, L.A.: The Concept of a Linguistic Variable and its Application to Approximate Reasoning I–II–III, Informa. Sciences, 8, 199–249, 301–357, 9, 43–80 (1975)

Soft Computing in Water Resources Management by Using OWA Operator

Mahdi Zarghami* and Ferenc Szidarovszky

Abstract. The Ordered Weighted Averaging (OWA) operator is an efficient multi criteria decision making (MCDM) method. This study introduces a new method to obtain the order weights of this operator. The new method is based on the combination of fuzzy quantifiers and neat OWA operators. Fuzzy quantifiers are applied for soft computing in modeling the social preferences (optimism degree of the decision maker, DM). In using neat operators, the ordering of the inputs is not needed resulting in better computation efficiency.

One of the frequently-used ways to control water shortages is inter-basin water transfer (IBWT). Efficient decision making on this subject is however a real challenge for the water institutions. These decisions should include multiple criteria, model uncertainty, and also the optimistic/pessimistic view of the decision makers. The theoretical results are illustrated by ranking four IBWT projects for the Zayanderud basin, Iran. The results demonstrate that by using the new method, more sensitive decisions can be obtained to deal with limited water resources.

The results of this study also show that this new method is more appropriate than the other traditional MCDM methods in systems engineering since it takes the optimism/pessimism nature of the DM into account in a quantifiable way. The comparison of the computational results with the current state of the projects shows the optimistic character of the DM. A sensitivity analysis illustrates how the rankings of the water projects depend on the optimism degree of the DMs.

Mahdi Zarghami
Faculty of Civil Engineering, University of Tabriz, Tabriz, I.R. Iran,
e-mail: zarghaami@gmail.com

Ferenc Szidarovszky
Systems and Industrial Engineering Department, University of Arizona,
Tucson, AZ, USA
e-mail: szidar@sie.arizona.edu

* Corresponding author.

R.R. Yager et al. (Eds.): Recent Developments in the OWA Operators, STUDFUZZ 265, pp. 269–279.
springerlink.com

1 Introduction

The increasing demand for water and the relative absence of new and less expensive resources are major challenges and should be considered in formulating the water resources management problems. These problems should include the social, economic, and environmental issues. In addition, there are many sources of uncertainty which makes the efficient decision making a complex problem.

Since the original work of Yager (1988) the OWA operator is used in many fields including water resources management problems. Table 1 summarizes the most important application being developed in this decade.

This chapter introduces a new version of OWA, which will be then applied to solve a practical problem of selecting the most appropriate water resources project. This new method will be then applied for a project selection problem. The study is organized as follows.

Table 1 Some applications of OWA in water resources management

Authors	Applications
Despic and Simonovic (2000)	Comparing the OWA with three other methods to select flood control measures in Manitoba, Canada
Yalcin and Akyurek (2004)	Mapping flood vulnerability in a basin in Turkey
McPhee and Yeh (2004)	Applying OWA in a multi-objective study to choose scenarios in aquifer management
Mysiak et al (2005)	A decision tool in the MULINO decision support system (DSS) for integrated water resources management
Makropoulos and Butler (2006)	Extending the OWA by applying it in GIS to produce prioritization maps for pipe replacement in a water distribution network
Fu et al. (2006)	Aggregating the possible climate change scenarios based on their probabilities by using the OWA approach
Malczewski (2006)	Using OWA with geographic information system (GIS) for multi-criteria evaluation for land-use suitability analysis in Canada
Zarghami et al. (2008)	Using the OWA operator in group decision making in a conflict among stakeholders in a watershed
Averna Valente and Vettorazzi (2008)	Integrating OWA with GIS to define the priority areas for forest conservation in a Brazilian river basin in order to increase the regional biodiversity
Stroppiana et al (2009)	To aggregate the anomaly scores of a set of contributing factors extracted from the analysis of historical time series, mostly of Earth observations data

The next section introduces the new method, called the Revised OWA. Then the multi-criteria case study of selecting IBWT projects will be described. The IBWT projects will be compared in the next section by using the new method. The last section concludes the chapter.

2 OWA Operator

Assume that the number of decision alternatives is m which are evaluated by n criteria. Let a_{ij} denote the evaluation of alternative i with respect to criterion j, then the overall goodness of this alternative can be characterized by the evaluation vector $X_i=(a_{i1}, a_{i2}, \ldots, a_{in})$. The comparison of these decision alternatives is based on a combined goodness measures. OWA can be used to aggregate the evaluations of each alternative with respect to the criteria. An n-dimensional OWA operator assigns a combined goodness measure

$$F_i(a_{i1}, a_{i2}, \ldots, a_{in}) = \sum_{j=1}^{n} w_j b_{ij} = w_1 b_{i1} + w_2 b_{i2} + \ldots + w_n b_{in}\ , \tag{1}$$

for each alternative i where $F: I^n \mapsto I$ with $I = [0,1]$, b_{ij} is the jth largest element in the set of $\{a_{i1}, a_{i2}, \ldots, a_{in}\}$ and w_j $(j = 1, 2, \ldots\ldots, n)$ are the order weights such that $w_j \geq 0$ and $\sum_{j=1}^{n} w_j = 1$. That is, the OWA operator is a convex linear combination of the b_{ij} values. Notice that the components of the input vector have been ordered before multiplying them by the order weights. The OWA method has a large variety by the different selections of the order weights. Order weights depend on the optimism degree (well known as Orness degree) of the decision maker (DM). The greater the weights at the beginning of the vector are, the higher is the optimism degree. Yager (1988) defined the optimism degree θ as

$$\theta = \frac{1}{n-1} \sum_{j=1}^{n} (n-j) w_j\ . \tag{2}$$

Xu (2005) gives a general overview of the different methods for determining the order weights. In the next section, we will introduce the new method, called the Revised OWA.

3 Revised OWA Operator

In common language we use many linguistic terms such as *most, few, many*, and *about half*. Zadeh (1983) recommended the modeling of these linguistic quantifiers by using fuzzy sets. In this chapter these linguistic inputs are modeled by Regular Increasing Monotonic (RIM) quantifiers. A RIM quantifier, Q, characterizes aggregation imperatives, in which higher satisfaction is obtained by including more objects. This quantifier has the following properties:

$$R(Q)=[0,\ 1],\quad Q(0)=0\ ,\quad Q(1)=1 \ \text{ and } Q(r_1)\geq Q(r_2) \quad \text{if} \quad r_1 \geq r_2. \tag{3}$$

Yager (1988) suggested obtaining the weights of an n-dimensional OWA operator as

$$w_j = Q(\frac{j}{n}) - Q(\frac{j-1}{n}), \quad j=1\ ,\ 2\ ,\ ...,n. \tag{4}$$

Notice first that the derivative of the fuzzy quantifier Q is as follows:

$$\frac{dQ}{dr} = \lim_{\Delta r \to 0} \frac{Q(r)-Q(r-\Delta r)}{\Delta r}. \tag{5}$$

In the special case when n is large we may select $\Delta r = 1/n$, and so

$$\frac{dQ}{dr} \approx \frac{Q(r)-Q(r-1/n)}{1/n}. \tag{6}$$

Yager (1993) evaluated the value of dQ/dr at $r = j/n$ by using equation (4) as

$$\left.\frac{dQ}{dr}\right|_{r=j/n} \approx \frac{Q(j/n)-Q((j-1)/n)}{1/n} = \frac{w_j}{1/n}$$

so

$$w_j \approx \left.\frac{1}{n}\frac{dQ}{dr}\right|_{r=j/n}. \tag{7}$$

These weights depend on only the order of the criteria. More sensitive weight selection can be obtained if the weights also depend on the evaluations of the criteria. So instead of using equation (7), we propose the following weight selection:

$$w_j = \left.\frac{1}{n}\frac{dQ}{dr}\right|_{r=1-b_j} \tag{8}$$

where $b_1 \geq b_2 \geq ... \geq b_n$. The reason of using the term (1-b_j) instead of b_j is due to the opposite ordering of the criteria in equation (7) in comparison to the ordering of the b_j values in the case of RIM quantifiers. These values however do not satisfy the necessary conditions of OWA weights since their sum usually differs from unity. After normalizing the w_j values in equation (8), the final weights are obtained as follows:

$$w_j = \frac{Q'(1-b_j)}{\sum_{l=1}^{n} Q'(1-b_l)}. \tag{9}$$

This method of weight selection is entitled Revised OWA (Zarghami and Szidarovszky, 2009), since it is based on the exact derivatives of the quantifier. The weights obtained by equation (9) satisfy all necessary conditions of the OWA weights. The Revised OWA operator with these weights and with any fuzzy quantifier is a neat operator since the combined goodness measure, F_i is independent of the ordering of the inputs:

$$F_i(a_{i1}, a_{i2}, ..., a_{in}) = \sum_{j=1}^{n} w_j b_{ij} = \sum_{j=1}^{n} \frac{Q'(1-b_{ij})}{\sum_{l=1}^{n} Q'(1-b_{il})} b_{ij} = \sum_{j=1}^{n} \frac{Q'(1-a_{ij})}{\sum_{l=1}^{n} Q'(1-a_{il})} a_{ij} = \frac{\sum_{j=1}^{n} Q'(1-a_{ij}) a_{ij}}{\sum_{l=1}^{n} Q'(1-a_{il})}. \tag{10}$$

An additional advantage of using neat OWA operators in comparison to the initial OWA is that in this case more attention is given to the circumstance of the problem (e.g. to the evaluation values b_{ij}). It is however a weakness of Revised OWA that the weights have to be calculated separately for each alternative.

4 IBWT Projects for Central Iran

The increasing water demand has caused an alarming decrease in the annual per capita water resources in Iran which is an arid country. The uneven distribution of the water resources across the country and the fast growth of the population and

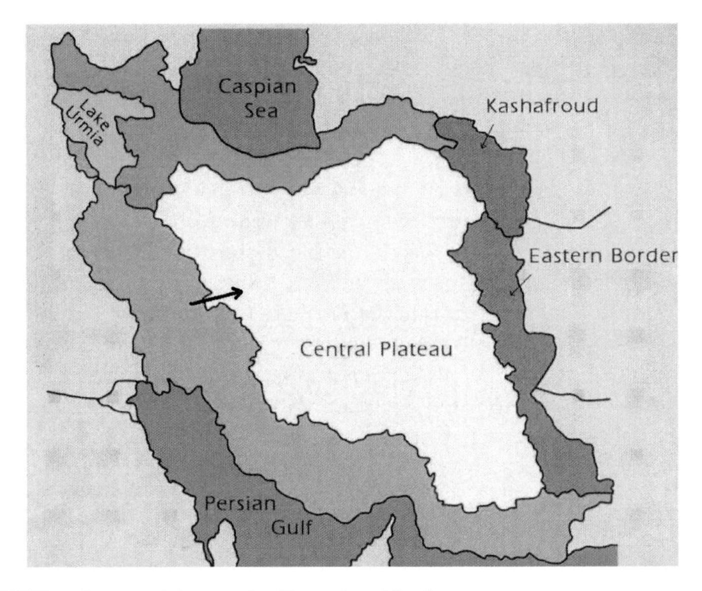

Fig. 1 IBWT projects positions to the Zayanderud basin

their water demands have led to the present water shortages in the major parts of the country, especially in the Central zone and in the Southeastern regions. The country is divided into six main hydrological basins as shown in Figure 1. The annual per capita water resources potentials (cubic meters) in the main basins are given in Figure 2 for the year 2000.

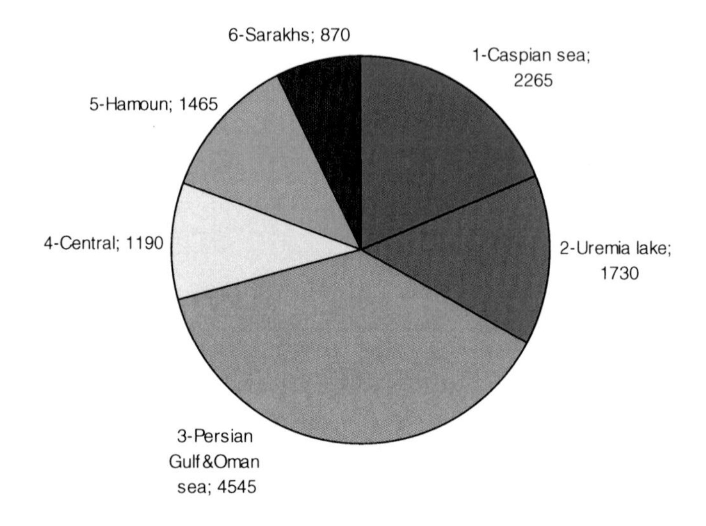

Fig. 2 Water resources potentials (m^3) per capita (year 2000) in the main basins of Iran

According to Figure 2, basin 3 (Persian Gulf and Oman Sea) has the highest amount of water resources (~4545 cubic meter/capita/year) while the neighboring basin 4 (Central) has only a quarter of that amount (~1190 cubic meter/capita/year), so it is facing a high degree of water shortage. The socio-economic life of the people in the Central basin can be significantly improved by water transfer from the neighboring basin, Large Karun, which is located in the Southwest corner of the country. It is the most important basin in Iran with respect to water potentials and the possibility of further water resources development. Part of its water potential is utilized inside the basin and transferred to other basins. The remaining water amount, around 20 to 25 billion cubic meters per year, is spilled and lost through outflow into the Persian Gulf. A part of the domestic and industrial wastewater returns to the ground water aquifers or to surface drainages. The assessment of the overall water quality of the system (Large Karun) shows that the IBWT projects from the upstream will worsen the water quality in the downstream. Due to the high potential of growth in the Karun basin, there are various concerns about water transfers to other basins, which generates conflicts among the stakeholders. Therefore it is essential to evaluate carefully any possible IBWT project from Large Karun before implementation.

In this study we focus on the IBWT projects to the Zayanderud basin. It is located West of the Central basin as shown in Figure 1. It has a population around four millions and in recent years, this basin is developed very extensively. In order to meet the increasing water demand in this basin, four IBWT alternative projects (Alt1: Gukan, Alt2: Cheshmelangan, Alt3: Kuhrang-III, and Alt4: Behestabad) have been developed to transfer water from the Large Karun River. These projects transfer water to the Zayanderud River, which passes through Isfahan, an important and historical city of the country. The city attracts more than one million domestic and foreign tourists every year.

The Isfahan Regional Water Company as DM wants to compare these IBWT projects. Before evaluating these projects, it is necessary to construct a general hierarchy of the criteria. In the first step, major watershed's plans of twenty countries were examined. Based on the state-of-the-art reviews and the national acts of Iran, a hierarchy of the criteria was then introduced (Zarghami et al 2007).

In this case study, only seven criteria were selected from the hierarchy, since some of them were irrelevant to the IBWT projects and also there was a lack of reliable data to evaluate the projects with respect to some others. The final seven criteria and the corresponding evaluation data are presented in Table 2. They were obtained by using a group of experts from the DM's company. The uncertainty of the data is taken into account by using linguistic variables, the followings were used: Very Low (VL), Low (L), Slightly Low (SL), Medium (M), Slightly High (SH), High (H), and Very High (VH).

Table 2. Evaluation matrix of the IBWT projects (adopted from Zarghami et al, 2007)

	Criteria						
	Allocation of Water to Prior Usages	Diversification of Financial Resources	Resettlement of People	Public Participation	Consistency with Policies	Benefit/Cost	Range of Environmental Impacts
	Weights of criteria						
Alternatives	VH	M	VH (Negative)	L	H	M	SL (Negative)
Alt1	SH	5.0	0.0	SH	H	1.5	L
Alt2	VH	0.0	0.0	M	VH	1.4	M
Alt3	VH	3.0	200.0	H	VH	1.1	SL
Alt4	VH	4.0	4000.0	VH	H	1.6	SH

5 Applying Revised OWA in Soft Ranking of the Alternatives

We can now determine the combined goodness measures of four IBWT projects by using the Revised OWA. The calculation procedure is as follows:

Step 1. At the beginning, consider the projects which are evaluated by using linguistic variables. The linguistic data were modeled by crisp numbers according to the uniform scale of {0.05, 0.20, 0.35, 0.50, 0.68, 0.80, 0.95} corresponding to the set {VL, L, SL, M, SH, H, VH}. Other scales could also be introduced based on non-uniform distributions.

Step 2. These evaluation numbers were then normalized into the unit interval [0, 1] as:

$$a_i = \begin{cases} \dfrac{A_i}{\underset{1\le j\le n}{Max}(A_j)} & \text{for positive criteria}, \\ \\ \dfrac{\underset{1\le j\le n}{Min}(A_j)}{A_i} & \text{for negative criteria}. \end{cases} \tag{11}$$

Step 3. In applying the original version of OWA, the criteria weights are considered to be equal, however in this case, they were different as it is shown in the first row of Table 2. These weights were multiplied by the normalized evaluations of the alternatives.

Step 4. The order weights were determined by using equation (9). Table 3 shows the results for Alt1.

Table 3 The OWA weights for Alt1 by using different quantifiers

	Fuzzy Quantifiers						
	All	*Most*	*Many*	*Half*	*Some*	*Few*	*At least one*
	Relevant optimism degree						
	0.001	0.091	0.333	0.500	0.667	0.909	0.999
w_1	0.000	0.071	0.199	0.143	0.095	0.056	0.047
w_2	0.000	0.012	0.162	0.143	0.105	0.067	0.058
w_3	0.000	0.000	0.100	0.143	0.134	0.104	0.094
w_4	1.000	0.910	0.264	0.143	0.083	0.043	0.036
w_5	0.000	0.000	0.015	0.143	0.344	0.562	0.615
w_6	0.000	0.007	0.153	0.143	0.109	0.071	0.062
w_7	0.000	0.000	0.107	0.143	0.130	0.098	0.088

Step 5. The combined goodness measures have been then calculated by using equation (10). The results are shown in Figure 3.

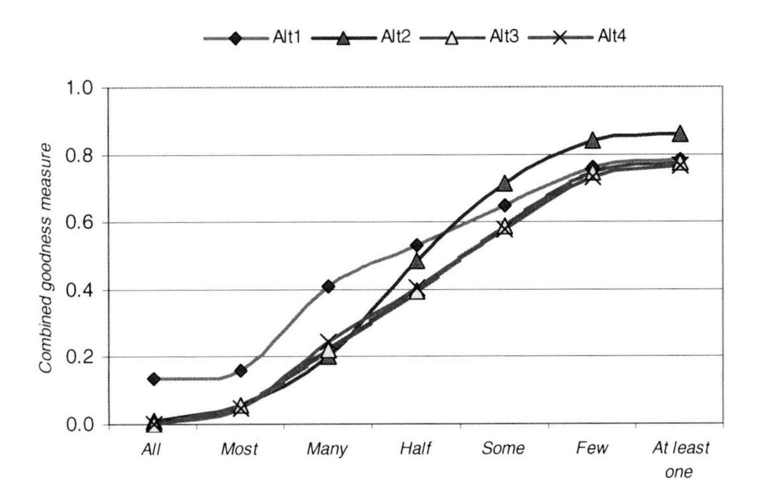

Fig. 3 The combined goodness measures of the IBWT projects

According to Figure 3, Alt2 is the most preferred project when the DM is optimistic (based on the quantifiers *At least, Few* and *Some*). Alt1 is however the most preferred project if the DM is neutral (by the quantifier *half*) or pessimistic (by the quantifiers *Many, Most* and *All*).

The corresponding ranks of the alternatives are shown in Table 4. The last row 'current state of project' reflects the previous decisions of the DM in which Alt2 is in operation (rank 1), Alt3 is under construction (rank 2), Alt1 is in the final study (rank 3) and Alt4 is under investigation (rank 4).

The most and the least preferred projects, according to the row 'current state of project', are the same as in the columns of *some*, *few* and *at least* which represent the optimistic view of the DM. Therefore, we can conclude that the DM was optimistic about the IBWT projects. Water managers are usually not risk-taking individuals. However in this case, the DM is water recipient and not water

Table 4 Ranking of the IBWT projects

Using Fuzzy quantifiers	*At least*	$A2 \succ A1 \succ A3 \succ A4$
	Few	$A2 \succ A1 \succ A3 \succ A4$
	Some	$A2 \succ A1 \succ A3 \succ A4$
	Half	$A1 \succ A2 \succ A4 \succ A3$
	Many	$A1 \succ A4 \succ A3 \succ A2$
	Most	$A1 \succ A3 \succ A2 \succ A4$
	All	$A1 \succ A2 \succ A3 \succ A4$
Current State of the projects		$A2 \succ A3 \succ A2 \succ A4$

supplier. He/she wants to bring as much water as possible to the Zayanderud basin, which explains the optimistic view.

However, the optimism degree of the DM is also subject to the national and local policies. If the DM felt to be pessimistic due to the risky conditions, the Alt1 project would have been the most preferred project. The ranks of the other projects also depend on the optimism degree. As an illustration, a sensitivity analysis was performed on the ranks of the alternatives due to changes of the optimism degree. The interval [0.01, 0.99] was selected with the increment of 0.03 and the entire procedure was repeated for all particular values of the optimism degree. Figures 4, as part of the results, illustrates how the most preferred alternative would change by the dynamic feature of the optimism degree.

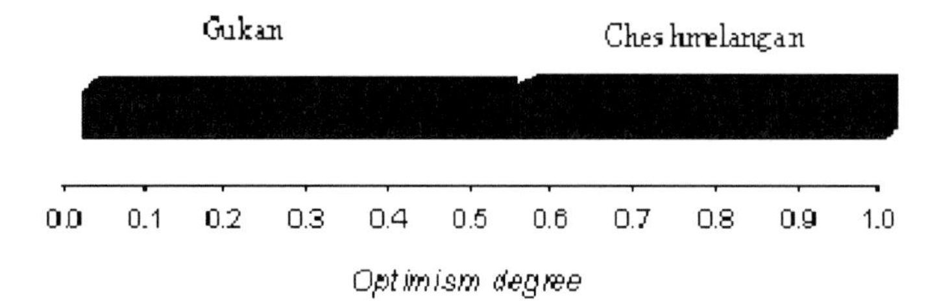

Fig. 4 Most preferred alternatives with different optimism degrees

According to results, the ranks of the projects are robust in the entire optimistic section ($\theta > 0.5$). However, in the pessimistic section their dependence on the optimism degree is not monotonic. Therefore, the precise knowledge of the optimism degree of the DM in the pessimistic region ($\theta < 0.5$) is very important for securing the safe and satisfactory decision.

6 Conclusions

The Revised OWA operator was introduced and applied successfully in the soft ranking of the IBWT projects for Zayanderud, Iran. The results of this case study show that this new method is more appropriate than the other traditional MADM methods since it reflects the optimism/pessimism nature of the DM by using a quantifiable method. The comparison of the obtained results with the current state of the projects shows the optimistic character of the DM.

Revised OWA benefits from fuzzy quantifiers to achieve a better characterization of the DM's satisfaction. It is therefore a context based model in which the ordering of the initial inputs is not required, so it is a neat operator. This new method therefore offers a more efficient way of computing the OWA weights. A sensitivity analysis illustrated the dependence of the rankings on the optimism degree of the DM.

Acknowledgments. The authors are grateful for the technical and financial supports of the University of Tabriz, Water Resources Management Company of Iran, Mahab Ghodss Consulting Engineers and the Isfahan Regional Water Company in Iran.

References

Averna Valente, R.O., Vettorazzi, C.A.: Definition of priority areas for forest conservation through the ordered weighted averaging method. Forest Ecology and Management 256(6), 1408–1417 (2008)

Despic, O., Simonovic, S.P.: Aggregation operators for soft decision making in water resources. Fuzzy Sets and Systems 115, 11–33 (2000)

Fu, G., Hall, J., Lawry, J.: Beyond probability: new methods for representing uncertainty in projections of future climate. Tyndall Centre for Climate Change Research Working Paper 75, `http://www.tyndall.ac.uk` (accessed 2006)

McPhee, J., Yeh, W.W.-G.: Multiobjective optimization for sustainable groundwater management in semiarid regions. Journal of Water Resources Planning and Management 130(6), 490–497 (2004)

Makropoulos, C.K., Butler, D.: Spatial ordered weighted averaging: incorporating spatially variable attitude towards risk in spatial multi-criteria decision-making. Environmental Modeling and Software 21(1), 69–84 (2006)

Malczewski, J.: Ordered weighted averaging with fuzzy quantifiers: GIS-based multicriteria evaluation for land-use suitability analysis. International Journal of Applied Earth Observation and Geoinformation 8(4), 270–277 (2006)

Mysiak, J., Giupponi, C., Rosato, P.: Towards the development of a decision support system for water resource management. Environmental Modelling & Software 20, 203–214 (2005)

Stroppiana, D., Boschetti, M., Brivio, P.A., Carrara, P., Bordogna, G.: A fuzzy anomaly indicator for environmental monitoring at continental scale. Ecological Indicators 9(1), 92–106 (2009)

Xu, Z.S.: An overview of methods for determining OWA weights. International Journal of Intelligent Systems 20, 843–865 (2005)

Yager, R.R.: On ordered weighted averaging aggregation operators in multi-criteria decision making. IEEE Transactions on Systems, Man and Cybernetics 18(1), 183–190 (1988)

Yager, R.: Families of OWA operators. Fuzzy Sets and Systems 59, 125–143 (1993)

Yalcin, G., Akyurek, Z.: Multiple criteria analysis for flood vulnerable areas. In: Proc. of 24th Annual ESRI International User Conference, San Diego, USA, August 9–13 (2004)

Zadeh, L.: A computational approach to fuzzy quantifiers in natural languages. Computers and Mathematics with Application 9, 149–184 (1983)

Zarghami, M., Ardakanian, R., Memariani, A.: Fuzzy multiple attribute decision making on inter-basin water transfers, case study: Transfers to Zayanderud basin in Iran. Water International 32(2), 280–293 (2007)

Zarghami, M., Ardakanian, R., Memariani, A., Szidarovszky, F.: Extended OWA operator for group decision making on water resources projects. Journal of Water Resources Planning and Management 134(3), 266–275 (2008)

Zarghami, M., Szidarovszky, F.: Revising the OWA operator for multi criteria decision making problems under uncertainty. European Journal of Operational Research 198(1), 259–265 (2009)

Combination of Similarity Measures in Ontology Matching Using the OWA Operator

Qiu Ji, Peter Haase, and Guilin Qi

Abstract. In this paper, we provide a novel solution for ontology matching by using the ordered weighted average (OWA) operator to aggregate multiple values obtained from different similarity measures. We have implemented the solution in the ontology matching system FOAM. Using the similarity measures in FOAM, we analyze the way to choose different OWA operators and compare our system with others.

1 Introduction

Ontology matching aims at identifying correspondences between the elements in multiple ontologies. Ontology matching has many application areas, such as data integration, data merging, and semantic search across heterogeneous data sources. So far, quite a number of ontology matching systems have been proposed. Good surveys of different approaches to the matching problem are provided in [9, 10].

It has been accepted that combining the values obtained by multiple similarity measures is a promising technique to obtain more accurate matching results than just using one similarity measure at a time. Usually, a simple *weighted average* is used as the aggregation operator where the weights can be obtained manually or by machine learning techniques. Obviously, it is difficult for a person to manually assign the weights to the similarity measures by experience. For the way based on machine learning techniques, rich data sets are needed to train the algorithms for obtaining useful weights.

To alleviate this problem, we investigate the use of the *Ordered Weighted Average* (OWA) [16] to aggregate the values obtained by individual similarity measures. It is noted that, a weight used by the OWA operator is associated not with a specific similarity measure, but instead with a specific *ordered position*. We have implemented

Qiu Ji · Peter Haase · Guilin Qi
Institute AIFB, University of Karlsruhe, D-76128 Karlsruhe, Germany
e-mail: qiji@aifb.uni-karlsruhe.de, pha@aifb.uni-karlsruhe.de
gqi@aifb.uni-karlsruhe.de

R.R. Yager et al. (Eds.): Recent Developments in the OWA Operators, STUDFUZZ 265, pp. 281–295.
springerlink.com

our solution in the ontology matching system FOAM[1]. There are two main reasons for integrating the OWA operator into FOAM:

1. FOAM provides many similarity measures according to various features of OWL[2] ontologies such as the feature of domain or range for an property.
2. The OWA operator is a powerful operator to aggregate multiple values, and there are many kinds of approaches to obtain OWA weights [5, 15, 6]. Particularly, the linguistic OWA operator [16] provides semantic explanations which can be understood by users easily.

This paper is organized as follows: In Section 2, we discuss the related work on the aggregation of the similarity measures in various ontology matching systems. In Section 3 we describe the background of the FOAM system and the OWA operator, and then we discuss the integration of the OWA operator into FOAM and the resulting problems. We evaluate the OWA operator as an aggregation operator in FOAM in Section 4. Finally, in Section 5 we conclude this paper and give an outlook to the future work.

2 Related Work

So far, various ontology matching systems have been developed by many researchers (see good surveys [9, 10]). They consider various kinds of information provided in the ontologies. To aggregate the values obtained by multiple individual similarity measures, many aggregation operators have been proposed. In the following we discuss those ontology matching systems which are most related to our work.

COMA [1] exploits *Max*, *Min*, *Average* and *Weighted* strategies for the aggregation operation. Where, the *Weighted* strategy computes a weighted sum of similarity values for the individual similarity measures which assigns relative weight to each similarity measure. *Average* is a special case of *Weighted* which considers each similarity measure equally important and returns the average similarity value over all similarity measures. *Max* and *Min* are two extreme cases which return the highest and lowest similarity value of any similarity measure respectively.

The study in paper [4] also adopts *Weighted* strategy to aggregate the similarity values obtained by all the features (i.e. similarity measures) that make the definition of an entity in an OWL-Lite ontology. The weight here is linked to entire descriptive aspects instead of particular similarity measures.

CMC [13] combines similarity values using weighted average based on credibility prediction. It needs to predict the accuracy of each similarity measure on the current matching task first by a manual rule or a machine learning method. Accordingly, different credibilities for the similarity measures are assigned. That is, for each similarity measure two matrices including the similarity matrix and the credibility matrix are provided. For each pair of entities to be compared, the credibilities in the

[1] http://ontoware.org/projects/map
[2] http://www.w3.org/TR/owl-features/

credibility matrices are used as weights to aggregate the similarity values obtained by different similarity measures into a combined one.

In the original version of FOAM, which implements the algorithms described in [2], both the manual way to assign weights and the automatic way to learn how to combine the similarity values are provided.

To sum up, *Weighted* is the most popular aggregation operator to combine multiple similarity values. The weight here is assigned to a particular similarity measure and can be obtained manually or by machine learning techniques. When it is not necessary or it is difficult to obtain weights for the similarity measures, only *Max*, *Min* and *Average* can be used. However, since each similarity measure performs differently under different conditions, these operators may be not enough to show various performance for complex situations.

In our previous work [8], the linguistic OWA operators are introduced to aggregate multiple similarity values for ontology matching. But there is no details about

1. How the performance of OWA operator behaves with more similarity measures and complex ontologies.
2. How to choose different aggregation operators for different purposes.
3. How the performance of the system compares with other ontology matching systems.

In this paper, we provide a new solution for ontology matching by integrating OWA operator into the FOAM system. Based on this solution, the questions above can be answered accordingly.

3 The OWA Operator for Ontology Matching

In this section, we first introduce some basic notions in ontology matching to be used throughout the paper. Then we will describe a specific ontology matching system FOAM in which we will integrate the OWA operator for similarity integration. After that, the OWA operator is introduced. Finally, the integration of the OWA operator into FOAM will be presented.

3.1 Basic Notions in Ontology Matching

We introduce here some basic notions given in [3], which will be used later in this paper. We assume the readers are familiar with ontologies, especially OWL ontologies.

Let O_1 and O_2 be two ontologies and Q be a function that defines sets of mappable elements $Q(O_1)$ and $Q(O_2)$. We call a 4-tuple $< e, e', r, \alpha >$ as a correspondence between O_1 and O_2 if

- $e \in Q(O_1)$ and $e' \in Q(O_2)$;
- r is a semantic relation which can be one of the semantic relations from the set $\{\equiv, \sqsubseteq, \sqsupseteq\}$;

- α is a confidence value (i.e. similarity value) from a suitable structure $< D, \leq >$. Usually, we adopt $D = [0.0, 1.0]$

A mapping between O_1 and O_2 consists of a set of correspondences between the pairs of entities which belong to $Q(O_1)$ and $Q(O_2)$ respectively.

3.2 Ontology Matching in FOAM

The FOAM system is based on a generic process for ontology matching, as described in [2]. Other ontology matching approaches can be described in terms of this process as well. Here, we only describe briefly the process to the extent that is necessary to understand how the role of similarity aggregation works within this process. Figure 1 illustrates the six main steps of the generic process.

Input:

Input for the process are two or more ontologies, which need to be matched with one another. Additionally, it is often possible to enter pre-known (manual) correspondences. They can help to improve the search for other correspondences.

1. Feature Engineering:

The role of feature engineering is to select relevant features of the ontology to describe a specific ontology entity, based on which the similarity with other entities will later be assessed. For instance, the matching process may only rely on a subset of OWL primitives. For each feature, a specific similarity value based on a corresponding similarity measure will be assigned.

2. Search Step Selection:

The derivation of the correspondences takes place in a search space of candidate correspondences. This step may choose to compute the similarity of certain candidate entity pairs and to ignore others in order to prune the search space [2].

3. Similarity Computation:

For a given description of two entities in a candidate correspondence, this step computes the similarity value of the entities using the selected features.

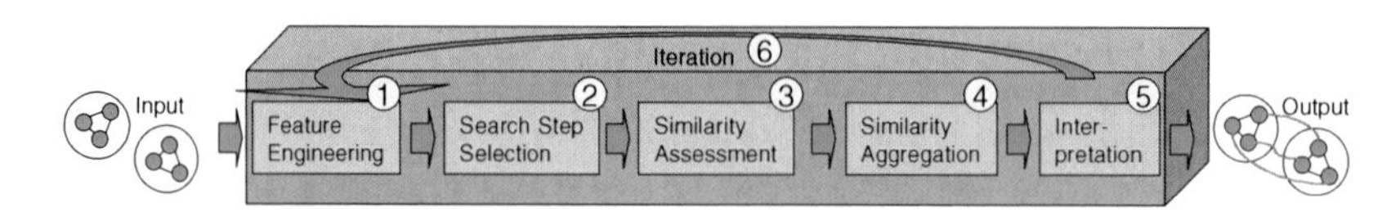

Fig. 1 Ontology matching process in FOAM

4. Similarity Aggregation:

In general, there may be several similarity values for a candidate pair of entities, e.g., one for the similarity of their labels and one for the similarity of their relationships (i.e. properties) to other entities. These different similarity values for one candidate pair have to be aggregated into a single aggregated similarity value. Often – as in the original FOAM system – a weighted average is used for the aggregation of similarity values.

5. Interpretation:

The interpretation finally uses individual or aggregated similarity values to derive correspondences between entities. A common approach is to use thresholds [1]: If the similarity value of two entities exceeds the threshold, the entities are mapped and they are considered as a correct correspondence.

6. Iteration:

The similarity of one entity pair influences the similarity of neighboring entity pairs. For example, if the individuals are equal, this affects the similarity of the concepts (i.e. the types of those individuals) and vice versa. Therefore, the matching process is repeated until no new correspondences are proposed or a fixed number (for our test, the maximal iteration is set 3) of iteration is reached.

Output:

The output is a representation of mapping including those correct correspondences (the correctness of a correspondence is determined in step 5) and possibly with additional confidence values (i.e. similarity values) based on the similarity of the entities.

3.3 The Ordered Weighted Average Operator

The ordered weighted averaging (OWA) operator is introduced in [16] to aggregate information. It has been used in a wide range of application areas, such as neural networks and fuzzy logic controllers.

Assume we are given a set of arguments $V_1 = (a_1, a_2, ..., a_n)$, $a_i \in [0,1]$, $1 \leq i \leq n$, and the weights for OWA operator $\mathrm{W} = (\mathrm{w}_1, ..., \mathrm{w}_n)$. After reordering the elements in V_1 in descending order, we mark it as $V_2 = (b_1, b_2, ..., b_n)$, where b_j is the j_{th} highest value in V_1. An OWA operator is a mapping function F from I^n to I, $I = [0,1]$:

$$\begin{aligned} F(a_1, a_2, ..., a_n) &= \textstyle\sum_{i=1}^{n} \mathrm{w}_i b_i \\ &= \mathrm{w}_1 b_1 + \mathrm{w}_2 b_2 + ... + \mathrm{w}_n b_n \end{aligned}$$

where $\mathrm{w}_i \in [0,1]$ and $\sum_{i=1}^{n} \mathrm{w}_i = 1$.

Note that a weight w_i is not associated with a particular argument a_i, but with a particular ordered position i of the arguments. That is w_i is the weight associated with the ith largest argument whichever component it is [16].

Obviously, determining the OWA weights w_i, $1 \leq i \leq n$ is a critical task. So far, quite a few approaches have been proposed. We adopt the linguistic quantifiers developed by Yager [16], since these quantifiers have semantics which can be accepted easily for users. They are defined as:

$$w_i = Q(i/n) - Q((i-1)/n), i = 1, 2, ..., n \tag{1}$$

where Q is a nondecreasing proportional fuzzy linguistic quantifier and is defined as the following:

$$Q(r) = \begin{cases} 0, & \text{if } r < a; \\ (r-a)/(b-a), & \text{if } a \leq r \leq b; \\ 1, & \text{if } r > b, \end{cases} \tag{2}$$

where $0 \leq a, b, r \leq 1$, a and b are the predefined thresholds. Obviously, the operators such as *Max*, *Min* and *Average* are three special cases of the OWA operator.

For some special linguistic operators like *at least half* which are used in the paper, we introduce their semantic interpretations within the application area of ontology matching to facilitate users to choose different operators for different tasks or purposes.

Assume there are n similarity measures m_1, m_2, ..., m_n. Each similarity measure can be regarded as a criteria, so the aggregation process is to form an overall decision by considering multiple criteria. For an entity pair (x, y), where x belongs to a source ontology and y belongs to a target ontology, $m_i(x, y)$ indicates the degree to which the entity pair (x, y) satisfies the criteria or similarity measure m_i. Actually, $m_i(x, y)$ is the similarity value between x and y obtained by similarity measure m_i, where i= 1, 2, ..., n.

1. Max: $Max(x, y) = Max\{m_1(x, y), m_2(x, y), ..., m_n(x, y)\}$, where *Max* means that (x, y) satisfies at least one of the similarity measures, i.e., it satisfies m_1 or m_2 or... or m_n.
2. Min: $Min(x, y) = Min\{m_1(x, y), m_2(x, y), ..., m_n(x, y)\}$. *Min* means that (x, y) satisfies all the similarity measures, that is to say, we are essentially requiring to satisfy m_1 and m_2 and... and m_n.
3. Average: *Average* means identity, which regards all similarity values equally.
4. At least half: This operator satisfies at least half similarity measures. Actually, it only considers the first half of similarity values after re-ordering them in the descending order.
5. Most: *Most* means most of the similarity measures is satisfied. Usually, this operator ignores some higher and lower similarity values, that is to give small weights on them, while paying more attention to the values in the middle of the input arguments after re-ordering them in the descending order.

6. As many as possible: It satisfies as many as possible similarity measures and is opposite to *at least half*. The second half of values after reordering is considered. So after an aggregation operation, the result obtained by *at least half* is always no less than that by *as many as possible*.

3.4 Integration of OWA into FOAM

In FOAM, originally a *weighted average* is used for the similarity aggregation step, which gives more importance to the similarity measure based on the labels of the entities to be compared. If there is no label available, the performance of ontology matching will become worse.

In our work, we use an OWA operator to combine similarity values obtained from multiple similarity measures. Obviously, the main advantage of our method is that weights are not fixed to these similarity measures, but assigned to the positions of these similarity values in a descending order. In this way, each similarity measure is treated equally. The second advantage is that, although the OWA weights can be obtained manually or by machine learning techniques, there are quite a few straightforward methods without data sets for training and too much preliminary knowledge. What the users need to do is to take several entity pairs to be compared as samples and observe the similarity values obtained by the similarity measures for each entity pair, or they can simply choose different linguistic OWA operators by their semantic interpretations.

To choose OWA operators by observation, we assume there are n individual similarity measures for a category of entities, which can be concept category, datatype property category, object property category or individual category. If most individual similarity measures could return m $(0 \leq m \leq n)$ similarity values sim_i above zero, where $0 \leq i \leq m$, and $0 < sim_i \leq 1$, then it is better to choose an OWA operator which can give more importance to most of the m highest values or all of them, while assigning lower or zero to other $n - m$ values. For the similarity measures we used in this paper, no more than half of the similarity measures return some similarity values above zero in most cases. Based on our experience, higher values are more reliable, but one should not rely on just one highest value. So it would be better to use *at least half* which only considers the half higher values, but not to use *Max* considering only one extreme value for each aggregation.

4 Evaluation and Discussion

4.1 Data Sets

We use the benchmarks which are provided by the OAEI campaign in 2008[3]. The test case of benchmarks includes 110 ontologies in OWL excluding ontology 102 as the corresponding reference mapping is empty. Ontology 101 is regarded as the

[3] http://oaei.ontologymatching.org/2008/

reference ontology, i.e., each ontology in the benchmarks, including ontology 101, will be matched against ontology 101. The benchmarks are divided into three groups marked as 1xx (ontology 101-104), 2xx (ontology 201-266) and 3xx (ontology 301-304). More details can be found on the website of OAEI 2008.

The goal of the benchmark series is to identify the areas in which each matching algorithm is strong and weak. In our experiment, the mappings are computed automatically without the participation of users. We also obey the rules to obtain mappings to compare with other systems in OAEI 2008. For example, for all tests based on our matching system we use the same parameters such as the threshold to determine the mapped entity pairs and maximal number of iterations in the matching process.

4.2 Evaluation Criteria

In order to compare the performance of different matching algorithms or systems, several evaluation criteria are used to give different views of the results. Except the standard measures such as *precision*, *recall* and *f-measure*, the harmonic mean measure is also used to compare our results with those provided by other ontology matching systems in OAEI 2008.

For the measures below, i indicates the ith test. $|R_i|$ refers to the number of correspondences in a reference mapping or golden standard which is manually created. $|P_i|$ is the number of all correspondences found automatically by the matching system. $|I_i|$ is the number of *correct correspondences* found by the matching system for test i, where the correct correspondences mean those correspondences found by a matching system are also in the reference mapping.

1. Precision (p):

$p_i = |I_i|/|P_i|$. It reflects the ratio of the correct correspondences among all correspondences discovered by the matching system.

2. Recall (r):

$r_i = |I_i|/|R_i|$ specifies the ratio of correct correspondences found by the matching system in comparison with total number of correspondences in the golden standard.

3. F-Measure (f):

$f_i = 2 * p_i * r_i/(p_i + r_i)$, which estimates the reliability of the match predictions [1].

4. Harmonic mean (H):

Harmonic mean[2] is an aggregation of standard measures such as *precision* and *recall*. Specifically, harmonic mean of *precision*, *recall* and F-Measure is defined respectively as followings:

$$H(p) = \sum_{i=1}^{n} |I_i| \; / \; \sum_{i=1}^{n} |P_i|$$
$$H(r) = \sum_{i=1}^{n} |I_i| \; / \; \sum_{i=1}^{n} |R_i|$$
$$H(f) = 2 * H(p) * H(r) \; / \; (H(p) + H(r))$$

Here, n is the number of the considered tests.

4.3 Similarity Measures

For our experiments, we used more than 20 similarity measures which are defined according to various characteristics of an OWL ontology. Usually, the entities in an ontology consist of the concepts, properties (which includes object properties and datatype properties) and individuals. Based on the classification of the entities, the similarity measures we used can be divided into the following categories:

- String-based similarity measures: This category computes the similarity between two entities using edit distance[4] technique by considering the string of a label or local name of a URI from an entity as the sequences of letters in an alphabet.
- Concept-based similarity measures: Such kind of measures regard the two concepts as similar if their super concepts, sub concepts, related properties or associated individuals are similar.
- Property-based similarity measures: Two properties will be similar if their super properties, sub properties, domains or ranges are similar.
- Individual-based similarity measures: Two individuals will be considered as similar if their type (i.e. the concept(s) that an individual belong to) or relevant properties are similar.

Here, we roughly described the similarity measures we used for our tests. More details can be found in the class of "ManualRuleSimple" in FOAM API which is open source.

4.4 Results and Discussion

We evaluate our algorithm from the following aspects. First of all, we evaluate the performance of various OWA operators based on the same similarity measures. Then we compare the OWA operator with weighted average operator. Finally, we compare our system using OWA operator with other systems.

4.4.1 The Performance of OWA Operators

In the first part of the evaluation, we compare the performance of the FOAM system when applying different OWA operators for the similarity aggregation.

---

[4] The edit distance between two strings is given by the minimum number of operations needed to transform one string into the other, where an operation is an insertion, deletion, or substitution of a single character.

Figure 2 shows the harmonic means (over the entire data set of benchmarks in OAEI 2008) of the precision, recall and f-measure for the different operators introduced in Section 3.

For these OWA operators, they assign the importance to different positions of the values to be aggregated in the descending order. For example, *max* considers the maximal value and *As many as possible* considers half of the smallest values. The first observation is that using *max* operator shows a poor performance, as this operator assigns all the weights to only one similarity value. The *As many as possible* operator assigns most weights to the similarity measures with lower similarity values and exhibits a very high precision, as it will return a match if half of the similarity measures with the smallest similarity values indicate a match. Obviously, this selectivity results in a low recall. We observe an increasing performance in terms of f-measure for the operators from *max*, *As many as possible*, *most*, *average*, to *at least half*. The best results in terms of f-measure are obtained for the *at least half* operator that assigns the weights to half of similarity values which are the highest ones.

However, it is worth noting that for different matching tasks, different operators may be appropriate. For example, if a high precision is required, an operator that assigns higher weights to the lower similarity values may be adequate, e.g. *as many as possible*. In any case, this selection can be performed easily based on the intuitive meaning of the lexical OWA operators, without any knowledge about the specific similarity measures.

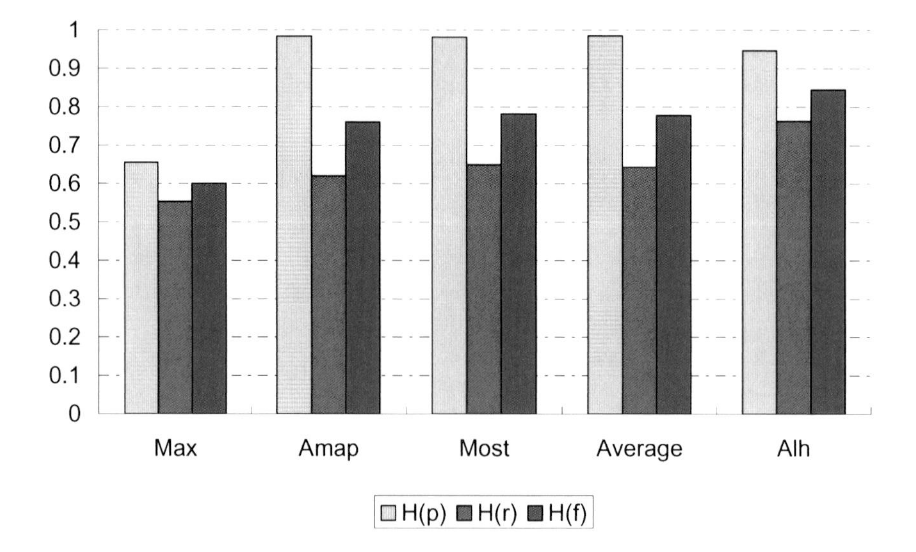

Fig. 2 The performance of different OWA combination methods

4.4.2 Comparison between OWA Operator and Weighted Average

In the second part of the evaluation, we compare the performance of the FOAM system by applying the OWA operator *at least half* with that applying the regular aggregation operator *weighted average* which was previously implemented in FOAM. Again, the comparison has been done based on exactly the same similarity measures.

In Figure 3, we only show the f-measure for ontologies from 248 to 304. For these ontologies, it is much harder and more challenging to find the correct correspondences than others in that these ontologies have quite different structure or lack annotation information like labels or comments comparing with the reference ontology 101. The number from 0 to 71 in X-axis indicates the ontology pair between the reference one 101 and an ontology from 248 to 304. For each correspondence between an ontology pair, one element is in the reference ontology 101, and another one is in the test ontology from 248 to 304. Two curves indicate the f-measure which are calculated by the *at least half* OWA operator and *weighted average* respectively.

We observe that in most cases, *at least half* outperforms *weighted average*. This is because ontologies from 248 to 266 derive from ontology 101 by changing the name of entities with different conventions, suppressing the comments, suppressing or flattening or expanding the concept hierarchy, and so on. For ontologies from 301 to 304, they are real-life ontologies which are created independently of the reference ontology 101. In such cases, some similarity measures will become useless if the corresponding features are not available. The *weighted average* is not flexible enough to deal with such cases since it gives weights to each similarity measure independent of the performance of the similarity measures. On the other hand, *at*

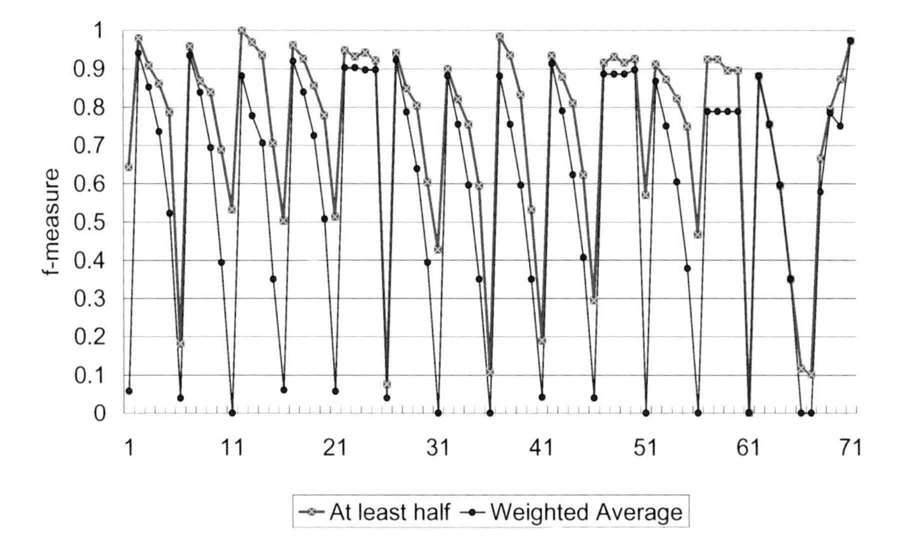

Fig. 3 The performance of *at least half* and *weighted average*

least half does not rely on particular similarity measures. Instead, it will assign the weights to that half of the similarity measures that perform best. On average, we observed an increase for the f-measure from 0.76 for the *weighted average* to 0.845 for the *at least half* OWA operator.

The benefits of using OWA aggregation operators are actually twofold: We do not need to assign weights to the individual similarity measures thus do not require any background knowledge about the similarity measures. At the same time we observe an improved performance.

4.4.3 Comparison with Other Ontology Matching Systems

Although the FOAM system did not participate in the official evaluation contest of OAEI in 2008, we give our results to compare with other systems which have participated, based on the same benchmarks as other systems and using the same evaluation measures. Since many systems attended the contest, we compare against the top eight systems.

Table 1 The comparison between FOAM (WA and OWA) and the top eight matching systems based on the benchmarks in OAEI 2008.

System	H(p)	H(r)	H(f)
aflood	0.97	0.71	0.820
aroma	0.95	0.70	0.806
ASMOV	0.95	0.86	0.903
CIDER	0.97	0.62	0.756
DSSim	0.97	0.67	0.793
Lily	0.97	0.88	0.923
RiMOM	0.96	0.84	0.896
SAMBO	0.99	0.58	0.731
FOAM-WA	0.95	0.63	0.760
FOAM-OWA	0.95	0.76	0.845

In Table 1, we show the harmonic means of precision and recall and f-measure to give an overview of the results. From the results shown in this table, we see that the matching systems ASMOV [7], Lily [14]and RiMOM [17] are the top three systems with respect to the harmonic mean of f-measure. For the three systems, the harmonic mean of f-measure achieves about 0.9 which are much better than all the others. Although the FOAM system – with either *weighted average* or OWA operator – can not beat with the top three systems, still it provides competitive performance. Especially when using the OWA aggregation operator, the FOAM system outperforms all the other systems except the top three ones.

Now we analyze why our system is outperformed by the top three systems. The main reason is that, we just use some simple and straight similarity measures and we are lacking of some heuristic algorithms to adapt the input parameters at runtime.

Take RiMOM system as an example. A key step of strategy selection is used. That is, if two ontologies have high label similarity, then the matching process will rely more on linguistic based strategies; while if the two ontologies have high structural similarity, they will employ similarity-propagation based strategies on them [11]. So RiMOM can perform better by using the flexible strategy selection.

Although FOAM-OWA is outperformed by ASMOV, Lily and RiROM, there is no big difference between ours and others. For example, the maximal difference between ours and the top system Lily is 0.078 regarding $\mathrm{H}(f)$. Besides, we still have quite a lot space to improve the performance of our system because of the flexibility of OWA operators and the support from the existing and ongoing theoretical and practical study of OWA operators. Furthermore, ours outperforms other systems except the top three systems. From Table 1, it can be seen that it is worth to integrate OWA operator into FOAM as the harmonic mean of recall has been improved a lot (i.e. from 0.63 to 0.76) when the same harmonic mean of precision (i.e. 0.95) is reached, comparing with FOAM-WA.

5 Conclusion and Future Work

It has been proved that, in most cases, combining the results of multiple similarity measures is a promising technique to get better results than just using one similarity measure at a time [1, 4, 2, 13]. In this paper, we integrated OWA aggregation operator into FOAM to provide a novel and promising solution for ontology matching. We summarize the answers for those questions given in Section 2.

1. **Test with more similarity measures and complex ontologies.** By testing more than 20 similarity measures and 110 ontology pairs, we can see that *At least half* operator outperforms other normal aggregation operators like *average* and *weighted average* in most cases.
2. **Choose different aggregation operators.** Generally, there are two ways to choose aggregation operators. One is according to their semantics explained in Section 3.4. Another way is according to various tasks which has been analyzed in our experiments in Section 4.4.1. For example, if a high precision is required, an operator that assigns higher weights to the lower similarity values may be adequate like *as many as possible*.
3. **Compare with other matching systems.** Although FOAM-OWA is outperformed by the top three systems ASMOV, Lily and RiROM, there is no big difference (e.g. the maximal difference is 0.078 regarding the harmonic mean of f-measure). As for the flexibility of OWA operators, there is still room for improving the performance of our system. Furthermore, our system outperforms others except the top three ones. We have shown that it is worth to integrate OWA operator into FOAM as the harmonic mean of recall has been improved when the same harmonic mean of precision is reached comparing with the previous version of FOAM (i.e. OWA-WA).

In the future work, we will extend our current work along two directions. One is the integration of machine learning techniques to obtain OWA weights when rich data sets related to the test ontologies are available. Another direction is combining OWA weights with the weights associated to the similarity measures using the techniques of the weighted OWA operator [12] when the similarity measures have different importance.

Acknowledgements. This work is partially supported by the EU in the IST project NeOn (IST-2006-027595, http://www.neon-project.org/web-content/).

References

1. Do, H., Rahm, E.: COMA - a system for flexible combination of schema matching approaches. In: Proceedings of 28th International Conference on Very Large Data Bases (VLDB 2002), pp. 610–621 (2002)
2. Ehrig, M., Staab, S.: QOM - quick ontology mapping. In: McIlraith, S.A., Plexousakis, D., van Harmelen, F. (eds.) ISWC 2004. LNCS, vol. 3298, pp. 683–697. Springer, Heidelberg (2004)
3. Euzenat, J., Shvaiko, P.: Ontology Matching. Springer, Heidelberg (2007)
4. Euzenat, J., Valtchev, P.: Similarity-based ontology alignment in OWL-Lite. In: Proceedings of 16th European Conference on Artificial Intelligence (ECAI 2004), pp. 333–337 (2004)
5. Filev, D., Yager, R.R.: On the issue of obtaining OWA operator weights. Fuzzy Sets and Systems 94(2), 157–169 (1998)
6. Fuller, R.: On obtaining OWA operator weights: a short survey of recent developments. In: Proceedings of 5th IEEE International Conference on Computational Cybernetics (ICCC 2007), pp. 241–244 (2007)
7. Jean-Mary, Y.R., Kabuka, M.R.: ASMOV: Results for OAEI 2008. In: Proceedings of Ontology Matching Workshop (OM 2008), pp. 132–139 (2008)
8. Ji, Q., Liu, W., Qi, G., Bell, D.A.: LCS: A linguistic combination system for ontology matching. In: Lang, J., Lin, F., Wang, J. (eds.) KSEM 2006. LNCS (LNAI), vol. 4092, pp. 176–189. Springer, Heidelberg (2006)
9. Rahm, E., Bernstein, P.: A survey of approaches to automatic schema matching. The International Journal on Very Large Data Bases 10(4), 334–350 (2001)
10. Shvaiko, P., Euzenat, J.: A survey of schema-based matching approaches. In: Spaccapietra, S. (ed.) Journal on Data Semantics IV. LNCS, vol. 3730, pp. 146–171. Springer, Heidelberg (2005)
11. Tang, J., Liang, B.-Y., Li, J., Wang, K.-H.: Risk minimization based ontology matching. In: Chi, C.-H., Lam, K.-Y. (eds.) AWCC 2004. LNCS, vol. 3309, pp. 469–480. Springer, Heidelberg (2004)
12. Torra, V.: The weighted OWA operator. International Journal of Intelligent Systems 12, 153–166 (1997)
13. Tu, K., Yu, Y.: CMC: Combining multiple schema-matching strategies based on credibility prediction. In: Zhou, L.-z., Ooi, B.-C., Meng, X. (eds.) DASFAA 2005. LNCS, vol. 3453, pp. 17–20. Springer, Heidelberg (2005)

14. Wang, P., Xu, B.: Lily: Ontology alignment results for OAEI 2008. In: Proceedings of Ontology Matching Workshop (OM 2008), pp. 167–175 (2008)
15. Xu, Z.: An overview of methods for determining OWA weights. International Journal of Intelligent Systems 20(8), 843–865 (2005)
16. Yager, R.R.: On ordered weighted averaging aggregation operators in multi-criteria decision making. IEEE Transactions on Systems, Man and Cybernetics 18(1), 183–190 (1988)
17. Zhang, X., Zhong, Q., Li, J., Tang, J.: RiMOM results for OAEI 2008. In: Proceedings of Ontology Matching Workshop (OM 2008), pp. 182–189 (2008)